PREFACE

This monograph is a collective work. The names appearing on the front cover are those of the people who worked on every chapter. But the contributions of others were also very important:

C. Risito for Chapters I, II and IV,

K. Peiffer for III, IV, VI, IX

R. J. Ballieu for I and IX,

Dang Chau Phien for VI and IX,

J. L. Corne for VII and VIII.

The idea of writing this book originated in a seminar held at the University of Louvain during the academic year 1971-72. Two years later, a first draft was completed. However, it was unsatisfactory mainly because it was excessively abstract and lacked examples. It was then decided to write it again, taking advantage of some remarks of the students to whom it had been partly addressed. The actual text is this second version.

The subject matter is stability theory in the general setting of ordinary differential equations using what is known as Liapunov's direct or second method. We concentrate our efforts on this method, not because we underrate those which appear more powerful in some circumstances, but because it is important enough, along with its modern developments, to justify the writing of an up-to-date monograph. Also excellent books exist concerning the other methods, as for example R. Bellman [1953] and W. A. Coppel [1965].

Liapunov's second method has the undeserved reputation of being mainly of theoretical interest, because auxiliary functions appear to be so difficult to construct. We feel this is the opinion of those people who have not really tried. Indeed, many mathematicians have tackled only theoretical problems. On the other hand, too many of those involved in applications are unaware of the useful theorems or are victims of the myth of the elusive Liapunov function. Our aim, in writing this book, has been twofold: to describe the present state of the most useful parts of the theory, and to appeal to the practical man with a wealth of applications taken from many varied fields.

Chapters I and II constitute an elementary self-contained treatment of stability theory. They should normally be read first. Almost every other chapter can be studied without further prerequisite, except that some definitions or propositions of Chapter VI are needed in Chapters VII, VIII, and IX. The whole of Chapter VI is used in Section IX.6.

We are also grateful to M. Everard, S. Spinacci and Kate MacDougall for their particularly expert typing of successive versions of the manuscript. Finally, it is a pleasure to acknowledge the financial support of this work by "Fonds National de la Recherche Scientifique".

Louvain-la-Neuve, October, 1975

N. Rouche
P. Habets
M. Laloy

Stability Theory by Liapunov's Direct Method

Students and External Readers | Staff & Research Students
DATE DUE FOR RETURN | DATE OF ISSUE

N.B. All books must be returned for the Annual Inspection in June

Any book which you borrow remains your responsibility until the loan slip is cancelled

App

Spr

Berlin

N. Rouche
P. Habets
M. Laloy

U.C.L.
Institut de Mathématique Pure et Appliquée
Chemin du Cyclotron 2
B-1348
Louvain-la-Neuve
Belgium

AMS Subject Classifications: 34D20, 93D05 (Primary), 34Dxx, 34H05

Library of Congress Cataloging in Publication Data

Rouche, Nicolas.
Stability theory by Liapunov's direct method.

(Applied mathematical sciences ; v. 22)
Bibliography: p.
Includes indexes.
1. Differential equations. 2. Stability.
3. Liapunov functions. I. Habets, P., 1943- joint author. II. Laloy, M., 1946- joint author.
III. Title. IV. Series.
QA1.A647 vol. 22 [QA372] 510'.8s [515'.352] 77-7285

Printed in the United States of America.

9 8 7 6 5 4 3 2 1

ISBN 0-387-90258-9 Springer-Verlag New York

ISBN 3-540-90258-9 Springer-Verlag Berlin Heidelberg

SOME NOTATIONS AND DEFINITIONS

This books requires a familiarity with some basic concepts from the theory of ordinary differential equations. As a general rule we have used symbols which are common place in mathematics. Let us however point out the following notations:

$\mathcal{R}$, the set of real numbers,

$\overline{\mathcal{R}}$, the extended real number system,

$a \geq 0$, a is a positive real number,

$a > 0$, a is a strictly positive real number,

$[a,b]$, closed interval,

$]a,b[$, open interval,

$(a|b)$ or $a^T b$, according to context, scalar product in $\mathcal{R}^n$,

$||x||$, norm of point x in $\mathcal{R}^n$,

$d(x,M) = \inf_{y \in M} ||x-y||$, distance from $x \in \mathcal{R}^n$ to $M \subset \mathcal{R}^n$,

$B_\varepsilon = \{x \in \mathcal{R}^n, ||x|| < \varepsilon\}$, open ball with center at the origin and radius $\varepsilon > 0$,

$B(a,\varepsilon) = \{x \in \mathcal{R}^n, ||x-a|| < \varepsilon\}$, open ball with center $a \in \mathcal{R}^n$ and radius $\varepsilon > 0$,

$B(M,\varepsilon) = \{x \in \mathcal{R}^n, d(x,M) < \varepsilon\}$, ε - neighborhood of the set $M \subset \mathcal{R}^n$,

$M_\varepsilon = B(M,\varepsilon) \cap \Omega$, ε - neighborhood of $M \in \mathcal{R}^n$ with respect to $\Omega \subset \mathcal{R}^n$,

E, unit $n \times n$ matrix,

$\dot{x} = \frac{dx}{dt}$, time derivative of the function $x: A \subset \mathcal{R} \to \mathcal{R}^n$,

$\frac{\partial f}{\partial x}$, jacobian matrix of the function $f: \mathcal{R}^n \to \mathcal{R}^m$, $x \to f(x)$,

J^+, see p. 7,

$\mathcal{K}$, see definition p. 12.

$\forall x$, universal quantifier; read "for all x" or "given x",

$\exists x$, existential quantifier; read "for some x" or "there exist x".

For general concepts on differential equations which are not defined in this text we refer to Ph. Hartman [1964], E. Coddington and N. Levinson [1965] or N. Rouche and J. Mawhin [1973]. The following definitions might be useful.

Let $A \subset \mathscr{R}$ and $f: A \to \mathscr{R}$, $x \to f(x)$ be a real valued function.

The function f is said to be:

increasing if $\forall x \in A$, $\forall y \in A$, $x < y$ implies $f(x) \leq f(y)$;
i.e., for all x and y in A, $x < y$ implies $f(x) \leq f(y)$.

strictly increasing if $\forall x \in A$, $\forall y \in A$, $x < y$ implies $f(x) < f(y)$,

decreasing if $\forall x \in A$, $\forall y \in A$, $x < y$ implies $f(x) \geq f(y)$,

strictly decreasing if $\forall x \in A$, $\forall y \in A$, $x < y$ implies $f(x) > f(y)$,

monotonic if it is increasing on A or decreasing on A.

Let $a \in \bar{A}$, the extended closure of A. Then the limit superior (upper limit) of f at a is

$$\limsup_{x \to a} f(x) = \inf_{\delta > 0}\{\sup\{f(x): x \in B(a,\delta),\ x \neq a\}\} \in \bar{\mathscr{R}}.$$

Similarly the limit inferior (lower limit) of f at a is

$$\liminf_{x \to a} f(x) = \sup_{\delta > 0}\{\inf\{f(x): x \in B(a,\delta),\ x \neq a\}\} \in \bar{\mathscr{R}}.$$

If $a \in A$, the function f is said to be lower semi-continuous at a if $\liminf_{x \to a} f(x) \geq f(a)$. If

$\limsup_{x \to a} f(x) \leq f(a)$, the function f is said to be <u>upper semi-continuous at</u> a. It is easy to verify that a function f is continuous at a if and only if it is lower and upper semi-continuous at a.

A function $V: \mathscr{R}^{1+n} \to \mathscr{R}$, $(t,x) \to V(t,x)$ is said to be <u>positive definite</u> (with respect to x) if there exists a function $a \in \mathscr{K}$ such that

(i) $V(t,0) = 0$

(ii) $V(t,x) \geq a(||x||)$.

If $-V$ is positive definite, the function V is said to be <u>negative definite</u> (with respect to x). If $V(t,0) = 0$ and $V(t,x) \geq 0$ the function V is said to be <u>positive semi-definite</u> (with respect to x). A function $V: \mathscr{R}^{1+n+m} \to \mathscr{R}$, $(t,x,y) \to V(t,x,y)$ is said to be <u>positive definite with respect to</u> x if for some function $a \in \mathscr{K}$

(i) $V(t,0,0) = 0$

(ii) $V(t,x,y) \geq a(||x||)$.

An important class of positive definite functions are the <u>positive quadratic forms</u>

$$V(x) = x^T A x$$

where A is a symmetric positive definite matrix (T denotes transpose).

TABLE OF CONTENTS

CHAPTER I

ELEMENTS OF STABILITY THEORY

The first two chapters are of an introductory character. Of the matters they exhibit, some have been known for a long time, others belong to the last fifteen years. Almost all will be considered over again in subsequent chapters, where the results will be extended or deepened. However, the next few pages are meant to give a fair idea of what stability and Liapunov's direct method are. Further, they should prove helpful to those concerned with simple practical applications. Of course, the rest of the book has been written to cope with less simple applications and, unfortunately or not, everyday practice proves how numerous they are ...

1. A First Glance at Stability Concepts

1.1. The English adjective "stable" originates from the Latin "stabilis", deriving itself from "stare", to stand. Its first acceptation is "standing firmly", "firmly established". A natural extension is "durable", not to mention the moral meaning "steady in purpose, constant". As it is, this concept of stability seems to be clear and of good use in everyday life. The layman might well wonder what reasons can be invoked to refine or complicate it. There are many, as we shall see.

Very early, the stability concept was specialized in mechanics to describe some type of equilibrium of a material particle or system. Consider for instance a particle subject to some forces and possessing an equilibrium point q_0. The equilibrium is called <u>stable</u> if, after any sufficiently small perturbations of its position and velocity, the particle remains forever arbitrarily near q_0, with arbitrarily small velocity. We shall not dwell on the well known example of a simple pendulum, whose lowest position, associated with zero velocity, is a stable equilibrium, whereas the highest one, also with zero velocity, is an unstable one.

Formulated in precise mathematical terms, this mechanical definition of stability was found useful in many situations, but inadequate in many others. This is why, with passing years, a host of other concepts have been introduced, each of them more or less related to the first definition and to the common sense meaning of stability. They were created either for definite technical or physical purposes, or for reasons of symmetry or completeness of the theory, or else to suit the fancy of their inventors. Later in this book (Chapter VI), we shall try, with much care, to separate the wheat from the chaff.

1.2. As contrasted with mechanical stability, the other concept known as Liapunov's stability has the following features: first, it pertains no more to a material particle (or the equations thereof), but to a general differential equation; second, it applies to a solution, i.e. not only to an equilibrium or critical point.

Let

$$\dot{x} = f(t,x), \tag{1.1}$$

where x and f are real n-vectors, t is the time (a real variable), f is defined on $\mathscr{R} \times \mathscr{R}^n$ and $\dot{x} = dx/dt$. We assume f smooth enough to ensure existence, uniqueness and continuous dependence of the solutions of the initial value problem associated with (1.1) over $\mathscr{R} \times \mathscr{R}^n$. For simplicity, we assume further that all solutions to be mentioned below exist for every $t \in \mathscr{R}$. Let $||\cdot||$ designate any norm on $\mathscr{R}^n$.

A solution $\bar{x}(t)$ of (1.1) is called <u>stable at</u> t_0, or, more precisely, <u>stable at</u> $t = t_0$ in the sense of A.M. Liapunov [1892] if, for every $\varepsilon > 0$, there is a $\delta > 0$ such that if $x(t)$ is any other solution with $||x(t_0) - \bar{x}(t_0)|| < \delta$, then $||x(t) - \bar{x}(t)|| < \varepsilon$ for all $t \geq t_0$. Otherwise, of course, $\bar{x}(t)$ is called <u>unstable at</u> t_0.

Thus, it turns out that stability at t_0 is nothing but continuous dependence of the solutions on $x_0 = x(t_0)$, uniform with respect to $t \in [t_0,\infty[$.

1.3. <u>Exercise</u>. Prove that stability at t_0 implies stability at any other initial time (usually with different values for δ).

Hint: use the fact that, if $x(t;t_0,x_0)$ is the solution passing through x_0 at t_0, then the mapping

$$x(t;t_0,\cdot): x_0 \to x(t;t_0,x_0)$$

is a homeomorphism; i.e., it and its inverse are one to one and continuous.

1.4. We may gain some geometrical insight into this stability concept by considering again a pendulum, whose equation is $\ddot{x} + \omega^2 \sin x = 0$, with x and $\omega \in \mathscr{R}$. This second order equation is equivalent to the first order system

$$\dot{x} = y$$
$$\dot{y} = -\omega^2 \sin x.$$

As is well known, the origin of the (x,y)-plane is a center, i.e. all the solutions starting near the origin form a family of non-intersecting closed orbits encircling the origin. Given $\varepsilon > 0$, consider an orbit entirely contained in the disk B_ε of radius ε with center at the origin. Further, choose any other disk B_δ of radius δ, contained in this orbit. Clearly, every solution starting in B_δ at any initial time remains in B_ε. This demonstrates stability of the origin for any initial time.

On the other hand however, any other solution corresponding to one of the closed orbits is unstable. In fact, the period of the solution varies with the orbit and two points of the (x,y)-plane, very close to each other at $t = t_0$, but belonging to different orbits, will appear in opposition after some time. This happens however small the difference between periods. But it remains that, in some sense, the orbits are closed to each other. Similar examples led to a new concept called <u>orbital stability</u>, to be discussed later in this book, in connection with the stability of sets of points.

1.5. To say a little more about possible variations on the theme of stability, notice that in the case of the pendulum, the equilibrium $x = y = 0$ is such that no neighbouring solution approaches it when $t \to \infty$, as it would do if some appropriate friction were present. In many practical situations, it is useful to require, besides mere Liapunov stability of a solution $\bar{x}(t)$, that all neighbouring solutions $x(t)$ tend to $\bar{x}(t)$ when $t \to \infty$. This leads to the notion of <u>asymptotic stability</u>.

1.6. Many other examples can illustrate the necessity of creating new specific concepts. The last one to be mentioned here will be borrowed from celestial mechanics. Following common sense, the solar system is called <u>stable</u> if it is "durable" (cf. 1.1), i.e. if none of its constituent bodies escapes to infinity, and further if no two such bodies meet each other. But the velocities are unbounded if and only if two bodies approach each other. Therefore, stability in this sense (it is called <u>Lagrange stability</u>), simply means that the coordinates and velocities of the bodies are bounded. Boundedness of solutions thus appears as a legitimate and natural type of stability.

In the next section, we introduce a small number of definitions, in fact the most widely used and studied.

2. Various Definitions of Stability and Attractivity

2.1. We presented above the concept of a stable solution $\bar{x}(t)$ for equation (1.1). If we replace x by a new variable $z = x - \bar{x}(t)$, then (1.1) becomes

$$\dot{z} = g(t,z) \equiv f(t,z + \bar{x}(t)) - f(t,\bar{x}(t)) \qquad (2.1)$$

where $g(t,0) = 0$ for every $t \in \mathscr{R}$. The origin is a critical point of (2.1) and stability of the solution $\bar{x}$ of (1.1) is equivalent to stability of this critical point for (2.1). Naturally, passing from (1.1) to (2.1) is not always possible, for $\bar{x}$ has to be explicitly known; nor is it always rewarding, for it often happens that (2.1) is more complicated than (1.1): for instance, when (1.1) is autonomous, (2.1) generally is not. Nevertheless, we shall, in this chapter and the next, concentrate on stability of critical points.

2.2. Equation considered, general hypotheses. Let us consider a continuous function

$$f\colon I \times \Omega \to \mathscr{R}^n, \ (t,x) \to f(t,x)$$

where $I =]\tau,\infty[$ for some $\tau \in \mathscr{R}$ or $\tau = -\infty$, and Ω is a domain (i.e. an open connected set) of $\mathscr{R}^n$, containing the origin. We assume that $f(t,0) = 0$ for every $t \in I$, so that for the differential equation

$$\dot{x} = f(t,x) \qquad (2.2)$$

the origin is an equilibrium or critical point. Further, let f be smooth enough in order that, through every $(t_0,x_0) \in I \times \Omega$, there passes one and only one solution of (2.2). We represent this solution by $x(t;t_0,x_0)$, thus displaying its dependence on initial conditions. By definition $x(t_0;t_0,x_0) = x_0$. For the right maximal interval where $x(\cdot;t_0,x_0)$ is defined, we write $J^+(t_0,x_0)$ or simply J^+ or $[t_0,\omega[$. Of course $J^+ \subset [t_0,\infty[$, but we do not assume $\omega = \infty$. Let us also recall that we write $B_\rho = \{x \in \mathscr{R}^n: ||x|| < \rho\}$.

In all definitions below, we use the logical quantifiers $\exists$ and $\forall$ in a systematic manner. This somewhat rigid way of presenting things is meant to avoid looseness in expression and ambiguities, a non-negligible danger in the manipulation of such delicate concepts.

2.3. Stability. The solution $x = 0$ of (2.2) is called

stable (A.M. Liapunov [1892]) if $(\forall\varepsilon > 0)(\forall t_0 \in I)(\exists\delta > 0)(\forall x_0 \in B_\delta)(\forall t \in J^+)$ $||x(t;t_0,x_0)|| < \varepsilon$; i.e., given $\varepsilon > 0$ and $t_0 \in I$, there is a $\delta > 0$ such that for all $x_0 \in B_\delta$ and $t \in J^+$ one has $||x(t;t_0,x_0)|| < \varepsilon$.

unstable if it is not stable, i.e. if $(\exists\varepsilon > 0)(\exists t_0 \in I)(\forall\delta > 0)(\exists x_0 \in B_\delta)(\exists t \in J^+)$ $||x(t;t_0,x_0)|| \geq \varepsilon$; i.e., for some $\varepsilon > 0$ and $t_0 \in I$ and each $\delta > 0$ there is an $x_0 \in B_\delta$ and a $t \in J^+$ such that $||x(t;t_0,x_0)|| \geq \varepsilon$.

uniformly stable (K.P. Persidski [1933]) if $(\forall\varepsilon > 0)(\exists\delta > 0)(\forall t_0 \in I)(\forall x_0 \in B_\delta)(\forall t \in J^+)$ $||x(t;t_0,x_0)|| < \varepsilon$; i.e., given $\varepsilon > 0$ there is a $\delta = \delta(\varepsilon)$ such that $||x(t;t_0,x_0)|| < \varepsilon$ for all $t_0 \in I$, all $||x_0|| < \delta$ and all $t \geq t_0$.

2.4. Remark. As already noticed, we did not presuppose $J^+ = [t_0,\infty[$. It follows that, in principle, any solution mentioned in the above definitions may cease to exist after some finite time. However, if $\overline{B}_\varepsilon \subset \Omega$, the solutions mentioned in the definitions of stability and uniform stability may be continued up to $+\infty$: in fact, they cannot approach the boundary of Ω. This is of course not the case for the solutions mentioned in the definition of instability.

2.5. <u>Attractivity</u>. The solution $x = 0$ of (2.2) is called <u>attractive</u> if $(\forall t_0 \in I)(\exists \eta > 0)(\forall \varepsilon > 0)(\forall x_0 \in B_\eta)$ $(\exists \sigma > 0,\ t_0 + \sigma \in J^+)(\forall t \geq t_0 + \sigma,\ t \in J^+)\quad ||x(t;t_0,x_0)|| < \varepsilon$; i.e., for each $t_0 \in I$ there is an $\eta = \eta(t_0)$, and for each $\varepsilon > 0$ and each $||x_0|| < \eta$ there is a $\sigma = \sigma(t_0,\varepsilon,x_0) > 0$ such that $t_0 + \sigma \in J^+$ and $||x(t;t_0,x_0)|| < \varepsilon$ for all $t \geq t_0 + \sigma$.

<u>equi-attractive</u> if $(\forall t_0 \in I)(\exists \eta > 0)(\forall \varepsilon > 0)(\exists \sigma > 0)(\forall x_0 \in B_\eta)$ $t_0 + \sigma \in J^+$ and $[(\forall t \geq t_0 + \sigma,\ t \in J^+)\quad ||x(t;t_0,x_0)|| < \varepsilon]$; i.e., for each $t_0 \in I$ there is an $\eta = \eta(t_0)$ and for each $\varepsilon > 0$ a $\sigma = \sigma(t_0,\varepsilon)$ such that $t_0 + \sigma \in J^+$ and $||x(t;t_0,x_0)|| < \varepsilon$ for all $||x_0|| < \eta$ and all $t \geq t_0 + \sigma$.

<u>uniformly attractive</u> if $(\exists \eta > 0)(\forall \varepsilon > 0)(\exists \sigma > 0)(\forall x_0 \in B_\eta)$ $(\forall t_0 \in I)\ t_0 + \sigma \in J^+$ and $[(\forall t \geq t_0 + \sigma,\ t \in J^+)\ ||x(t;t_0,x_0)|| < \varepsilon]$; i.e., for some $\eta > 0$ and each $\varepsilon > 0$ there is a $\sigma = \sigma(\varepsilon) > 0$ such that $t_0 + \sigma \in J^+$ and $||x(t;t_0,x_0)|| < \varepsilon$ for all $||x_0|| < \eta$, all $t_0 \in I$ and all $t \geq t_0 + \sigma$.

2.6 <u>Remark</u>. As in Remark 2.4, if $\overline{B}_\varepsilon \subset \Omega$, the solutions mentioned in all three definitions above exist over $[t_0,\infty[$. Thus, in the definition of attractivity, all solutions starting from B_η approach the origin as $t \to \infty$. In the case of equi-attractivity, they tend to 0 uniformly with respect to $x_0 \in B_\eta$, whereas in the case of uniform attractivity, they tend to 0 uniformly with respect to $x_0 \in B_\eta$ and $t_0 \in I$.

2.7. Attractivity does not imply stability! For an example pertaining to an autonomous system in $\mathscr{R}^2$, see R.E. Vinograd [1957], also presented in W. Hahn [1967]. The orbits are not the same, but similar to those of Figure 1.1, where γ is a curve separating bounded and unbounded orbits. The origin is unstable, in spite of the fact that every solution tends to it as $t \to \infty$.

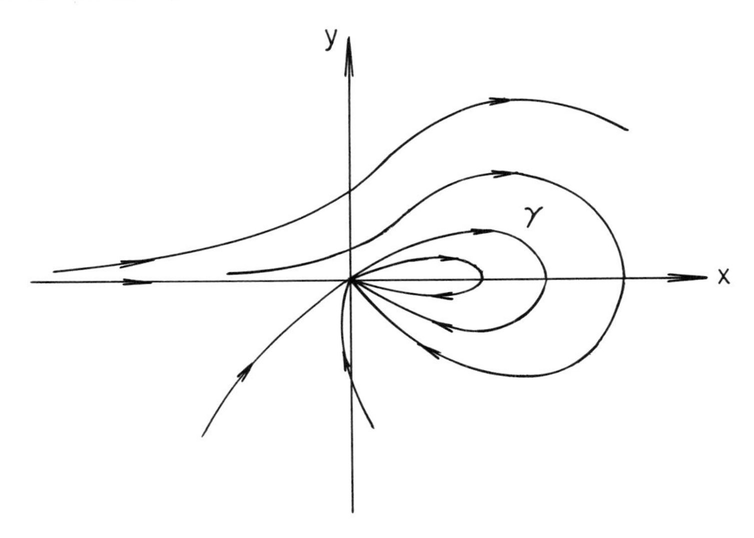

Figure 1.1. The origin is unstable and attractive

2.8. Exercise (J.L. Massera [1949]). For a scalar equation $\dot{x} = f(t,x)$ $(x,f \in \mathscr{R})$, attractivity of the origin implies equi-attractivity. Hint: use the uniqueness of the solutions.

2.9. Exercise. For Equation (2.2), equi-attractivity of the origin implies its stability.

2.10. The region of attraction of the origin at time t_0 is the set

$$A(t_0) = \{x_0 \in \Omega: x(t;t_0,x_0) \to 0, t \to \infty\}.$$

If $A(t_0)$ does not depend on t_0, we say that the region of attraction is uniform. Moreover, if $\Omega = \mathscr{R}^n = A(t_0)$ for every t_0, the origin is said to be globally attractive. Some kind of uniformity in t_0 and x_0 can also be added to global attractivity. The origin is said to be uniformly globally attractive if, for any $\eta > 0$, any $t_0 \in I$ and any $x_0 \in B_\eta$, $x(t;t_0,x_0) \to 0$ as $t \to \infty$, uniformly with respect to t_0 and x_0. This concept is obtained from uniform attractivity (Section 2.5) by replacing $(\exists\eta > 0)$ by $(\forall\eta > 0)$.

2.11. Asymptotic stability. The solution $x = 0$ of (2.2) is called

asymptotically stable (A.M. Liapunov [1892]) if it is stable and attractive;

equi-asymptotically stable (J.L. Massera [1949]) if it is stable and equi-attractive; (Cf. Exercise 2.9).

uniformly asymptotically stable (I.G. Malkin [1954]), if it is uniformly stable and uniformly attractive;

globally asymptotically stable (E.A. Barbashin and N.N. Krasovski [1952]) if it is stable and globally attractive;

uniformly globally asymptotically stable (E.A. Barbashin and N.N. Krasovski [1952]), if it is uniformly stable and uniformly globally attractive.

2.12. Exercise. For the equation $\dot{x} = -x$, with $x \in \mathscr{R}$, the origin is uniformly globally asymptotically stable.

2.13. Exercise. Consider, for $x \in \mathscr{R}$ and $t > 0$, the differential equation defined by

$$\dot{x} = -\frac{x}{t} \quad \text{for} \quad -1 \leq xt \leq 1,$$

$$\dot{x} = \frac{x}{t} - \frac{2}{t^2} \quad \text{for} \quad 1 < xt,$$

$$\dot{x} = \frac{x}{t} + \frac{2}{t^2} \quad \text{for} \quad xt < -1.$$

Prove that the origin is equi-asymptotically stable, but not uniformly nor globally asymptotically stable.

For more examples, see for instance J.L. Massera [1949] or W. Hahn [1967]. The following theorem is important.

2.14. Theorem. Let $f(t,x)$ in Equation (2.2) be independent of t or periodic in t. Then, stability of the origin implies uniform stability, and asymptotic stability implies uniform asymptotic stability.

For the proof, see T. Yoshizawa [1966], pp. 30-31. Notice, by the way, that there is no such theorem for attractivity or uniform attractivity: indeed, uniform attractivity implies stability (see 2.9) and therefore attractivity would imply stability, which is not the case, even for autonomous systems (see 2.7).

3. Auxiliary Functions

3.1. What is known as Liapunov's direct or second* method for the study of stability, makes an essential use of

---

*The first method rests on the consideration of some explicit representation of the solutions, in particular by infinite series.

auxiliary functions, also called Liapunov or Liapunovlike functions. Those of the simplest type are C^1 functions

$$V: I \times \Omega \to \mathscr{R}, \ (t,x) \to V(t,x)$$

where I and Ω are as in Section 2.2.

If $x(t)$ is some solution of Equation (2.2), the derivative $\dot{V}^*(t)$ of the time function $V^*(t) = V(t,x(t))$ exists and is given by

$$\dot{V}^*(t) = (\frac{\partial V}{\partial x}(t,x(t)) \mid f(t,x(t))) + \frac{\partial V}{\partial t}(t,x(t)). \tag{3.1}$$

Then, if one introduces the function

$$\dot{V}: I \times \Omega \to \mathscr{R}, \ (t,x) \to \dot{V}(t,x)$$

defined by $\dot{V}(t,x) = (\frac{\partial V}{\partial x}(t,x) \mid f(t,x)) + \frac{\partial V}{\partial t}(t,x)$, it follows that

$$\dot{V}^*(t) = \dot{V}(t,x(t)).$$

Hence, one finds out that computing $\dot{V}^*(t)$ at some given time t does not require a knowledge of the solution $x(t)$, but only the value of x at t. For simplicity, we shall often write $V(t)$ for $V^*(t)$, accepting the ambiguity of using the same symbol for V as a function of t and a function of t and x. In the same way, we shall use $\dot{V}(t)$ instead of $\dot{V}^*(t)$.

3.2. Another useful type of function is the so-called function of class $\mathscr{K}$ in the sense of W. Hahn [1967], i.e., a function $a: \mathscr{R}^+ \to \mathscr{R}^+$, continuous, strictly increasing, with $a(0) = 0$. For such a function, we shall usually write $a \in \mathscr{K}$.

3.3. Consider now a function $V(t,x)$ defined as above, but such that $V(t,0) = 0$ and, for some function $a \in \mathcal{K}$ and every $(t,x) \in I \times \Omega$: $V(t,x) \geq a(||x||)$. Such a function is said to be positive definite on Ω. Stating that $V(t,x)$ is positive definite, i.e., without mentioning Ω, means that for some open neighborhood $\Omega' \subset \Omega$ of the origin, V is positive definite on Ω'.

3.4. A well known necessary and sufficient condition for a function $V(t,x)$ to be positive definite on Ω is that there exists a continuous function $V^*: \overline{\Omega} \to \mathcal{R}$ such that $V^*(0) = 0$, $V^*(x) > 0$ for $x \subset \overline{\Omega}$, $x \neq 0$ and further that $V(t,0) = 0$ and $(\forall (t,x) \in I \times \Omega) V(t,x) \geq V^*(x)$.

We assume that the reader knows how to recognize a positive definite quadratic form in x, for instance by using Sylvester's criterion. For all these notions, we refer to N. Rouche and J. Mawhin [1973], vol. II.

4. Stability and Partial Stability

4.1. Theorems 4.2 and 4.3 to follow pertain to Equation (2.2) and to the accompanying general hypotheses.

4.2. Theorem (A.M. Liapunov [1892]). If there exists a $\mathcal{C}^1$ function $V: I \times \Omega \to \mathcal{R}$ such that, for some $a \in \mathcal{K}$ and every $(t,x) \in I \times \Omega$:

(i) $V(t,x) \geq a(||x||)$; $V(t,0) = 0$;

(ii) $\dot{V}(t,x) \leq 0$;

then, the origin is stable.

Proof. Let $t_0 \in I$ and $\varepsilon > 0$ be given. As V is continuous and $V(t_0,0) = 0$, there is a $\delta = \delta(t_0,\varepsilon) > 0$ such that $V(t_0,x_0) < a(\varepsilon)$ for every $x_0 \in B_\delta$. Writing $x(t)$ in place of $x(t;t_0,x_0)$ and using (ii), one gets for every $x_0 \in B_\delta$ and every $t \in J^+$:

$$a(||x(t)||) \leq V(t,x(t)) \leq V(t_0,x_0) < a(\varepsilon).$$

Since $a \in \mathscr{K}$, one obtains that $||x(t)|| < \varepsilon$. Q.E.D.

4.3. Theorem (K.P. Persidski [1933]). If one adds to the hypotheses of Theorem 4.2 that for some $b \in \mathscr{K}$ and every $(t,x) \in I \times \Omega$: $V(t,x) \leq b(||x||)$, then the origin is uniformly stable.

Indeed, in this case, δ can be chosen to be independent of t_0.

4.4. The concept of partial stability. Let $n > 0$ and $m > 0$ be two integers, and consider two continuous functions $f\colon I \times \Omega \times \mathscr{R}^m \to \mathscr{R}^n$, $g\colon I \times \Omega \times \mathscr{R}^m \to \mathscr{R}^m$, where I is defined as above and Ω is a domain of $\mathscr{R}^n$, containing the origin. We assume that $f(t,0,0) = 0$ and $g(t,0,0) = 0$ for every $t \in I$ and further that f and g are smooth enough in order that, through every point of $I \times \Omega \times \mathscr{R}^m$, there passes one and only one solution of the differential system

$$\begin{aligned} \dot{x} &= f(t,x,y), \\ \dot{y} &= g(t,x,y). \end{aligned} \tag{4.1}$$

To shorten our notation, we shall write z for the vector $(x,y) \in \mathscr{R}^{n+m}$ and also $z(t;t_0,z_0) = (x(t;t_0,z_0), y(t;t_0,z_0))$ for the solution of (4.1) starting from z_0 at t_0. The following definition of partial stability was given by

V.V. Rumiantsev [1957]: the solution $z = 0$ of (4.1) is <u>stable with respect to</u> x if

$$(\forall \varepsilon > 0)(\forall t_0 \in I)(\exists \delta > 0)(\forall z_0 \in B_\delta)(\forall t \in J^+)\ ||x(t;t_0,z_0)|| < \varepsilon;$$

i.e., given $\varepsilon > 0$ and $t_0 \in I$ there exists $\delta > 0$ such that $||x(t;t_0,z_0)|| < \varepsilon$ for all $z_0 \in B_\delta$ and all $t \in J^+$.

<u>Uniform stability with respect to</u> x is defined in the same way, following the example of Section 2.3 above.

Remark 2.4 is no longer true if partial stability replaces stability, because the domain defined by $||x|| < \varepsilon$ is unbounded in the (x,y)-space and a solution, even if it remains in this domain, can escape to infinity in a finite time: consider for example, the trivial system $\dot{x} = -x$, $\dot{y} = y^2$ for $x,y \in \mathscr{R}$.

4.5. <u>Theorem</u> (V.V. Rumiantsev [1957]). If there exists a $\mathscr{C}^1$ function $V: I \times \Omega \times \mathscr{R}^m \to \mathscr{R}$ such that, for some $a \in \mathscr{K}$ and every $(t,x,y) \in I \times \Omega \times \mathscr{R}^m$:

(i) $V(t,x,y) \geq a(||x||)$, $V(t,0,0) = 0$;

(ii) $\dot{V}(t,x,y) \leq 0$;

then the origin $z = 0$ is stable with respect to x.

Moreover, if for some $b \in \mathscr{K}$ and every $(t,x,y) \in I \times \Omega \times \mathscr{R}^m$:

(iii) $V(t,x,y) \leq b(||x|| + ||y||)$;

then the origin is uniformly stable in x.

The proof is left as an exercise.

4.6. <u>A first example</u>. Consider, for $x \in \mathscr{R}^n$, the linear differential equation

$$\dot{x} = (D(t) + A(t))x$$

where D and A are $n \times n$ matrices, being continuous functions of t on $I =]\tau,\infty[$, D diagonal and A skew-symmetric. Choosing $V(t,x) = (x,x)$ (where (x,x) is a scalar product), we get $\dot{V}(t,x) = 2(x,D(t)x)$. If the elements of D are ≤ 0 for every $t \in I$, then $\dot{V} \leq 0$ and, following Theorem 4.3, the origin is uniformly stable.

4.7. Stability of some steady rotations of a rigid body (cf. N.G. Chetaev [1955]). Consider a rigid body with a fixed point 0 in some inertial frame of reference, and no exterior forces applied whatsoever. Let A,B and C be the principal moments of inertia of the body with respect to 0, and ω its (vectorial) angular velocity in the inertial frame. The so-called Euler equations for the components p,q and r of ω in the principal axes of inertia read (cf., e.g., H. Cabannes [1966])

$$\begin{aligned} A\dot{p} &= (B-C)qr, \\ B\dot{q} &= (C-A)rp, \\ C\dot{r} &= (A-B)pq. \end{aligned} \tag{4.2}$$

The steady rotations around the first axis correspond to the critical point $p = p_0$, $q = 0$, $r = 0$. Using new variables x,y,z defined by $x = p - p_0$, $y = q$, $z = r$, we shift the critical point to the origin:

$$\begin{aligned} \dot{x} &= \frac{B-C}{A} yz, \\ \dot{y} &= \frac{C-A}{B} (p_0+x)z, \\ \dot{z} &= \frac{A-B}{C} (p_0+x)y. \end{aligned} \tag{4.3}$$

If $A < B \leq C$, the steady rotation is around the largest axis of the ellipsoid of inertia. An auxiliary function suiting Theorem 4.2 is

$$V = B(B-A)y^2 + C(C-A)z^2 + [By^2 + Cz^2 + A(x^2+2xp_0)]^2 \quad (4.4)$$

which is a first integral, therefore such that $\dot{V} = 0$. It follows that the origin is stable for (4.3), and even uniformly stable since the system is autonomous.

On the other hand, if $A > B \geq C$, one obtains a similar result using the first integral

$$V = B(A-B)y^2 + C(A-C)z^2 + [By^2+ + Cz^2 + A(x^2+2xp_0)]^2. \quad (4.5)$$

Therefore, the steady rotations of the body around the largest and the smallest axes of its ellipsoid of inertia are stable with respect to p,q,r. The auxiliary functions (4.4) and (4.5) are combinations of the integrals of energy and of moment of momenta. A general method for constructing such combinations will be presented in Chapter IV.

4.8. <u>Stability of a glider</u>. Consider a glider which might as well be a hovering bird, and suppose its plane of symmetry coincides at any moment with a vertical plane in an inertial frame of reference. Let v be the velocity of its center of inertia G and θ the angle between this velocity and a horizontal line. The longitudinal "axis" of the glider is assumed to make a constant angle α with v. Let m be the mass of the glider and g the acceleration of gravity. Using notations which are usual in aerodynamics, let $C_D(\alpha)$ and $C_L(\alpha)$ be the coefficients of drag and lift respectively. The equations for v and θ read:

$$m\dot{v} = -mg \sin\theta - C_D(\alpha)v^2,$$

$$mv\dot{\theta} = -mg \cos\theta + C_L(\alpha)v^2.$$

Putting $v_0^2 = mg/C_L$, $\tau = gt/v_0$, $y = v/v_0$ and $a = C_D/C_L$, we transform the equations into

$$\frac{dy}{d\tau} = -\sin\theta - ay^2,$$
$$\frac{d\theta}{d\tau} = (-\cos\theta + y^2)/y. \tag{4.6}$$

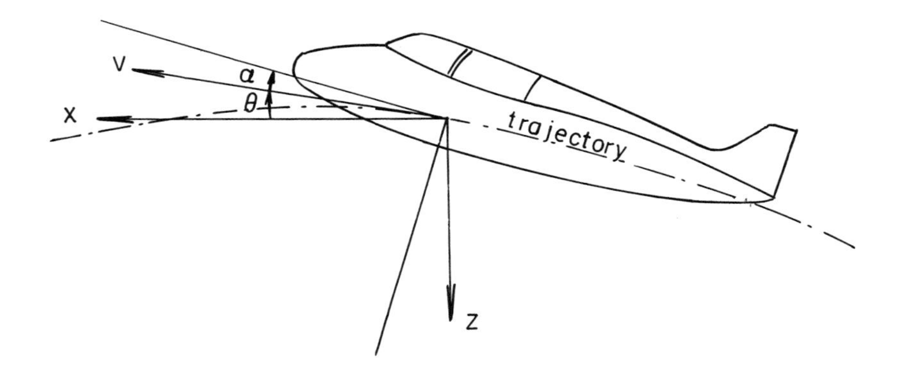

Figure 1.2. The glider

We have introduced a non-vanishing drag only for future reference (see Section 6.18). For more complete equations, including an equation of momentum, cf. N.G. Chetaev [1955].

If $a = 0$, the above equations have the critical points $y_0 = 1$, $\theta_0 = 2k\pi$ for k an integer, all of them corresponding to one and the same horizontal flight with constant velocity. Let us therefore concentrate on $y_0 = 1$, $\theta_0 = 0$. It is easily verified that the function

$$V(y,\theta) = \frac{y^3}{3} - y\cos\theta + \frac{2}{3}$$

is a first integral for the equations (4.6) with $a = 0$. Further, in some neighborhood of $(1,0)$, $V(y,\theta) > 0$ except that $V(1,0) = 0$. Therefore, if the critical point $(1,0)$ is shifted to the origin by a suitable translation of the axes, V expressed in the new variables satisfies the hypotheses of Theorem 4.3 (remember Section 3.4) and uniform stability is thus proved. For more details on this problem, see N.E. Zhukovski [1891], F.W. Lanchester [1908], B. Etkin [1959], and Section 6.18 below.

5. Instability

We go back to the general setting of Section 2.2 and prove now some sufficient conditions for instability.

5.1. Theorem (N.G. Chetaev [1934]). If there exist $t_0 \in I$, $\varepsilon > 0$ (with $\overline{B}_\varepsilon \subset \Omega$), an open set $\Psi \in B_\varepsilon$ and a $\mathscr{C}^1$ function $V\colon [t_0,\infty[\times B_\varepsilon \to \mathscr{R}$ such that, on $[t_0,\infty[\times \Psi$:

(i) $0 < V(t,x) \leq k < \infty$, for some k;

(ii) $\dot{V}(t,x) \geq a(V(t,x))$, for some $a \in \mathscr{K}$;

if further

(iii) the origin of the x-space belongs to $\partial\Psi$;

(iv) $V(t,x) = 0$ on $[t_0,\infty[\times (\partial\Psi \cap B_\varepsilon)$;

then, the origin is unstable.

Proof. In fact, due to (iii), there exists for every $\delta > 0$ an $x_0 \in \Psi \cap B_\delta$ such that $V(t_0,x_0) > 0$ (see (i)). Consider now $x(t;t_0,x_0)$, abbreviated as $x(t)$. As long as $x(t) \in \Psi$, one gets, owing to (i) and (ii), that for every $t \in J^+$:

$$k \geq V(t,x(t)) = V(t_0,x_0) + \int_{t_0}^{t} \dot{V}(\tau,x(\tau))d\tau$$
$$\geq V(t_0,x_0) + a(V(t_0,x_0))(t-t_0).$$

Therefore, $x(t)$ must leave Ψ after some time. But because of (iv), it cannot leave Ψ through $\partial\Psi \in B_\varepsilon$. Therefore, $x(t)$ leaves B_ε. Q.E.D.

Geometrically speaking, we have proved there exist a sequence $\{x_{0i}: x_{0i} \in \Psi, x_{0i} \to 0\}$ and a family of paths* γ_i connecting an initial point x_{0i} to some terminal point in $\partial B_\varepsilon, \gamma_i$ being included in Ψ except for its terminal point.

5.2. The two following corollaries of 5.1 were in fact established long before 5.1 was known.

5.3. <u>Corollary</u> (Liapunov's first theorem on instability (1892]). If there exist $t_0 \in I$, $\varepsilon > 0$ (with $\overline{B}_\varepsilon \subset \Omega$), an open set $\Psi \subset B_\varepsilon$ and $\mathscr{C}^1$ function $V\colon [t_0,\infty[\times B_\varepsilon \to \mathscr{R}$ such that, on $[t_0,\infty[\times \Psi$:

(i) $0 < V(t,x) \leq b(||x||)$, for some $b \in \mathscr{K}$;

(ii) $\dot{V}(t,x) \geq a(||x||)$, for some $a \in \mathscr{K}$;

if further:

(iii) the origin of the x-space belongs to $\partial\Psi$;

(iv) $V(t,x) = 0$ on $[t_0,\infty[\times (\partial\Psi \cap B_\varepsilon)$;

then, the origin is unstable.

5.4. <u>Corollary</u> (Liapunov's second theorem on instability [1892]). If, in Corollary 5.3, (ii) is replaced by

(ii) $\dot{V}(t,x) = cV(t,x) + W(t,x)$ on $[t_0,\infty[\times \Psi$,

---

*If f is a continuous function on an interval $[a,b]$ into $\mathscr{R}^n$, by definition $f([a,b])$ is a path.

where $c > 0$ and $W: [t_0,\infty[\times \Psi \to \mathcal{R}$ is continuous and ≥ 0, then the origin is unstable.

5.5. If the differential equation is autonomous and if V depends on x only, then (i) and (ii) in Chetaev's Theorem can be simplified to:

(i) $V(x) > 0$ on Ψ;

(ii) $\dot{V}(x) > 0$ on Ψ.

5.6. Instability of some steady rotations of a rigid body (cf. N.G. Chetaev [1955]). Let us consider the steady rotations of a rigid body around the intermediate axis of its ellipsoid of inertia (cf. 4.7). The equations are still (4.3). If $V(x,y,z) = yz$, one computes

$$\dot{V} = (p_0 + x)\left(\frac{C-A}{B} z^2 + \frac{A-B}{C} y^2\right).$$

The orientation of a principal axis of inertia is at our disposal. Therefore, we may choose $p_0 > 0$. Further, if two at least of the quantities A,B,C are unequal, we may decide that either $C > A \geq B$ or $C \geq A > B$. Then we choose

$$\Psi = \{(x,y,z): x^2 + y^2 + z^2 < \varepsilon^2, \ y > 0, \ z > 0\}$$

with ε small enough in order that $p_0 + x > 0$ on Ψ. All the hypotheses of Theorem 5.1 are verified and the origin is unstable: the steady rotations around the intermediate axis (or as the case may be, the equatorial axis) are unstable with respect to (p,q,r).

5.7. Instability proved by using the first approximation. The main use of Corollary 5.4 is to help proving instability by consideration of the linear approximation. This is a useful way of looking at many applications. Suppose Equation (2.2) is particularized as

$$\dot{x} = Ax + g(t,x) \tag{5.1}$$

where A is an $n \times n$ real matrix and $Ax + g(t,x)$ has all the properties required from $f(t,x)$ in Section 2.2. Then, the following theorem holds true.

5.8. Theorem (A.M. Liapunov [1892]). If at least one eigenvalue of A has strictly positive real part and if

$$\frac{||g(t,x)||}{||x||} \to 0 \quad \text{as} \quad x \to 0 \tag{5.2}$$

uniformly for $t \in I$, then the origin is unstable for (5.1).

Proof. The following lemma will be used and is stated here without proof (see, e.g. N. Rouche and J. Mawhin [1973]).

Lemma. If at least one eigenvalue of the $n \times n$ real matrix A has strictly positive real part, then to every positive definite quadratic form $U(x)$ for $x \in \mathscr{R}^n$, there corresponds a quadratic form $V(x)$ and a constant $c > 0$, such that

$$\left(\frac{\partial V}{\partial x}\middle| \, Ax\right) = cV + U \tag{5.3}$$

and $V(x) > 0$ for some x.

Now using (5.3), one computes

$$\dot{V} = cV + U + \left(\frac{\partial V}{\partial x}\middle| \, g(t,x)\right).$$

(5.2) shows that

$$W = U + \left(\frac{\partial V}{\partial x}\middle| \, g(t,x)\right)$$

is positive definite in some neighborhood of the origin.

Therefore, by Corollary 5.4, the origin is unstable for (5.1). Q.E.D.

5.9. <u>More about the steady rotations of a rigid body.</u> If we restrict ourselves to the case where $C > A > B$, Theorem 5.8 yields another proof of the result established in Section 5.6. The linear part of Equations (4.3) admits of a strictly positive real root equal to

$$\left[p_0^2 \frac{(C-A)(A-B)}{BC}\right]^{1/2}$$

and the origin is unstable. Notice that Theorem 5.8, as contrasted with Theorem 5.1, yields no instability result when either $C > A = B$ or $C = A > B$, for in such cases, all three relevant eigenvalues vanish.

5.10. <u>Watt's governor.</u> As Watt's governor is a well-known device, it will suffice to present it as in Figure 1.3 to define our principal symbols.

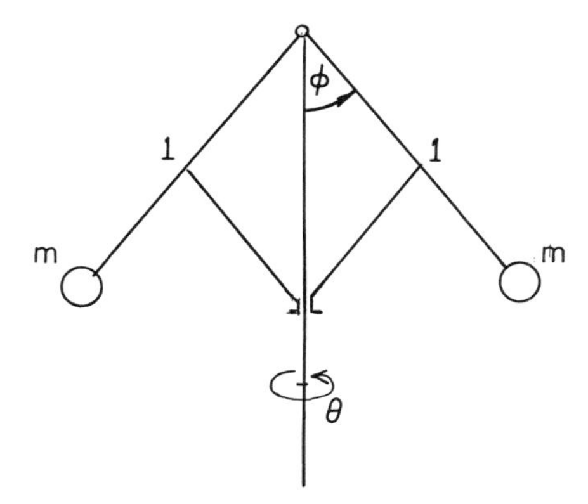

Figure 1.3. Watt's governor

The angles ϕ and θ correspond to the two degrees of freedom. If friction is disregarded, the equations (see e.g. T. Levi-Civita and U. Amaldi [1922-27], Vol. 2, Part 1) read

$$\frac{d}{dt}(\mathcal{J}\dot{\theta}) = -k(\phi-\phi_0),$$
$$\frac{d}{dt}(\mathcal{J}'\dot{\phi}) - \frac{1}{2}\dot{\theta}^2\frac{\partial\mathcal{J}}{\partial\phi} = \frac{\partial U}{\partial\phi}$$

where $\mathcal{J} = C + 2\,ml^2\sin^2\phi$, $\mathcal{J}' = 2\,ml^2$, $U = 2\,mgl\cos\phi$ and C, l, m, g and k are positive constants.

The steady motion $\phi = \phi_0$, $\dot{\phi} = 0$, $\dot{\theta}^2 = \dot{\theta}_0^2 = g/l\cos\phi_0$ is unstable, as can be established using Theorem 5.8. In fact, the eigenvalue equation for the linear part of the equations is

$$\left(\frac{C}{2\,ml^2} + \sin^2\phi_0\right)\lambda^3 + \dot{\theta}_0^2\sin^2\phi_0\left(1 + 3\cos^2\phi_0 + \frac{C}{2\,ml^2}\right)\lambda$$
$$+ \frac{k}{2\,ml^2}\dot{\theta}_0\sin 2\,\phi_0 = 0. \tag{5.4}$$

When $\dot{\theta}_0 < 0$, then $\Delta(0) < 0$, whereas $\Delta(\lambda) \to \infty$ when $\lambda \to \infty$: there is a strictly positive eigenvalue. Analogously, when $\dot{\theta}_0 > 0$, there is a strictly negative eigenvalue. But $\lambda_1 + \lambda_2 + \lambda_3 = 0$ and thus at least one eigenvalue has a strictly positive real part. We conclude therefore that the steady motion being considered is unstable. For more details on Watt's governor and the use of friction to stabilize its steady motions, we refer to L.S. Pontryagin [1961].

5.11. A wealth of further illustrations of Chetaev's Theorem 5.1 will appear in Chapter III.

6. Asymptotic Stability

6.1. In this section, we still consider equation (2.2) and the corresponding hypotheses. Uniform asymptotic stability has been studied long before (simple) asymptotic stability. The first theorem which was proved corresponds to thesis (b) of the following statement and is due to A.M. Liapunov [1892]. Thesis (a) gives an interesting estimate of the region of attraction $A(t_0)$.

6.2. Theorem. Suppose there exists a $\mathscr{C}^1$ function $V: I \times \Omega \to \mathscr{R}$ such that, for some functions $a, b, c \in \mathscr{K}$ and every $(t,x) \in I \times \Omega$:

(i) $a(||x||) \leq V(t,x) \leq b(||x||)$;

(ii) $\dot{V}(t,x) \leq -c(||x||)$

Choosing $\alpha > 0$ such that $\overline{B}_\alpha \subset \Omega$, let us put for every $t \in I$

$$V^{-1}_{t,\alpha} = \{x \in \Omega\colon V(t,x) \leq a(\alpha)\}.$$

Then

(a) for any $t_0 \in I$ and any $x_0 \in V^{-1}_{t_0,\alpha}$: $x(t;t_0,x_0) \to 0$ uniformly in t_0,x_0 when $t \to \infty$;

(b) the origin is uniformly asymptotically stable.

Proof. (a) We choose an $\alpha > 0$ such that $\overline{B}_\alpha \subset \Omega$ and deduce from (i) that for every $t \in I$

$$V^{-1}_{t,\alpha} \subset \overline{B}_\alpha \subset \Omega. \tag{6.1}$$

For any $t_0 \in I$ and $x_0 \in V^{-1}_{t_0,\alpha}$, it follows from (ii) that

$$x(t) \in V^{-1}_{t,\alpha}$$

for any $t \in J^+$, and therefore from (6.1) that $x(t)$ cannot approach the boundary of Ω. Hence $J^+(t_0,x_0) = [t_0,\infty[$.

For any $\varepsilon > 0$, let us choose $\eta > 0$ such that $b(\eta) < a(\varepsilon)$ and also choose a σ larger than $b(\alpha)/c(\eta)$. Now $||x(t)||$ cannot be larger than η for every $t \in [t_0,t_0+\sigma]$, for if this were the case, one should obtain for $t = t_0 + \sigma$:

$$V(t,x(t)) \leq V(t_0,x_0) - \int_{t_0}^{t} c(||x(s)||)ds \leq b(\alpha) - c(\eta)\sigma < 0,$$

which contradicts (i). Therefore, there exists a $t_1 \in [t_0,t_0+\sigma]$ such that $b(||x(t_1)||) \leq b(\eta) < a(\varepsilon)$, and since V is decreasing, one obtains for $t \geq t_0 + \sigma$

$$a(||x(t)||) \leq V(t,x(t)) \leq V(t_1,x(t_1)) < b(||x(t_1)||) < a(\varepsilon).$$

Therefore, for $t \geq t_0 + \sigma$: $||x(t)|| < \varepsilon$ and part (a) of the thesis is proved.

(b) Uniform stability of the origin follows from Theorem 4.3. Further, for any $\delta > 0$ such that $b(\delta) \leq a(\alpha)$:

$$B_\delta \subset V^{-1}_{t_0,\alpha}.$$

This shows that the attractivity is uniform. Q.E.D.

Several remarks on this theorem can prove helpful.

6.3. It might be important in practical cases to obtain an upper estimate of the time needed by the solutions to reach a given ε. It follows from the above proof that for $x_0 \in V^{-1}_{t_0,\alpha}$, one can choose any number larger than $\sigma = b(\alpha)/c(\eta)$, η being known as soon as ε is.

6.4. Assumptions (i) and (ii) are equivalent to (i) and

(ii)' $\dot{V}(t,x) \leq -c'(V(t,x))$ for some $c' \in \mathscr{K}$.

6.5. The existence of a $\mathscr{C}^1$ function $V(t,x)$ such that, for some a and $c \in \mathscr{K}$ and every $(t,x) \in I \times \Omega$:

(i) $V(t,x) \geq a(||x||)$; $V(t,0) = 0$;

(ii) $\dot{V}(t,x) \leq -c(||x||)$;

does not imply uniform asymptotic stability, nor even asymptotic stability! This is shown by the following counter-example, borrowed from J.L. Massera [1949].

Let $g: [0,\infty[\to \mathscr{R}$ be a $\mathscr{C}^1$ function coinciding with e^{-t} except at some peaks where it reaches the value 1. Figure 1.4. is a diagram of $g^2(t)$. There is one peak for each integer value of t. The width of the peak corresponding to abcissa n is supposed to be smaller than $(1/2)^n$.

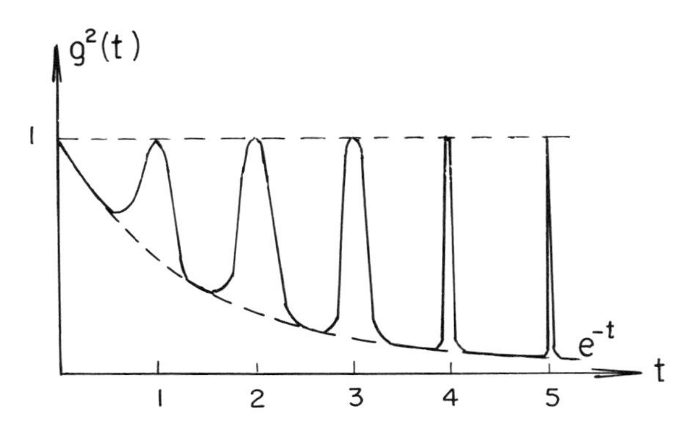

Figure 1.4. The function $g^2(t)$

Consider now the following differential equation

$$\dot{x} = \frac{\dot{g}(t)}{g(t)} x.$$

The form of its general solution, namely

$$x(t) = \frac{g(t)}{g(t_0)} x_0,$$

shows obviously that the origin is not asymptotically stable for this equation. However, if we choose the auxiliary function

$$V(t,x) = \frac{x^2}{g^2(t)} [3 - \int_0^t g^2(\tau)d\tau],$$

we see first that $V(t,x) \geq x^2$, since

$$\int_0^\infty g^2(\tau)d\tau < \int_0^\infty e^{-\tau}d\tau + \sum_{1 \leq n < \infty} (\tfrac{1}{2})^n = 2,$$

and secondly that $\dot{V}(t,x) = -x^2$, as is easily computed.

6.6. Thus, we have established that hypotheses (i) and (ii) of Section 6.5 do not imply asymptotic stability. Are there nevertheless any conclusions which can be drawn out of them? To answer this question, let us introduce a new definition: the solution $x = 0$ of (2.2) is called <u>weakly attractive</u> if $(\forall t_0 \in I)(\exists \eta > 0)(\forall x_0 \in B_\eta)(\exists \{t_i\} \subset J^+, t_i \to \omega$ as $i \to \infty)$ $x(t_i) \to 0$ as $i \to \infty$.

6.7. <u>Exercise</u>. The hypotheses of Section 6.5, imply that the origin is weakly attractive.

6.8. <u>Exercise</u>. By the way, the following property will be useful in the sequel: weak attractivity along with uniform stability imply attractivity.

6.9. Corollary to Theorem 6.2. (E.A. Barbashin and N.N. Krasovski [1952]). The origin is uniformly globally asymptotically stable if the assumptions of Theorem 6.2 are satisfied for $\Omega = \mathscr{R}^n$ and

$$a(r) \to \infty \quad \text{as} \quad r \to \infty. \tag{6.2}$$

6.10. Exercise. Corollary 6.9 remains true if (6.2) is replaced by

$$\lim_{r\to\infty} a(r) = \lim_{r\to\infty} b(r).$$

6.11. An RLC circuit with parametric excitation. To illustrate Theorem 6.2, let us consider the scalar equation

$$\ddot{x} + a\dot{x} + b(t)x = 0 \tag{6.3}$$

where $a > 0$, $b = b_0(1 + \varepsilon f(t))$ with $b_0 \geq 0$, and $f(t)$ is a bounded function from $\mathscr{R} \to \mathscr{R}$. This equation can be interpreted as representing an RLC circuit with time-varying capacitance, or a mechanical oscillator with viscous friction and a time-varying spring parameter. Equation (6.3) is equivalent to

$$\dot{x} = y,$$
$$\dot{y} = -ay - b(t)x.$$

The auxiliary function

$$V(x,y) = \frac{1}{2}\left(y + \frac{ax}{2}\right)^2 + \left(\frac{a^2}{4} + b_0\right)\frac{x^2}{2}$$

is positive definite. According to Theorem 6.2, the origin will be uniformly asymptotically stable if the time derivative

$$\dot{V}(t,x,y) = -\frac{a}{2}y^2 - (b-b_0)xy - \frac{ab}{2}x^2$$

is negative definite. This happens, following Sylvester's criterion, if, for some α

$$\varepsilon^2 b_0 f(t)^2 - a^2(1+\varepsilon f(t)) \leq -\alpha < 0. \tag{6.4}$$

This condition is satisfied for any small enough ε, a result which can be interpreted as follows in the language of electrical engineers. There are two opposing forces at work: a parametric excitation proportional to ε and a load, the damping force $a\dot{x}$. Satisfying (6.4) amounts to choosing the resistance, designated by a, large enough for the load to absorb all the energy provided by the excitation. In this case, the origin is asymptotically stable. In the opposite case, i.e. if the load is not large enough, one may expect the energy balance of the system to increase and the origin to become unstable. Of course, this is but a heuristic view of the problem.

6.12. A damped pendulum. Consider the damped pendulum described by the scalar equation

$$\ddot{x} + \dot{x} + \sin x = 0,$$

for which, as everyone knows, the origin is asymptotically stable. To check this by Liapunov's method, one might think it natural, at first sight, to choose the total energy

$$V_1 = \frac{\dot{x}^2}{2} + (1 - \cos x)$$

as an auxiliary function. This is not a good choice however, for the time derivative $\dot{V}_1 = -\dot{x}^2$ is not negative definite: it proves stability, not asymptotic stability. Therefore, a "natural" choice may not always fit Theorem 6.2. Finding a suitable auxiliary function is often a matter of habit and feeling. For instance

$$V_2 = \dot{x}^2 + (\dot{x}+x)^2 + 4(1 - \cos x)$$

is such that

$$\dot{V}_2 = -2(\dot{x}^2 + x \sin x)$$

and can be used to prove uniform asymptotic stability of the origin. It could hardly be said however, that V_2 has any physical interpretation, not that it is "natural" in any sense!

As it is often very difficult to exhibit an auxiliary function whose time derivative is negative definite, an alternative way of proving asymptotic stability will be to work out some more elaborate theorems, allowing one to use functions like V_1 whose derivative is only ≤ 0, but of course along with some more information. Much effort will be made in this direction in subsequent sections.

6.13. Asymptotic stability proved by using the first approximation. Theorem 6.2 can be used to prove asymptotic stability by consideration of the linear approximation. Suppose Equation (2.2) is particularized as $\dot{x} = Ax + g(t,x)$, where A is an $n \times n$ real matrix and $Ax + g(t,x)$ has all the properties required from $f(t,x)$ in Section 2.2. Then the following theorem holds true.

6.14. Theorem (A.M. Liapunov [1892]). If all eigenvalues of A have strictly negative real parts and if

$$\frac{||g(t,x)||}{||x||} \to 0 \quad \text{as} \quad x \to 0,$$

uniformly for $t \in I$, then the origin is uniformly asymptotically stable.

The proof consists in exhibiting a quadratic form in x, which, considered as an auxiliary function, satisfies all the hypotheses of Theorem 6.2. This quadratic form is built up using the following lemma.

6.15. Lemma (A.M. Liapunov [1892]). If all eigenvalues of the $n \times n$ real matrix A have strictly negative real parts, then to every negative definite quadratic form $U(x)$ for $x \in \mathscr{R}^n$, there corresponds one and only one quadratic form $V(x)$ which is positive definite and such that

$$\left(\frac{\partial V}{\partial x}, Ax\right) = U.$$

(See e.g. N. Rouche and J. Mawhin [1973]).

Theorem 6.14 has no immediate extension to the case where the matrix A is a continuous function of t. There exist counter-examples proving that the origin can be unstable for a linear equation $\dot{x} = A(t)x$ even though, for any t, the real part of every eigenvalue of $A(t)$ is strictly negative. Cf. Exercise V.9.1.

6.16. Exercise. Prove Theorem 6.14.

6.17. Exercise. Let $A(t)$ be an $n \times n$ real matrix, continuous function of $t \in I$ and periodic of period $T > 0$. Suppose the second member of the differential equation $\dot{x} = A(t)x + g(t,x)$ has, as above, all the properties required from $f(t,x)$ in Section 2.2. If all the characteristic exponents of $\dot{x} = A(t)x$ have strictly negative real parts, and if

$$\frac{||g(t,x)||}{||x||} \to 0 \quad \text{as} \quad x \to 0$$

uniformly for $t \in I$, then the origin is uniformly asymptotically stable. The latter condition, along with the existence of a characteristic exponent with strictly positive real part, implies instability of the origin. <u>Hint</u>: use the existence of a differentiable, regular, periodic matrix $S(t)$ such that $S(t)^{-1}(A(t)S(t) - \dot{S}(t))$ is constant. Concerning linear periodic differential equations and characteristic exponents, see e.g. N. Rouche and J. Mawhin [1973].

6.18. <u>Asymptotic stability of a glider</u>. Let us come back to the problem of the glider, already presented in Section 4.8. We consider here the case of a motion with non-vanishing drag, i.e. described by the Equations (4.6). These equations admit of the critical point

$$y_0 = (1+a^2)^{-1/4}, \quad \theta_0 = -\tan^{-1}a,$$

corresponding to a rectilinear down motion at constant velocity. Without loss of generality, we assume that $0 > \theta_0 > -\pi/2$. Transferring the origin to the critical point by the change of variables $y = y_0 + y_1$, $\theta = \theta_0 + \theta_1$ and computing the terms of first order in y_1, θ_1, we obtain for the linear variational equation

$$\dot{y}_1 = -\frac{2a}{(1+a^2)^{1/4}} y_1 - \frac{1}{(1+a^2)^{1/2}} \theta_1,$$

$$\dot{\theta}_1 = 2y_1 - \frac{a}{(1+a^2)^{1/4}} \theta_1.$$

It is readily verified that the eigenvalues of the second

member have strictly negative real parts. Therefore, according to 6.14, the critical point is asymptotically stable.

6.19. Theorems like 6.14 and 6.17 are very useful. They have a drawback, however, as compared to other theorems using an auxiliary function: they do not yield any estimate of the region of attraction.

6.20. _Exercise_ (J.L. Massera [1949], see also H.A. Antosiewicz [1958]). If we replace in Theorem 6.2, Hypotheses (ii) by

(ii)' there exist a function $U: I \times \Omega \to \mathscr{R}$ and a function $c \in \mathscr{K}$ such that

$$U(t,x) \geq c(||x||), \quad U(t,0) = 0,$$

and for any ρ_1, ρ_2 with $0 < \rho_1 < \rho_2$

$$\dot{V}(t,x) + U(t,x) \to 0 \quad \text{as} \quad t \to \infty.$$

uniformly on $\rho_1 \leq ||x|| \leq \rho_2$, then the origin is equi-asymptotically stable.

6.21. _Exercise_ (H.A. Antosiewicz [1958]). If the origin is uniformly stable and if there exists a $\mathscr{C}^1$ function $V: I \times \Omega \to \mathscr{R}$ such that, for some functions $a, c \in \mathscr{K}$ and every $(t,x) \in I \times \Omega$:

(i) $V(t,x) \geq a(||x||)$; $V(t,0) = 0$;

(ii) $\dot{V}(t,x) \leq -c(||x||)$;

then the origin is equi-asymptotically stable.

6.22. Let us now introduce for the first time in this book (but it will occur very often in the sequel!) a theorem which makes use of two auxiliary functions.

6.23. Theorem (L. Salvadori [1972]). Suppose there exist two $\mathscr{C}^1$ functions $V: I \times \Omega \to \mathscr{R}$ and $W: I \times \Omega \to \mathscr{R}$ such that, for some functions $a, b, c \in \mathscr{K}$ and every $(t,x) \in I \times \Omega$:

(i) $V(t,x) \geq a(||x||)$; $V(t,0) = 0$;

(ii) $W(t,x) \geq b(||x||)$; $W(t,0) = 0$;

(iii) $\dot{V}(t,x) \leq -c(W(t,x))$;

(iv) $\dot{W}(t,x)$ is bounded from below or from above.

Choosing $\alpha > 0$ such that $\overline{B}_\alpha \subset \Omega$, we put for any $t \in I$

$$V_{t,\alpha}^{-1} = \{x \in \Omega: V(t,x) \leq a(\alpha)\}.$$

Then

(a) the region of attraction $A(t_0) \supset V_{t,\alpha}^{-1}$;

(b) the origin is asymptotically stable.

Proof. (a) We choose an $\alpha > 0$ such that $\overline{B}_\alpha \subset \Omega$ and deduce, as in Theorem 6.2, that for any $t_0 \in I$ and $x_0 \in V_{t_0,\alpha}^{-1}$: $J^+(t_0,x_0) = [t_0,\infty[$. Let us now prove that $W(t,x(t)) \to 0$ as $t \to \infty$. If this were not the case, two mutually exclusive behaviors would be possible for W, and we shall rule them out one after the other.

First there might exist a $\sigma > 0$ and a $k > 0$ such that, for every $t \geq t_0 + \sigma$: $W(t,x(t)) \geq k > 0$. But then it would follow from (iii) that $\dot{V}(t,x(t)) \leq -c(k)$ and therefore, that $V(t,x(t)) \to -\infty$ as $t \to \infty$, which contradicts (i).

Secondly, there might exist two increasing sequences $\{t_i\},\{t_i'\}$ such that $t_i \to \infty$ as $i \to \infty$ and for every

$i = 1,2,\ldots$: $t_i < t_i' < t_{i+1}$, and further such that, for some $k > 0$:

$$W(t_i,x(t_i)) = k/2, \quad W(t_i',x(t_i')) = k$$
$$k/2 < W(t,x(t)) < k \quad \text{for every} \quad t \in]t_i,t_i'[. \tag{6.5}$$

Of course, one might as well have written

$$W(t_i,x(t_i)) = k, \; W(t_i',x(t_i')) = k/2. \tag{6.6}$$

According to whether $\dot{W}$ is bounded from above or from below, we use (6.5) or (6.6) in much the same way for both cases. So let us suppose there is an $M > 0$ such that $\dot{W}(t,x) \leq M$. It is clear from (6.5) that $t_i' - t_i \geq k/2\,M$. Using (iii), we get

$$\begin{aligned} V(t_n',x(t_n')) &\leq V(t_0,x(t_0)) + \sum_{1\leq i\leq n} \int_{t_i}^{t_i'} \dot{V}(s,x(s))ds \\ &\leq V(t_0,x(t_0)) - n\, c\left(\frac{k}{2}\right)\frac{k}{2\,M} . \end{aligned}$$

The last member becomes negative for n large enough, and this again contradicts (i). The reasoning would be similar for $\dot{W}$ bounded from below. Thus, $W(t,x(t)) \to 0$ as $t \to \infty$ and (ii) shows that $x(t) \to 0$ as $t \to \infty$.

(b) Part (b) of the thesis is immediate: stability derives from Theorem 4.2; attractivity results from the fact that, for every t_0: $V_{t_0,\alpha}^{-1}$ is a neighborhood of the origin. Q.E.D.

6.24. The function W in Theorem 6.23 can be particularized in several ways. For instance, we obtain Corollary 6.25 by identifying $W(t,x)$ with $(x|x)$ and Corollary 6.26 by putting $W(t,x) = V(t,x)$. By the way, these two corollaries were known long before Theorem 6.23.

6.25. <u>Corollary</u> (M. Marachkov [1940]). Suppose there exists a $\mathscr{C}^1$ function $V: I \times \Omega \to \mathscr{R}$ such that, for some functions $a, c \in \mathscr{K}$ and every $(t,x) \in I \times \Omega$:

(i) $V(t,x) \geq a(||x||)$; $V(t,0) = 0$;

(ii) $\dot{V}(t,x) \leq -c((x|x))$.

If, moreover, $f(t,x)$ is bounded on $I \times \Omega$, then

(a) for every α such that $\overline{B}_\alpha \subset \Omega$, the region of attraction $A(t_0) \supset V^{-1}_{t_0,\alpha}$;

(b) the origin is asymptotically stable.

There is no restriction of generality to suppose that Ω is a bounded set. Then the bound on $f(t,x)$ enables one to prove that the time derivative of $(x|x)$ along the solutions is bounded.

6.26. <u>Corollary</u> (J.L. Massera [1956]). Suppose there exists a $\mathscr{C}^1$ function $V: I \times \Omega \to \mathscr{R}$ such that, for some functions $a, c \in \mathscr{K}$ and every $(t,x) \in I \times \Omega$:

(i) $V(t,x) \geq a(||x||)$; $V(t,0) = 0$;

(ii) $\dot{V}(t,x) \leq -c(V(t,x))$;

then

(a) for every $\alpha > 0$ such that $\overline{B}_\alpha \subset \Omega$, the region of attraction $A(t_0) \supset V^{-1}_{t_0,\alpha}$;

(b) the origin is asymptotically stable.

One may even prove that the origin is equi-asymptotically stable.

6.27. <u>A pendulum with time-varying friction</u>. Let us generalize Example 6.12 and consider the pendulum with variable friction described by the equation

$$\ddot{x} + h(t)\dot{x} + \sin x = 0, \tag{6.7}$$

where h is a $\mathscr{C}^1$ function from $I \to \mathscr{R}$. We are looking for some hypotheses, as mild as possible, concerning $h(t)$ and which will entail asymptotic stability of the origin of the phase plane. Let us try, as an auxiliary function

$$V(t,x,\dot{x}) = \frac{(\dot{x} + a \sin x)^2}{2} + b(t)(1 - \cos x),$$

which is the sum of a quadratic function

$$\frac{(\dot{x} + ax)^2}{2} + b(t)\,\frac{x^2}{2}$$

and of terms of order at least 4 in $x,\dot{x}$. If we put $b(t) = 1 + ah(t) - a^2$, we obtain

$$\begin{aligned}\dot{V}(t,x,\dot{x}) &= -(h(t) - a \cos x)\dot{x}^2 - a \sin^2 x + ah'(t)(1 - \cos x)\\ &\qquad - a^2(1 - \cos x)\dot{x} \sin x\\ &= -(h(t) - a)\dot{x}^2 - a(2-h'(t))(1 - \cos x) + \mathscr{O}_3\end{aligned}$$

where $\mathscr{O}_3$ contains only terms of order at least 3 in $x,\dot{x}$, all independent of t.

Now, if there exist two constants $\alpha > a$ and $\beta < 2$ such that

(i) $h(t) \geq \alpha > a > 0$;

(ii) $h'(t) \leq \beta < 2$;

then V and $-\dot{V}$ are positive definite, as can be readily established. Further, let us define

$$W(x,\dot{x}) = \frac{\dot{x}^2}{2} + (1 - \cos x)$$

the derivative of which, namely

$$\dot{W}(t,x,\dot{x}) = -h(t)\dot{x}^2 \leq 0,$$

is bounded from above. Asymptotic stability of the origin follows then from Theorem 6.23.

Notice that Corollary 6.25 could not be used to get this result, because of the requirement that $f(t,x)$ be bounded and therefore that a bound should have been assumed for $h(t)$. Corollary 6.26 could not be used either, at least using the function $V(t,x,\dot{x})$ chosen above, because satisfying hypothesis (ii) of this corollary would also require further hypotheses on $h(t)$. In fact, consider a point of Ω where $\dot{x} = 0$ and a $h(t)$ approaching ∞ while $h'(t)$ is bounded from below: hypothesis (ii) of Corollary 6.26 would be violated.

Notice further that, although condition (ii) of this section may appear somewhat odd, for instance from a physicist's point of view, it remains that (i) alone would not entail asymptotic stability, as we shall prove presently by way of a counter-example. Consider first (after J.K. Hale [1969]) the similar problem

$$\ddot{y} + (2+e^t)\dot{y} + y = 0 \qquad (6.8)$$

which, for any a, admits the solution $y = a(1+e^{-t})$. The orbit of such a solution is contained in the line of equation $y + \dot{y} = a$. Of course, the origin of the phase plane is not asymptotically stable.

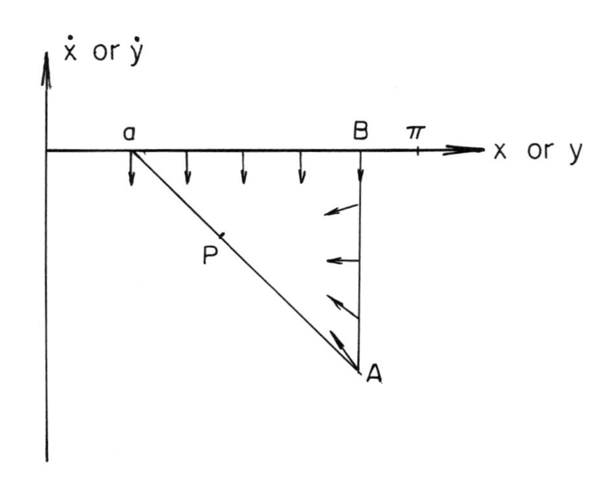

Figure 1.5.
The phase plane for equations (6.7) and (6.8)

Let us now come back to Equation (6.7) with $h = 2 + e^t$ and let us, for some a, draw the line $x + \dot{x} = a$ in its phase plane, as shown in Figure 1.5. The phase velocity, i.e., the velocity of the point representing a solution in the $(x,\dot{x})$-plane, is $(\dot{x},\ddot{x})$ or, using (6.7)

$$(\dot{x},\ -(2+e^t)\dot{x} - \sin x). \tag{6.9}$$

The analogous velocity for (6.8) would be

$$(\dot{y},\ -(2+e^t)\dot{y} - y). \tag{6.10}$$

Consider now a triangle such as aAB, with AB parallel to the $\dot{x}$ axis. It is readily seen that at any point of the side AB and for any t, the phase velocity (6.9) enters the triangle: this is because $\dot{x} < 0$. Similarly, for any t and any point of the side aB, this velocity enters the triangle if B has been chosen at the left of the point of abcissa π. Consider now two solutions, one $x(t)$ of (6.7), the other $y(t)$ of (6.8), both starting from A at some time t_0. Comparing (6.9) and (6.10) shows that the solution of (6.7) enters the triangle at A. But this solution cannot come out of the triangle, neither by crossing AB, nor by crossing aB. Suppose now that having left A, it meets aA for the first time at time τ. Let P be the corresponding point of aA. Let τ' be the time at which $y(t)$ reaches P. Observe that $\tau' < \tau$, since

$$\tau - t_0 = \int_{\Gamma_x} \frac{dx}{\dot{x}} > \int_{\Gamma_y} \frac{dy}{\dot{y}} = \tau' - t_0$$

where Γ_x is the orbit of $x(t)$ from A to P while Γ_y is the segment AP which is also the orbit of $y(t)$ from A to P. Therefore, we may write

$$\ddot{x}(\tau) = -(2+e^{\tau})\dot{x}(\tau) - \sin x(\tau) > -(2+e^{\tau'})\dot{y}(\tau') - y(\tau') = \ddot{y}(\tau').$$

We conclude that at point P, $x(t)$ would enter the triangle with a nonvanishing velocity, which is absurd.

We sum up our results: for $h(t) = 2 + e^t$ (a function satisfying condition (i)), for any a and any point on $x + \dot{x} = a$, we have found a solution of (6.7) remaining forever in the triangle aBA. Thus, condition (i) alone cannot ensure asymptotic stability of the origin for (6.7).

6.28. Exercise (L. Salvadori [1972]). Suppose there exist two $\mathscr{C}^1$ functions $V: I \times \Omega \to \mathscr{R}$ and $W: I \times \Omega \to \mathscr{R}$ such that, for some $a \in \mathscr{R}$, some functions $b, c \in \mathscr{K}$ and every $(t,x) \in I \times \Omega$:

(i) $V(t,x) \geq a$;

(ii) $W(t,x) \geq b(||x||)$; $W(t,0) = 0$;

(iii) $\dot{V}(t,x) \leq -c(W(t,x))$;

(iv) $\dot{W}(t,x) \leq 0$;

then the origin is equi-asymptotically stable for equation (2.2).

The two following exercises yield improvements of Corollaries 6.25 and 6.26. In the first one (of Marachkov), the hypotheses can be weakened while proving the same thesis. In the second one (of Massera), the thesis can be made stronger without modification of the hypotheses.

6.29. Exercise. Use Exercise 6.28 to prove that the hypotheses of Corollary 6.26 imply equi-asymptotic stability of the origin.

6.30. Exercise (J.R. Haddock [1972], A.S. Oziraner [1972]). Show that in Corollary 6.25, Hypothesis (i) can be replaced by: for every $(t,x) \in I \times \Omega$, $V(t,0) = 0$ and $V(t,x) \geq 0$.

6.31. Exercise (N.G. Chetaev [1955]). Suppose there exist a $\mathscr{C}^1$ function $V\colon I \times \Omega \to \mathscr{R}$ and a continuous function $k\colon I \to \mathscr{R}$ such that for some function $a \in \mathscr{K}$ and every $(t,x) \in I \times \Omega$:

(i) $V(t,x) \geq k(t)a(||x||)$; $V(t,0) = 0$;

(ii) $\dot{V}(t,x) \leq 0$;

(iii) $k(t) > 0$; $k(t) \to \infty$ as $t \to \infty$;

then

(a) the region of attraction $A(t_0) \supset V^{-1}_{t_0,\alpha}$;

(b) the origin is equi-asymptotically stable.

6.32. <u>Partial asymptotic stability</u>. Let us get back to the Equations (4.1) along with the accompanying hypotheses as presented in Section 4.4. Here we suppose further that all solutions of (4.1) exist on $[t_0,\infty[$. The solution $z = 0$ of (4.1) is said to be <u>uniformly asymptotically stable with respect to</u> x, if it is uniformly stable with respect to x and if $(\exists\eta > 0)(\forall\varepsilon > 0)(\exists\sigma > 0)(\forall z_0: ||z_0|| < \eta)$ $(\forall t_0 \in I)(\forall t \geq t_0 + \sigma)$ $||x(t;t_0,z_0)|| < \varepsilon$. <u>Asymptotic stability with respect to</u> x is defined similarly (V.V. Rumiantsev [1957]).

6.33. <u>Theorem</u>. Suppose there exists a $\mathscr{C}^1$ function $V: I \times \Omega \times \mathscr{R}^m \to \mathscr{R}$ such that, for some functions a, b, $c \in \mathscr{K}$ and every $(t,z) \in I \times \Omega \times \mathscr{R}^m$:

(i) $a(||x||) \leq V(t,z) \leq b(||x||)$;

(ii) $\dot{V}(t,z) \leq -c(||x||)$.

Then

(a) for any $\alpha > 0$ and any $(t_0,z_0) \in I \times [(B_\alpha \cap \Omega) \times \mathscr{R}^m]$, $x(t;t_0,z_0) \to 0$ uniformly in t_0, z_0 when $t \to \infty$;

(b) the origin is uniformly asymptotically stable with respect to x.

The proof is along the same lines as for Theorem 6.2. The theorem corresponding to thesis (b) appears in N. Rouche and K. Peiffer [1967]. See also V.V. Rumiantsev [1957] and [1971]. From this last author [1970], we borrow the following statement.

6.34. Exercise. Define $u \in \mathscr{R}^{n+k}$, $0 \leq k \leq m$, as a vector containing all components of x and k components of y. Suppose there exists a $\mathscr{C}^1$ function $V\colon I \times \Omega \times \mathscr{R}^m \to \mathscr{R}$ such that, for some functions $a,b,c \in \mathscr{K}$ and every $(t,z) \in I \times \Omega \times \mathscr{R}^m$:

(i) $a(||x||) \leq V(t,z) \leq b(||u||)$;

(ii) $\dot{V}(t,z) \leq -c(||u||)$;

then the origin is uniformly asymptotically stable with respect to x.

6.35. Exercise (A. Halanay [1963]). Suppose there exists a $\mathscr{C}^1$ function $V\colon I \times \Omega \times \mathscr{R}^m \to \mathscr{R}$ such that, for some functions a, $c \in \mathscr{K}$ and every $(t,z) \in I \times \Omega \times \mathscr{R}^m$:

(i) $V(t,z) \geq a(||x||)$; $V(t,0) = 0$;

(ii) $\dot{V}(t,z) \leq -c(V(t,z))$;

then the origin is asymptotically stable with respect to x.

7. Converse Theorems

7.1. In most theorems given so far, the existence of a Liapunov-like function is assumed. The question arises whether such a function actually exists: given some stability or attractivity property of the origin, can we build up an auxiliary function suiting the corresponding theorem? The answer is contained in the so-called converse theorems. Three comments are in order here:

(1) First, most converse theorems are proved by actually constructing a suitable auxiliary function. Unfortunately, this construction assumes almost always a knowledge of the solutions of the differential equation. This is why converse theorems give usually no clue to the practical search for Liapunov-like functions. In this respect, all they show is that this search, however difficult, is not hopeless, and this is a little more than nothing.

(2) When the existence of such and such a function is necessary and sufficient for some stability property, one knows of course that any other sufficient condition will imply the first one. But this observation should not prevent the mathematician from looking for other sufficient conditions which might happen to be more practical, easier to apply. Many theorems in the rest of this book, and in particular those utilizing several auxiliary functions, present this type of interest.

(3) Sometimes a stability property of a system can be investigated by considering first a simplified system. Let us suppose that the stability of the latter can be easily established. Then we can deduce from a converse theorem the existence of a suitable auxiliary function. Under appropriate assumptions, it might also be a good auxiliary function for the original system and prove the desired stability property. This way of thinking is typical of total stability (see Section II.4) and of the proofs of stability using the first approximation, but is used in several other circumstances (cf. F.C. Hoppensteadt [1966]).

Hereafter, we state without proof the converses of

three important theorems given above. The general setting is again as in Section 2.2. For proofs and further results, the reader is referred to I.G. Malkin [1952], N.N. Krasovski [1959], T. Yoshizawa [1966] and W. Hahn [1967].

7.2. The converses of Theorems 4.2 and 4.3 have been obtained by K.P. Persidski [1933] and J. Kurzweil [1955] respectively.

Theorem. If the function f is $\mathscr{C}^1$ and if the origin is stable, there exist a neighborhood $\Psi \subset \Omega$ of the origin and a $\mathscr{C}^1$ function $V: I \times \Psi \to \mathscr{R}$ such that, for some $a \in \mathscr{K}$ and every $(t,x) \in I \times \Psi$:

(a) $V(t,x) \geq a(||x||)$, $V(t,0) = 0$;

(b) $\dot{V}(t,x) \leq 0$.

7.3. Theorem. If the function f is $\mathscr{C}^1$ and if the origin is uniformly stable, there exist a neighborhood $\Psi \subset \Omega$ of the origin and a $\mathscr{C}^1$ function $V: I \times \Psi \to \mathscr{R}$ such that, for some $a \in \mathscr{K}$ and $b \in \mathscr{K}$, and every $(t,x) \in I \times \Psi$:

(a) $b(||x||) \geq V(t,x) \geq a(||x||)$, $V(t,0) = 0$;

(b) $\dot{V}(t,x) \leq 0$.

7.4. J.L. Massera, [1949] and [1956], was the first to obtain a converse of Liapunov's theorem on uniform asymptotic stability (cf. 6.2). Before stating it, let us recall that the function $f(t,x)$ defined in Section 2.2 is said to be locally lipschitzian in x uniformly with respect to t if, for every point of Ω, there exist a neighborhood N of this point, $N \subset \Omega$, and a constant k such that, for any $t \in I$ and any pair of points x_1, $x_2 \in N$, one gets:

$||f(t,x_1) - f(t,x_2)|| \leq k||x_1 - x_2||$. The same function is said to be <u>lipschitzian in</u> x <u>uniformly with respect to</u> t if there exists a constant k such that, for any $t \in I$ and any pair of points $x_1, x_2 \in \Omega$, one gets: $||f(t,x_1) - f(t,x_2)|| \leq k||x_1 - x_2||$.

<u>Theorem</u>. If the function f is, on $I \times \Omega$, locally lipschitzian in x uniformly with respect to t and if the origin is uniformly asymptotically stable, there exist a neighborhood $\Psi \subset \Omega$ of the origin and a function $V: I \times \Psi \to \mathscr{R}$ possessing partial derivatives in t and x of arbitrary order, such that for some functions $a, b, c \in \mathscr{K}$ and every $(t,x) \in I \times \Psi$:

(a) $a(||x||) \leq V(t,x) \leq b(||x||)$;

(b) $\dot{V}(t,x) \leq -c(||x||)$.

<u>Corollary</u>. If f is, on $I \times \Omega$, lipschitzian in x uniformly with respect to t, V can be chosen such that all partial derivatives of any order of V are bounded on $I \times \Psi$, with the same bound for all of them. If f is, on $I \times \Omega$, independent of t or periodic in t, V can be chosen independent of t or periodic in t respectively.

8. Bibliographical Note

Several references have been given in the text, where we have made an effort to mention the author of each theorem. Although it would be interesting to write in detail the early story of stability theory, for lack of space we refrain from doing so. Let us content ourselves here with the mention of a few introductory texts, the contents of which might usefully precede or complement the present one.

I.G. Malkin [1952] contains a very explicit and careful treatment of the basic theory (without use of vector notations, which renders the text a bit lengthy). N.G. Chetaev [1955], although somewhat old-fashioned now, is still interesting, in particular because it contains many mechanical applications, a too rare feature indeed! W. Hahn [1959] was amongst the first western authors to write on Liapunov's second method. The first four chapters of his book constitute a very good and rigorous introduction to the subject. They are accompanied by an extensive bibliography. In a second book [1967], he gathered more, more varied, and up to date material. H.A. Antosiewicz [1958] is a survey of the theory at the end of the fifties, the instability theorems being excluded. It is written in a beautifully concise style and should be highly recommended. J.P. LaSalle and S. Lefschetz [1961] is interesting for the beginner: the exposition is definitely elementary, being in particular limited to autonomous equations. Finally, many textbooks on differential equations devote some chapter to stability theory, for instance A. Halanay [1963], C. Corduneanu [1971], N. Rouche and J. Mawhin [1973], Vol. II, H.W. Knobloch and F. Kappel [1974] and several others.

CHAPTER II

SIMPLE TOPICS IN STABILITY THEORY

This chapter is varied in character. It is meant to complete the elementary view of stability theory which we began to present in Chapter I. Sections 1 and 2 examine in different ways what conclusions can be drawn from the knowledge of a positive definite auxiliary function $V(t,x)$ possessing a derivative which is only smaller than or equal to zero. Section 3 is an elementary presentation of the so-called comparison method. All these matters will be investigated further in later chapters. Section 4 deals briefly with total stability, i.e. stability with respect to perturbations not only of the initial conditions, but also of the second member of the differential equation. The main theorem proved here will be needed in Chapter IV. Section 5 is but a glance at the "frequency method", introduced here because of its intrinsic importance and for the sake of completeness, but considered later as outside the scope of the rest of the book. The last section is an example of circuit theory showing that continuously differentiable auxiliary functions are not always satisfactory, and thus justifying the fact that, in subsequent chapters, the required regularity conditions will be somewhat weakened.

1. Theorems of E.A. Barbashin and N.N. Krasovski for Autonomous and Periodic Systems

1.1. As already observed in Section I.6.12, it is sometimes difficult and time consuming to find an auxiliary function with negative definite time derivative. In this section, we endeavor to prove asymptotic stability, global as well as local, and instability, by using a $V(t,x)$ whose time derivative is only smaller than or equal to zero, which is a much more common situation. Of course, we shall have to compensate by introducing a new hypothesis: namely that the set M where $\dot{V}(t,x) = 0$ contains no complete trajectory. Inasmuch as this condition is often realized and easily recognized, the theorems to come are amongst the most useful ones. In particular, as we shall see in several instances, they allow using the total energy as an auxiliary function in mechanical problems with dissipation. They have a drawback however, in that they pertain to autonomous and periodic equations only.

1.2. Our general hypotheses in this section will be those of Section I.2.2, but we assume further that $f(t,x)$ is periodic with respect to t, for some period $T > 0$.

1.3. Theorem (N.N. Krasovski [1959]). Suppose there exists a $\mathscr{C}^1$ function $V\colon I \times \Omega \to \mathscr{R}$, periodic in t with period T, such that for some function $a \in \mathscr{K}$ and every $(t,x) \in I \times \Omega$:

(i) $V(t,x) \geq a(||x||)$; $V(t,0) = 0$;

(ii) $\dot{V}(t,x) \leq 0$; we put $M = \{(t,x) \in I \times \Omega: \dot{V}(t,x) = 0\}$;

(iii) except for the origin, M contains no complete positive semi-trajectory.

Choosing $\alpha > 0$ such that $\overline{B}_\alpha \subset \Omega$, let us put for every $t \in I$:

$$V_{t,\alpha}^{-1} = \{x \in \Omega: V(t,x) \leq a(\alpha)\}.$$

Then

(a) the region of attraction $A(t_0) \supset V_{t_0,\alpha}^{-1}$;

(b) the origin is uniformly asymptotically stable.

If (i) is replaced by

(i)' $(\exists t_0 \in I)(\forall \delta > 0)(\exists x_0 \in B_\delta)$ $V(t_0,x_0) < 0$; $V(t,0) = 0$;

then the origin is unstable.

Proof. To prove the first part of this statement, observe first that the origin is stable, by Theorem I.4.2 and uniformly stable by I.2.14. Let us choose $\alpha > 0$ such that $\overline{B}_\alpha \subset \Omega$ and $t_0 \in I$. We deduce as in I.6.2 that every solution starting at some $x_0 \in V_{t_0,\alpha}^{-1}$ is defined over $[t_0,\infty[$. We prove now that such a solution $x(t;t_0,x_0) \to 0$ as $t \to \infty$. Due to uniform stability (cf. I.6.8), it suffices to prove that $(\forall \delta > 0)(\exists t \geq t_0)$ $||x(t;t_0,x_0)|| < \delta$. If this were not true, there would be an $\overline{x}_0 \in V_{t_0,\alpha}^{-1}$ and a $\delta > 0$ such that

$$(\forall t \geq t_0) \qquad \delta \leq ||x(t;t_0,\overline{x}_0)|| \leq \alpha. \tag{1.1}$$

But then, the sequence $x_0^k = x(t_0+kT;t_0,\bar{x}_0)$ for $k = 1,2,\ldots$ has a cluster point x_0^* and, for some subsequence (also written $\{x_0^k\}$), $x_0^k \to x_0^*$ as $k \to \infty$. As a function of t, $V(t,x(t;t_0,\bar{x}_0))$ is decreasing and bounded from below: therefore, it tends to a limit as $t \to \infty$. But, V being continuous and periodic in t, the limit is obtained thus:

$$\begin{aligned}\lim_{t\to\infty} V(t,x(t;t_0,\bar{x}_0)) &= \lim_{k\to\infty} V(t_0+kT,x(t_0+kT;t_0,\bar{x}_0)) \\ &= \lim_{k\to\infty} V(t_0,x_0^k) = V(t_0,x_0^*).\end{aligned} \tag{1.2}$$

Considering now $x(t;t_0,x_0^*)$, we deduce from (iii) the existence of a $t^* > t_0$ such that $\dot{V}(t^*,x(t^*;t_0,x_0^*)) < 0$ and therefore that

$$V(t^*,x(t^*;t_0,x_0^*)) \neq V(t_0,x_0^*). \tag{1.3}$$

The periodicity of $f(t,x)$ yields

$$\begin{aligned}x(t^*;t_0,x_0^k) &= x(t^*+kT;t_0+kT,x_0^k) \\ &= x(t^*+kT;t_0+kT,x(t_0+kT;t_0,\bar{x}_0)) = x(t^*+kT;t_0,\bar{x}_0).\end{aligned}$$

Due to (1.2), (1.3) and the periodicity of $V(t,x)$, we get finally the following contradiction

$$\begin{aligned}V(t_0,x_0^*) &= \lim_{k\to\infty} V(t^*+kT,x(t^*+kT;t_0,\bar{x}_0)) \\ &= \lim_{k\to\infty} V(t^*,x(t^*;t_0,x_0^k)) = V(t^*,x(t^*;t_0,x_0^*)) \neq V(t_0,x_0^*).\end{aligned}$$

Part (a) of the thesis is proved and, because $V_{t_0,\alpha}^{-1}$ is a neighborhood of the origin, this point is asymptotically stable. Then it is uniformly asymptotically stable by Theorem I.2.14.

The part of the thesis concerning instability is proved <u>ab absurdo</u>. If the origin were stable, then for any $\alpha > 0$ with $\overline{B}_\alpha \subset \Omega$, one should have

$$(\exists \eta > 0)(\forall x_0 \in B_\eta)(\forall t_0 \in I)(\forall t \geq t_0) \quad ||x(t;t_0,x_0)|| < \alpha.$$

But (i)' shows that for some $t_0 \in I$ and $\overline{x}_0 \in B_\eta$: $V(t_0,\overline{x}_0) < 0$. In that case, for some $\delta > 0$ and every $t \geq t_0$: $||x(t;t_0,\overline{x}_0)|| \geq \delta$ because V is continuous, periodic and $V(t,0) = 0$. But then (1.1) is verified and the expected contradiction is obtained as above. Therefore, the origin is unstable. Q.E.D.

1.4. <u>Remarks</u>. (1) Theorem 1.3 cannot be extended to general nonautonomous systems, as is shown by the equation $\dot{x} = -p(t)x$ for $x \in \mathscr{R}$, with $p(t) > 0$ for $t \in [0,\infty[$ and $\int_0^\infty p(t)dt < \infty$. The origin is not asymptotically stable, whereas $V = x^2$ satisfies Hypotheses (i) to (iii) of Theorem 1.3 (cf. V.M. Matrosov $[1962]_1$).

(2) To prove instability, Hypothesis (iii) can be replaced by the following one:

(iii)' The set $M \cap \{(t,x): V(t,x) < 0\}$ contains no complete positive semi-trajectory.

In the Corollary to follow, the general setting is as in Theorem 1.3, with the exception that $\Omega = \mathscr{R}^n$.

1.5. <u>Corollary</u>. In the hypotheses of Theorem 1.3, if one assumes further that $\Omega = \mathscr{R}^n$ and that $a(r) \to \infty$ as $r \to \infty$, the origin is globally asymptotically stable.

Proof. Indeed, due to the fact that $a(r) \to \infty$, for every $x_0 \in \mathscr{R}^n$ there is an $\alpha > 0$ such that $x_0 \in V^{-1}_{t_0,\alpha}$, and thus $x(t;t_0,x_0) \to 0$ as $t \to \infty$. Q.E.D.

1.6. Several examples illustrate hereafter Theorems 1.3 and Corollary 1.5. Another example, pertaining to the asymptotic stability of a mechanical equilibrium appears in Section III.6.

1.7. Exercise. A simple pendulum with a constant torque L applied to it and some viscous friction admits of the equation, for $x \in \mathscr{R}$: $\ddot{x} + a\dot{x} + \omega^2 \sin x = L$, where $a > 0$, ω and L are real quantities. One assumes that $|L| < \omega^2$. What about the stability behavior of the equilibrium positions? (see E.A. Barbashin [1967]).

1.8. A transistor oscillator. One of the simplest types of transistor oscillators is the based tuned circuit shown in Figure 2.1. It is studied in all elementary books on electronic engineering and its equation has been investigated

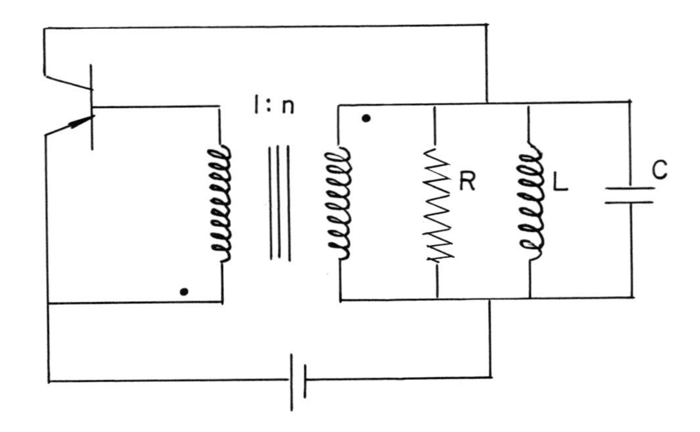

Figure 2.1. A transistor oscillator

by many mathematicians since Van der Pol. Of course, in the days of Van der Pol, it was the equation of a vacuum tube oscillator! Although the purpose of this circuit is to generate electrical oscillations, we shall limit ourselves here to exhibit the stability properties of its rest state: in case of instability, it generates oscillations, in case of asymptotic stability it does not.

It consists of a linear RLC circuit connected as shown between emitter and collector, but with an ideal transformer in the base circuit. We refer to specialized books for the derivation of the equation, which reads

$$\ddot{x} + a\dot{x} + \omega^2 x = g(\dot{x}) \tag{1.4}$$

where x is the current through the inductor L,

$$a = \frac{1}{RC} > 0, \ \omega^2 = \frac{1}{LC} > 0 \quad \text{and} \quad g: \mathscr{R} \to \mathscr{R}$$

is a $\mathscr{C}^1$ function, depending on the characteristics of the transistor. In fact g is strictly increasing and $g(0) = 0$.

There is a unique equilibrium at $x = \dot{x} = 0$. The derivative of the auxiliary function

$$V(x,\dot{x}) = \frac{1}{2}\dot{x}^2 + \frac{1}{2}\omega^2 x^2$$

is computed easily and reads

$$\dot{V}(x,\dot{x}) = \dot{x}g(\dot{x}) - a\dot{x}^2 = [g'(0) - a]\dot{x}^2 + h(\dot{x})$$

where $h(\dot{x})/\dot{x} \to 0$ as $\dot{x} \to 0$. Accordingly as $g'(0) < a$ or $g'(0) > a$, $\dot{V}$ is negative or positive definite, but with respect to $\dot{x}$ only: $\dot{V}$ vanishes if and only if $\dot{x} = 0$. But, as Equation (1.4) shows and except at the origin, $\dot{x} = 0$

implies $\ddot{x} \neq 0$. Therefore, there is no complete positive semi-trajectory in the set M of the Theorem 1.3. We conclude that the origin is unstable or asymptotically stable according to whether $g'(0) > a$ or $g'(0) < a$. The functions to consider in applying Theorem 1.3 are V for asymptotic stability and $-V$ for instability.

1.9. Example of an equation of third order (E.A. Barbashin and N.N. Krasovski [1952], and E.A. Barbashin [1967]). Let f and ϕ be two functions from $\mathscr{R} \to \mathscr{R}$, f of class $\mathscr{C}^1$ with $f(0) = 0$, ϕ continuous with $\phi(0) = 0$, and let $a > 0$. The equation

$$\dddot{x} + a\ddot{x} + \phi(\dot{x}) + f(x) = 0 \tag{1.5}$$

arises in several problems of control theory and in the dynamics of systems containing a gyroscopic pendulum, like the monorail (cf. H. Leipholz [1968]). Putting $y = \dot{x}$ and $z = \dot{y} + ay$ transforms (1.5) into

$$\begin{aligned} \dot{x} &= y, \\ \dot{y} &= z - ay, \\ \dot{z} &= -\phi(y) - f(x). \end{aligned} \tag{1.6}$$

Let us look for some conditions which would entail global asymptotic stability of the origin $x = y = z = 0$.

The time derivative of the function

$$V(x,y,z) = aF(x) + f(x)y + \Phi(y) + \frac{1}{2}z^2$$

where

$$F(x) = \int_0^x f(\xi)d\xi, \qquad \Phi(y) = \int_0^y \phi(\eta)d\eta,$$

reads

$$\dot{V}(x,y) = [f'(x)y - a\phi(y)]y.$$

Let us put

$$W(x,y) = aF(x) + f(x)y + \Phi(y)$$

and assume the following:

(i) $f(x)x > 0$ for $x \neq 0$;

(ii) $[a\phi(y) - yf'(x)]y > 0$ for any x, if $y \neq 0$;

(iii) $W(x,y) \to \infty$ as $x^2 + y^2 \to \infty$.

Because of (iii), to prove that $V(x,y,z)$ is positive definite, we only have to prove that $W(x,y,z) > 0$ for any $(x,y) \neq (0,0)$. The way to prove that, consists in considering first the integral

$$\int_0^y [a\phi(\eta) - \eta f'(x)]d\eta$$

which, because of (ii), is strictly positive for any x, if $y \neq 0$. Multiplying by $f(x)$ and integrating with respect to x, one gets

$$\int_0^x f(\xi)\left[\int_0^y [a\phi(\eta) - \eta f'(\xi)]d\eta\right]d\xi = aF(x)\Phi(y) - \frac{y^2}{4} f^2(x) \quad (1.7)$$

a function which, due to (i) and the way it was built up, is strictly positive for every $(x,y) \neq (0,0)$. Observing then that

$$W(x,y) = \frac{(2\Phi(y) + yf(x))^2}{4\Phi(y)} + \frac{4aF(x)\Phi(y) - y^2f^2(x)}{4\Phi(y)},$$

and because $\Phi(y) > 0$ for $y \neq 0$, we see that effectively $W(x,y) > 0$ for any $(x,y) \neq (0,0)$. That $\Phi(y) > 0$ results from the second member of (1.7) being positive.

As far as $\dot{V}$ is concerned, it is obviously ≤ 0 and vanishes if and only if $y = 0$. But the plane $y = 0$ in the (x,y,z)-space contains, besides the origin, no complete orbit. In fact, for an orbit in this plane, one would obtain identically $y = \dot{y} = 0$. The second equation (1.6) yields then identically $z = \dot{z} = 0$. Finally, the third equation (1.6) gives $f(x) = 0$, which is equivalent to $x = 0$. All the hypotheses of Corollary 1.5 being satisfied, the origin is globally asymptotically stable.

1.10. <u>Exercise</u> (J.P. LaSalle and S. Lefschetz [1961]). Consider the equation

$$\dddot{x} + f(\dot{x})\ddot{x} + a\dot{x} + bx = 0,$$

where $a > 0$, $b > 0$ and f is a continuous function from $\mathcal{R} \to \mathcal{R}$. Find some conditions on f in order that the origin $x = y = z = 0$ be globally asymptotically stable for the equivalent system

$$\dot{x} = y, \quad \dot{y} = z, \quad \dot{z} = -f(y)z - ay - bx.$$

<u>Hint</u>: the following auxiliary function may be used:

$$V(x,y,z) = \frac{a}{2} z^2 + byz + b \int_0^y f(\eta)\eta d\eta + \frac{1}{2} (bx+ay)^2.$$

1.11. <u>An example from chemical kinetics in biology</u>. Genetic control mechanisms in bacteria can be described (cf. R. Rosen [1970]). using a primary gene product X, an enzymatic protein Y determined by this gene and a metabolite M produced by the metabolic activity of the protein. Let x,y and m be the concentrations of these products. A first model of the genetic mechanism corresponds to the following equations

$$\dot{x} = \frac{K}{m} - \alpha,$$
$$\dot{y} = Ax - \beta,$$
$$\dot{m} = Cy - \gamma m,$$

where K, A, C, α, β and γ are positive constants. The first equation expresses the fact that the metabolite M is an inhibitor for the primary gene product. A common assumption is that m varies slowly, in such a way that $\dot{m}$ can be approximated by zero. In this case, the equations become (B. Goodwin [1963]):

$$\dot{x} = \frac{K\gamma}{Cy} - \alpha,$$
$$\dot{y} = Ax - \beta,$$

and the stability of the equilibrium

$$x = \beta/A, \; y = K\gamma/C\alpha$$

follows, by Theorem I.4.2, from the existence of the first integral

$$V(x,y) = -\frac{K\gamma}{C}\ln y + \alpha y + \frac{A}{2}x^2 - \beta x$$

which is such that $V(x,y) - V(\beta/A, K\gamma/C\alpha)$ is positive definite in some neighborhood of the equilibrium (prove this !).

A criticism to Goodwin's equations is that, the influence of x being considered apart, y disappears at a constant rate, whereas it is more plausible to assume that it disappears at a rate proportional to its concentration.

In that case, the equations become

$$\dot{x} = \frac{K\gamma}{Cy} - \alpha,$$
$$\dot{y} = Ax - \beta y,$$

and the equilibrium

$$x = \frac{\beta K\gamma}{AC\alpha}, \quad y = \frac{K\gamma}{C\alpha}$$

is asymptotically stable as follows, by Theorem 1.3, from the existence of the auxiliary function

$$V(x,y) = -\frac{K\gamma}{C} \ln y + \alpha y + \frac{1}{2A}(Ax-\beta y)^2$$

whose derivative reads

$$\dot{V}(x,y) = -\frac{\beta}{A}(Ax-\beta y)^2.$$

2. A Theorem of V.M. Matrosov on Asymptotic Stability

2.1. Section I was entirely devoted to proving asymptotic stability or instability through the use of an auxiliary function $V(t,x)$ whose time derivative $\dot{V}(t,x)$ is not negative definite, but only smaller than or equal to zero. The corresponding theorems were restricted to the autonomous or periodic case. In this Section, we present a theorem of V.M. Matrosov dealing with the general nonautonomous case. The crucial condition that the set M where $\dot{V}(t,x) = 0$ contains no complete positive semi-trajectory can no longer be used as such. The difficulty is overcome by introducing a second auxiliary function defined in some appropriate neighborhood of M and possessing adequate properties in order that the solutions do not spend too much time near M.

2.2. We consider again Equation I(2.2), and the general hypotheses associated with it. As the proof of the theorem itself is somewhat intricate, it will be rewarding to prove first two lemmas.

2.3. Lemma (lower bound for the transit time of a solution). If t_1 and t_2 are in I, with $t_1 < t_2$, if a solution $x(t)$ is defined on $[t_1,t_2]$ and if, for every $(t,x) \in \Omega$: $||f(t,x)|| \leq A$ where A is a constant, then $||x(t_1) - x(t_2)|| \geq r > 0$ implies that $t_2 - t_1 \geq r/A$.

Proof. In fact

$$||x(t_2) - x(t_1)|| = \left|\left| \int_{t_1}^{t_2} f(\tau,x(\tau))d\tau \right|\right| \leq$$

$$\leq \int_{t_1}^{t_2} ||f(\tau,x(\tau))||d\tau \leq A(t_2-t_1). \qquad \text{Q.E.D.}$$

2.4. Lemma (upper bound for the escape time of a solution). If there exist a $\mathscr{C}^1$ function $W: I \times \Omega \to \mathscr{R}$, a function $c \in \mathscr{K}$, a set $E \subset \Omega$ and a constant $L > 0$ such that, for every $(t,x) \in I \times \Omega$:

(i) $|W(t,x)| < L$;

(ii) $\max(d(x,E),|\dot{W}(t,x)|) \geq c(||x||)$;

then, for any $\eta > 0$, a solution $x(t)$ cannot remain in the set

$$U = \{x \in \Omega: c(||x||) \geq \eta, d(x,E) \leq \eta\}$$

on a period of duration $2L/\eta$.

Proof. Suppose $x(t) \in U$ for every t in some closed interval $[t_1,t_2]$. It follows then from (ii) that

$$|\dot{W}(t,x(t))| \geq \eta \quad \text{for} \quad t \in [t_1,t_2].$$

Using (i), one gets successively

$$2L > |W(t_2,x(t_2))| + |W(t_1,x(t_1))| \geq |W(t_2,x(t_2)) - W(t_1,x(t_1))|$$

$$= \left|\int_{t_1}^{t_2} \dot{W}(\tau,x(\tau))d\tau\right| = \int_{t_1}^{t_2} |\dot{W}(\tau,x(\tau))|d\tau \geq \eta(t_2-t_1),$$

which yields the expected result. We are able to write that the absolute value of the integral is equal to the integral of the absolute value because $\dot{W}$ is continuous and therefore does not change sign on $[t_1,t_2]$. Q.E.D.

2.5. Theorem (V.M. Matrosov [1962]$_1$). Let there exist two $\mathscr{C}^1$ functions $V: I \times \Omega \to \mathscr{R}$, $W: I \times \Omega \to \mathscr{R}$, a $\mathscr{C}^0$ function $V^*: \Omega \to \mathscr{R}$, three functions $a, b, c \in \mathscr{K}$ and two constants $A > 0$ and $L > 0$ such that, for every $(t,x) \in I \times \Omega$:

(i) $a(||x||) \leq V(t,x) \leq b(||x||)$;

(ii) $\dot{V}(t,x) \leq V^*(x) \leq 0$; put $E = \{x \in \Omega: V^*(x) = 0\}$;

(iii) $|W(t,x)| < L$;

(iv) $\max(d(x,E), |\dot{W}(t,x)|) \geq c(||x||)$;

(v) $||f(t,x)|| < A$;

choosing $\alpha > 0$ such that $\overline{B}_\alpha \subset \Omega$, let us put for every $t \in I$

$$V^{-1}_{t,\alpha} = \{x \in \Omega: V(t,x) \leq a(\alpha)\}.$$

Then

(a) for any $t_0 \in I$ and any $x_0 \in V^{-1}_{t_0,\alpha}$: $x(t;t_0,x_0) \to 0$ uniformly in t_0, x_0, when $t \to \infty$;

(b) the origin is uniformly asymptotically stable.

Proof. The reasoning begins as in Theorem I.6.2. We choose an $\alpha > 0$ such that $\overline{B}_\alpha \subset \Omega$ and deduce from (i) that for every $t \in I$

$$V^{-1}_{t,\alpha} \subset \Omega. \tag{2.1}$$

For any $t_0 \in I$ and $x_0 \in V^{-1}_{t_0,\alpha}$, it follows from (ii) that

$$x(t) \in V^{-1}_{t,\alpha}$$

for any $t \in J^+$, and therefore from (2.1) that $x(t)$ cannot approach the boundary of Ω. Hence $J^+(t_0,x_0) = [t_0,\infty[$. For any $\varepsilon > 0$, choose any $\eta > 0$ such that $b(c^{-1}(\eta)) < a(\varepsilon)$. Further, let us consider any $(t_0,x_0) \in I \times V^{-1}_{t_0}$, and suppose ab absurdo that the corresponding solution $x(t)$ remains for all $t \geq t_0$ in the union of the sets

$$U = \{x \in \overline{B}_\alpha : c(||x||) \geq \eta, d(x,E) \leq \eta\},$$

$$U^* = \{x \in \overline{B}_\alpha : c(||x||) \geq \eta, d(x,E) \geq \eta/2\}.$$

Then, on any subinterval $[\tau_1,\tau_2] \subset [t_0,\infty[$, with $\tau_2 - \tau_1 = 2L/\eta$, one would have two possibilities:

(α) either $x(t) \in U^*$ for $t \in [\tau_1,\tau_2]$; then if we write $V(t)$ for $V(t,x(t))$ and $\dot{V}(t)$ for $\dot{V}(t,x(t))$, we get

$$V(\tau_2) - V(\tau_1) = \int_{\tau_1}^{\tau_2} \dot{V}(\tau)d\tau \leq \int_{\tau_1}^{\tau_2} V^*(x(\tau))d\tau \leq -2\ell L/\eta$$

where $\ell = \inf\{|V^*(x)|: x \in U^*\}$;

(β) or there is a $t \in [\tau_1, \tau_2]$ such that $x(t) \in U \backslash U^*$. But it follows from Lemma 2.4 that there is a $t \in [\tau_1, \tau_2]$ for which $x(t) \notin U$. At the first mentioned of these instants $d(x,E) < \eta/2$, while at the second $d(x,E) > \eta$. Since $x(t)$ is continuous, there exist two instants τ_1', $\tau_2' \in [\tau_1, \tau_2]$ such that

$$d(x(\tau_1'),E) = \eta/2, \quad d(x(\tau_2'),E) = \eta,$$

and, if we suppose for definiteness that $\tau_1' < \tau_2'$,

$$d(x(t),E) \geq \eta/2 \quad \text{for} \quad t \in [\tau_1', \tau_2'].$$

We deduce from Lemma 2.3 that $|\tau_2' - \tau_1'| \geq \eta/2A$. Thus, as $V(t)$ is monotonic decreasing

$$V(\tau_2) - V(\tau_1) \leq V(\tau_2') - V(\tau_1') = \int_{\tau_1'}^{\tau_2'} \dot{V}(\tau)d\tau \leq -\ell\eta/2A.$$

In either case

$$V(\tau_2) - V(\tau_1) \leq -\min[2\ell L/\eta, \ell\eta/2A].$$

In order to juxtapose several intervals like $[\tau_1, \tau_2]$, let us put, for some integer $k \geq 1$,

$$t_i = t_0 + 2iL/\eta \qquad 1 \leq i \leq k.$$

Then

$$V(t_k) = V(t_0) + \sum_{i=0}^{k-1} (V(t_{i+1}) - V(t_i))$$

and further,

$$V(t_k) \leq a(\alpha) - k \min[2\ell L/\eta, \ell\eta/2A].$$

Of course, if k is chosen large enough, this inequality contradicts (i). Noticing that k, determined in this way, is independent of (t_0,x_0), we obtain the following: there exists a $\sigma > 0$ and, for every $(t_0,x_0) \in I \times V^{-1}_{t_0,\alpha}$, a $t_1 \in [t_0,t_0+\sigma]$ such that $||x(t_1)|| \leq c^{-1}(\eta)$. The rest of the proof is a trivial transposition from Theorem I.6.2. Q.E.D.

2.6. <u>The pendulum with time varying friction</u>. Let us consider again a pendulum with time varying friction described by the equation

$$\ddot{x} + h(t)\dot{x} + \sin x = 0.$$

Using the total energy

$$V = \frac{\dot{x}^2}{2} + (1-\cos x)$$

as auxiliary function, and in view of obtaining the largest possible region $V^{-1}_{t_0,\alpha}$, it seems appropriate to choose $\Omega = \{(x,\dot{x}): (x^2+\dot{x}^2)^{1/2} < \pi - \varepsilon\}$ for some $\varepsilon > 0$ arbitrarily small. One can then verify that the function $a(r)$ of class $\mathcal{K}$ of Theorem 2.5 can be chosen as $2r^2/\pi^2$. The existence of the function b is obvious and it would be of no use to specify it further. The derivative of V reads $\dot{V} = -h(t)\dot{x}^2$. If we suppose that there exist two constants k_1 and k_2 such that $k_1 \geq h(t) \geq k_2 > 0$, we may choose $V^* = -k_2\dot{x}^2$, in such a way that $E = \{(x,\dot{x}) \in \Omega: \dot{x} = 0\}$. Let us use $W = \dot{x}$. This function is bounded on Ω and further, there exists a function $c \in \mathcal{K}$ such that

$$\max(d((x,\dot{x}),E),|\dot{W}|) = \max(|\dot{x}|,|\ddot{x}|)$$

$$= \max(|\dot{x}|,|h(t)\dot{x} + \sin x|) \geq c[(x^2+\dot{x}^2)^{1/2}].$$

This can be proved as follows. First

$$|h(t)\dot{x} + \sin x| \geq h(t)|\dot{x}| - |\sin x| \geq k_2|\dot{x}| - |\sin x|$$

and

$$|h(t)\dot{x} + \sin x| \geq |\sin x| - h(t)|\dot{x}| \geq |\sin x| - k_1|\dot{x}|.$$

Hence

$$\max(|\dot{x}|,|h(t)\dot{x} + \sin x|) \geq$$

$$\geq \max(|\dot{x}|,k_2|\dot{x}| - |\sin x|,|\sin x| - k_1|\dot{x}|).$$

The function $F(x,\dot{x}) = \max(|\dot{x}|,k_2|\dot{x}| - |\sin x|,|\sin x| - k_1|\dot{x}|)$ is continuous. Further it is > 0 for any $(x,\dot{x}) \in \Omega$, $(x,\dot{x}) \neq 0$. Let us write

$$c^*(r) = \inf\{F(x,\dot{x}),\ 0 \leq r < (x^2+\dot{x}^2)^{1/2} \leq \pi - \varepsilon\}.$$

This function is monotonic increasing, and vanishes only for $r = 0$. As is well known (see e.g., N. Rouche and J. Mawhin [1973]), there exists a function $c \in \mathscr{K}$ such that for every r: $c(r) < c^*(r)$. Hence, the desired result.

All the hypotheses of Theorem 2.5 being verified, th origin is uniformly asymptotically stable. Further, for any $\alpha < \pi$,

$$V^{-1}_{t_0,\alpha} = \{(x,\dot{x}) : \frac{\dot{x}^2}{2} + (1-\cos x) \leq \frac{2\alpha^2}{\pi^2}\} \subset A(t_0).$$

This result extends in some way those of Examples I.6.12 and I.6.27. However, we know from I.6.27 that the condition $h(t) \leq k_1$ is not necessary for asymptotic stability, a fact which illustrates one of the limitations of Theorem 2.5.

2.7. Asymptotic stability of a gyroscope (V.M. Matrosov [1962]$_1$). Consider a symmetric rigid body with a fixed point 0 in some inertial frame of reference 0XYZ, where 0Z is a vertical axis oriented upward. Let 0xyz be a system of axes along the principal directions of inertia of the body at 0 and fixed in the body. With suitable hypotheses about friction forces and the driving torque along 0z, the equations of motion read

$$A\dot{p} = (A-C)qr + mgz_0\gamma_2 - \partial R/\partial p, \tag{2.2}$$

$$A\dot{q} = (C-A)pr - mgz_0\gamma_1 - \partial R/\partial q, \tag{2.3}$$

$$C\dot{r} = M(t,r), \tag{2.4}$$

$$\dot{\gamma}_1 = r\gamma_2 - q\gamma_3, \tag{2.5}$$

$$\dot{\gamma}_2 = p\gamma_3 - r\gamma_1, \tag{2.6}$$

$$\dot{\gamma}_3 = q\gamma_1 - p\gamma_2, \tag{2.7}$$

$$\gamma_1^2 + \gamma_2^2 + \gamma_3^2 = 1. \tag{2.8}$$

In these equations A,A,C are the principal moments of inertia of the body at 0; p,q,r are the components in 0xyz of the instantaneous rotation of the body in the inertial frame; $R = R(p,q)$ is a positive definite homogeneous form of degree $k \geq 2$, with coefficients continuous and bounded functions of t, and $M(t,r)$ is the resultant torque around 0z of the friction and driving forces. Further, m is the mass of the body, g the acceleration of gravity, z_0 the z-coordinate of the center of inertia of the body and finally $\gamma_1,\gamma_2,\gamma_3$ are the components in the inertial frame of the unit vector along 0z. We shall suppose hereafter that $z_0 < 0$.

We assume that equation (2.4) possesses a solution

$r(t)$ which is a bounded function of t. Replacing r by $r(t)$ in the other equations and replacing also γ_3 by its value taken from (2.8), we get the following four equations

$$A\dot{p} = (A-C)qr(t) + mgz_0\gamma_2 - \partial R/\partial p,$$
$$A\dot{q} = (C-A)pr(t) - mgz_0\gamma_1 - \partial R/\partial q,$$
$$\dot{\gamma}_1 = r(t)\gamma_2 - q\sqrt{1 - \gamma_1^2 - \gamma_2^2},$$
$$\dot{\gamma}_2 = p\sqrt{1 - \gamma_1^2 - \gamma_2^2} - r(t)\gamma_1.$$

We choose the plus sign for the square root in order for the equilibrium $p = q = \gamma_1 = \gamma_2 = 0$ of these equations to correspond to $\gamma_3 = 1$, and therefore to the $0z$ axis pointing upwards. Remembering that $z_0 < 0$, we see that the center of inertia at equilibrium is below 0.

As auxiliary functions suiting Theorem 2.5, we choose

$$V = \frac{1}{2} A(p^2+q^2) - \frac{1}{2} mgz_0(\gamma_1^2 + \gamma_2^2 + (1 - \sqrt{1 - \gamma_1^2 - \gamma_2^2})^2),$$

$$W = A(p\gamma_2 - q\gamma_1).$$

One computes easily that

$$\dot{V} = -p\frac{\partial R}{\partial p} - q\frac{\partial R}{\partial q} = -kR$$

and, as $\dot{V}$ is negative definite with respect to p and q, one has $E = \{(p,q,\gamma_1,\gamma_2): p = q = 0\} \cap \Omega$, where Ω is an appropriate neighborhood of the equilibrium. One also computes

$$\dot{W} = mgz_0(\gamma_1^2+\gamma_2^2) - Cr(t)(q\gamma_2+p\gamma_1) + \gamma_1\frac{\partial R}{\partial q} - \gamma_2\frac{\partial R}{\partial p} + A(p^2+q^2)\sqrt{1 - \gamma_1^2 - \gamma_2^2}.$$

Due to the boundedness properties of $r(t)$ and of the

coefficients of R, one recognizes that hypothesis (iv) of Theorem 2.5 is satisfied. All the other hypotheses being verified also, the equilibrium is uniformly asymptotically stable. Notice that V.M. Matrosov $[1962]_1$ proves further the instability of the equilibrium when $z_0 > 0$.

2.8. Exercise (A Lienard type equation). Let $f(y)$ and $g(x)$ be continuous functions from $\mathscr{R} \to \mathscr{R}$ and $h(t,x,y)$ be a continuous function from $\mathscr{R}^3 \to \mathscr{R}$. For the equation

$$\ddot{x} + h(t,x,\dot{x})\dot{x} + f(\dot{x})g(x) = 0,$$

which is equivalent to the system

$$\dot{x} = y,$$
$$\dot{y} = -h(t,x,y)y - f(y)g(x),$$

the origin is uniformly asymptotically stable if

(i) there exist continuous functions $k_1(x,y)$ and $k_2(x,y)$ from $\mathscr{R}^2 \to \mathscr{R}$ such that for $y \neq 0$

$$k_1(x,y) \geq h(t,x,y) \geq k_2(x,y) > 0;$$

(ii) $f(y) > 0$ for every y;

(iii) $g(x)x > 0$ for every $x \neq 0$ and $g(0) = 0$;

further, the origin is uniformly globally asymptotically stable if

(iv) $\lim_{|x|\to\infty} \int_0^x g(\xi)d\xi = \infty$ and $\lim_{|y|\to\infty} \int_0^y \frac{\eta d\eta}{f(\eta)} = \infty.$

Hint: .the auxiliary functions

$$V(x,y) = \int_0^y \frac{\eta d\eta}{f(\eta)} + \int_0^x g(\xi)d\xi \tag{2.9}$$

and

$$W = \int_0^y \frac{d\eta}{f(\eta)}$$

can be used successfully in this problem. By the way, it is interesting to observe how V has been constructed, using a separation of variables. One tries the form

$$V(x,y) = G(x) + \Phi(y).$$

Its time derivative reads

$$\dot{V}(x,y) = G'(x)y - \Phi'(y)[h(t,x,y)y + f(y)g(x)].$$

In view of simplifying $\dot{V}$, one tries then to determine G and Φ in order that $G'(x)y - \Phi'(y)f(y)g(x)$ vanishes identically. By separation of variables, one gets

$$\frac{G'(x)}{g(x)} = \frac{\Phi'(y)f(y)}{y}$$

and, of course, both members have to be equal to one and the same constant. Choosing the constant equal to 1, one obtains (2.9).

2.9. Example of a chemical reactor. Consider a continuously agitated tank reactor where a first order irreversible reaction

$$A \to B$$

takes place (cf. R.B. Warden, R. Aris and N.R. Amundson [1964]). Let V be the volume of the vessel into which a stream of reactant A flows at a constant rate q, with constant concentration C_0 and temperature T_0. The vessel is stirred so that the concentration C and temperature T of the reactant A are constant throughout the volume. A mass balance for reactant A yields the equation

$$V \frac{dC}{dt} = q(C_0 - C) - VF(C,T) \qquad (2.10)$$

where $F(C,T)$ is the reaction rate. A heat balance gives

$$VC_p\rho \frac{dT}{dt} = qC_p\rho(T_0 - T) - VU^*(T) + (-\Delta H)VF(C,T) \qquad (2.11)$$

where $U^*(T)$ is the heat removed to the coolant, $-\Delta H$ the heat evolved per mole of consumed A, C_p the specific heat of the reactant and ρ its density.

Let us introduce the normalized variables

$$\xi = \frac{C}{C_0}, \quad \eta = C_p\rho T/(-\Delta H)C_0, \quad \tau = tq/V,$$

and let us put further

$$G = VF/qC_0 \quad \text{and} \quad U = VU^*/qC_0(-\Delta H).$$

But generally,

$$F = kC \exp(-\frac{E}{RT})$$

where k, E and R are appropriate constants. Hence, we may put

$$G = \xi K(\eta)$$

where $K(\eta) > 0$. Equations (2.10) and (2.11) become

$$\dot{\xi} = 1 - \xi(1+K(\eta))$$

$$\dot{\eta} = \eta_0 - \eta + \xi K(\eta) - U(\eta)$$

where the dot represents from now on, the derivation with respect to τ.

Let us study the stability of a constant regime $\xi = \xi_s$, $\eta = \eta_s$ through the use of the auxiliary function

$$V(\xi,\eta) = \frac{1}{2}[\xi - \xi_s + P(\eta)]^2 + Q(\eta)$$

with

$$P(\eta) = \int_{\eta_s}^{\eta} p(\eta)d\eta, \quad Q(\eta) = \int_{\eta_s}^{\eta} q(\eta)d\eta.$$

The functions p and q are left undefined at this stage. One computes that

$$\begin{aligned}
\dot{V}(\xi,\eta) &= (\xi-\xi_s)^2[-1 - K(\eta) + p(\eta)K(\eta)] \\
&+ (\xi-\xi_s)[\xi_s(K(\eta_s) - K(\eta)) - P(\eta)(1+K(\eta)) \\
&+ p(\eta)\{\eta_s - \eta + \xi_s(K(\eta) - K(\eta_s)) + U(\eta_s) - U(\eta) + P(\eta)K(\eta)\} \\
&+ q(\eta)K(\eta)] \\
&+ P(\eta)[\xi_s(K(\eta_s) - K(\eta)) + p(\eta)\{\eta_s - \eta + \xi_s(K(\eta) - K(\eta_s)) \\
&+ U(\eta_s) - U(\eta)\}] \\
&+ q(\eta)[\eta_s - \eta + \xi_s(K(\eta) - K(\eta_s)) + U(\eta_s) - U(\eta)].
\end{aligned}$$

Let us now choose p and q such that $\dot{V}$ does not depend on ξ any more, by putting

$$\begin{aligned}
p(\eta) &= \frac{1 + K(\eta)}{K(\eta)} > 0, \\
q(\eta) &= -\frac{1}{K(\eta)}[\xi_s(K(\eta_s) - K(\eta)) + p(\eta)\{\eta_s - \eta + \xi_s(K(\eta) - K(\eta_s)) \\
&\quad + U(\eta_s) - U(\eta)\}].
\end{aligned}$$

Then

$$\dot{V} = -q(\eta)K(\eta)H(\eta)$$

where

$$H(\eta) = P(\eta) - \frac{\eta_s - \eta + \xi_s(K(\eta) - K(\eta_s)) + U(\eta_s) - U(\eta)}{K(\eta)}.$$

It follows from Theorem I.4.2 that the constant regime ξ_s, η_s is stable if, in some neighborhood of (ξ_s, η_s),

(i) V is positive definite, i.e. $q'(\eta_s) > 0$;

(ii) $\dot{V}$ is negative, i.e. $H'(\eta_s) > 0$.

It can be shown that these conditions are nothing but Routh-Hurwitz conditions for the linearized system. Observe that under Hypothesis (ii), $\dot{V}$ is negative semi-definite in the (ξ,η)-space, because it is independent of ξ.

Theorem 2.5 shows that, under the same hypothesis, the stability is asymptotic since $\dot{V}(\xi,\eta) = 0$ if and only if $\eta = \eta_s$, whereas, on the set

$$E = \{(\xi,\eta): \eta = \eta_s,\ \xi \neq \xi_s\},$$

the auxiliary function $W(\xi,\eta) = \eta - \eta_s$ is such that

$$\dot{W}(\xi,\eta) = (\xi-\xi_s)K(\eta_s) \neq 0.$$

3. Introduction to the Comparison Method

3.1. In this section, we still consider Equation I(2.2), with no change of the general hypotheses. For convenience, let us rewrite this equation here:

$$\dot{x} = f(t,x). \tag{3.1}$$

Suppose that there exists a scalar differential equation

$$\dot{u} = \omega(t,u) \tag{3.2}$$

with a critical point at the origin $u = 0$, and some known stability properties. The so-called comparison method studies the relationship which should exist between (3.1) and (3.2) in order that the stability properties of (3.2) entail the corresponding properties for (3.1). An appropriate relationship is obtained by using a differential inequality due to T. Wazewski [1950], involving an auxiliary function which, although positive definite, will not be required to have a total time derivative smaller than or equal to zero. This is the kind of generalization of the classical stability and asymptotic stability theorems which we now set out to study.

3.2. Preliminary remark. It is customary to have the second member of a differential equation defined on some open subset of the (t,x)-space. This is to avoid needless intricacies in the computation of derivatives $\dot{x}(t)$. In this section however, it will prove helpful to define the second member $\omega(t,u)$ of (3.2) on $I \times \mathscr{R}^+$, where $\mathscr{R}^+ = [0,\infty[$. Of course, the reader may always imagine any continuous extension of ω in the region $u < 0$. Nevertheless, when it will be supposed that (3.2) admits of $u = 0$ as a critical point and that this point is stable, or asymptotically stable, positive perturbations only will have to be considered.

The following lemma is due to T. Wazewski [1950]. Uniqueness of the solutions of the comparison equation is assumed here for convenience. But this condition is not necessary: cf. Chapter IX of this book, where the comparison technique is dealt with over again in a more general setting.

3.3. Lemma. Let $\omega\colon I \times \mathscr{R}^+ \to \mathscr{R}^+$, $(t,u) \to \omega(t,u)$ be continuous and such that (3.2) has a unique solution $u(t;t_0,u_0)$ through any $(t_0,u_0) \in I \times \mathscr{R}^+$. Let $[t_0,\beta[$ be the maximal future interval where $u(t;t_0,u_0)$ is defined. Let $v\colon [t_0,\beta[\to \mathscr{R}$ be such that

(i) $v(t_0) \leq u_0$;

(ii) $\dot{v}(t) \leq \omega(t,v(t))$ on $[t_0,\beta[$;

then $v(t) \leq u(t)$ on $[t_0,\beta[$.

For the proof, see Chapter IX.

3.4. Theorem (C. Corduneanu [1960]). Suppose there exists a function ω as described in Lemma 3.3, with $\omega(t,0) = 0$, and a $\mathscr{C}^1$ function $V: I \times \Omega \to \mathscr{R}$, such that for some function $a \in \mathscr{K}$ and every $(t,x) \in I \times \Omega$:

(i) $V(t,x) \geq a(||x||)$, $V(t,0) = 0$,

(ii) $\dot{V}(t,x) \leq \omega(t,V(t,x))$;

then

(a) stability of $u = 0$ implies stability of $x = 0$;

(b) asymptotic stability of $u = 0$ implies equi-asymptotic stability of $x = 0$;

if moreover, for some function $b \in \mathscr{K}$ and every $(t,x) \in I \times \Omega$:

(iii) $V(t,x) \leq b(||x||)$;

then

(c) uniform stability of $u = 0$ implies uniform stability of $x = 0$;

(d) uniform asymptotic stability of $u = 0$ implies uniform asymptotic stability of $x = 0$.

Proof. Because of (i), (ii) and the lemma, we have for any $(t_0,x_0) \in I \times \Omega$ and any $t \geq t_0$ where $u(t;t_0,V(t_0,x_0))$ and $x(t;t_0,x_0)$ are defined, that

$$a(||x(t)||) \leq V(t,x(t)) \leq u(t;t_0,V(t_0,x_0)). \tag{3.3}$$

(a) The origin $u = 0$ being stable, for any $t_0 \in I$ and $\varepsilon > 0$, there exists a $\delta^*(t_0,\varepsilon) > 0$ such that for arbitrary $u_0 < \delta^*$ and $t \geq t_0$, $u(t;t_0,u_0) < a(\varepsilon)$. Notice, in passing, that $u(t;t_0,u_0)$ is defined over $[t_0,\infty[$. For continuity reasons, there exists a $\delta(t_0,\varepsilon)$ such that $||x_0|| < \delta$ implies $u_0 = V(t_0,x_0) < \delta^*$. But then one deduces from (3.3) that $||x(t;t_0,x_0)|| < \varepsilon$ for any $t \geq t_0$, and thus $x = 0$ is stable.

(b) Since (3.2) is a scalar equation, the assumed asymptotic stability of $u = 0$ implies equi-asymptotic stability (cf. I.2.8). Therefore, for any given $t_0 \in I$, there exists an $\eta^*(t_0) > 0$ and for any $\varepsilon > 0$, a $\sigma(t_0,\varepsilon) > 0$, such that $u_0 < \eta^*$ and $t \geq t_0 + \sigma$ imply that $u(t;t_0,u_0) < a(\varepsilon)$. But now, again for continuity reasons, there exists an $\eta(t_0) > 0$ such that $||x_0|| < \eta$ implies that $u_0 = V(t_0,x_0) < \eta^*$. But then one gets from (3.3) that $x_0 \in B_\eta$ and $t \geq t_0 + \sigma$ imply that $x(t;t_0,x_0) \in B_\varepsilon$, and therefore the origin is equi-attractive.

(c) To prove thesis (c), one has only to observe that in the proof a), δ^* and δ can be chosen independent of t_0.

(d) One merely has to observe that in the proof of (b), η^* and σ can be chosen independent of t_0, and further that the same is true for η. Q.E.D.

3.5. Particular cases. For $\omega(t,u) \equiv 0$, Theorem 3.4(a) is reduced to Liapunov's Theorem I.4.2 on stability, whereas 3.4(c) yields Persidski's Theorem I.4.3 on uniform stability.

Choosing $\omega(t,u) = -c(u)$ for some function $c \in \mathscr{K}$, we observe that for the equation $\dot{u} = -c(u)$, the origin $u = 0$ is uniformly asymptotically stable. Because we have assumed uniqueness for the solutions of (3.2), c has to be chosen such that for $u_0 > 0$,

$$\int_0^{u_0} \frac{du}{c(u)} = \infty. \tag{3.4}$$

Hypothesis (ii) reads here $\dot{V}(t,x) \leq -c(V(t,x))$. Therefore, 3.4(b) is reduced to Massera's Theorem I.6.26, and 3.4(d) to Liapunov's Theorem I.6.2. To establish the latter point, observe that $\dot{V}(t,x) \leq -c[a(||x||)]$ and that $c \circ a$ is a function of class $\mathscr{K}$.

Because uniqueness has been assumed throughout for the solutions of differential equations in hand, condition (3.4) is no restriction whatsoever. Indeed, if in the above mentioned theorems of Liapunov and Massera, the function c was chosen such as to let the integral in (3.4) converge, uniqueness would be violated for the solutions of $\dot{x} = f(t,x)$ (prove this!).

Various ways of choosing the function $\omega(t,u)$ are illustrated in the following exercises.

3.6. Exercise. If $c \in \mathscr{K}$ and if $\phi\colon I \to \mathscr{R}^+$ is continuous, prove that the origin $u = 0$ is uniformly stable for the equation $\dot{u} = -\phi(t)c(u)$. If, moreover

$$\int^{\infty} \phi(t)\,dt = \infty,$$

the origin is equi-asymptotically stable.

3.7. Exercise. If $\lambda\colon I \to \mathscr{R}$ is a continuous function, the equation $\dot{u} = \lambda(t)u$ has a critical point at the origin, which is stable, uniformly stable or equi-asymptotically stable according to

$$(\forall t_0 \in I)(\exists A > 0)(\forall t \geq t_0) \int_{t_0}^{t} \lambda(s)ds \leq A,$$

or

$$(\exists A > 0)(\forall t_0 \in I)(\forall t \geq t_0) \int_{t_0}^{t} \lambda(s)ds \leq A,$$

or

$$(\forall t_0 \in I) \int_{t_0}^{t} \lambda(s)ds \to -\infty \quad \text{as} \quad t \to \infty.$$

The following example illustrates the second of these three possibilities.

3.8. <u>Example</u>. Consider the differential equation

$$\dot{x} = (E \sin t + A(t,x))x \tag{3.5}$$

where E is a unit $n \times n$ identity matrix and $A(t,x)$ an $n \times n$ matrix which is skew symmetric and satisfies appropriate regularity conditions on some domain $I \times \Omega$. The auxiliary function $V(x) = x_1^2 + x_2^2 + \ldots + x_n^2$ has a time derivative $\dot{V}(t,x) = 2V(x) \sin t$ which is not sign constant. But the solution $u = 0$ of the scalar comparison equation $\dot{u} = 2u \sin t$ is uniformly stable. This entails uniform stability for $x = 0$ and Equation (3.5). Compare this result with I.4.6.

The following exercise makes use of the comparison method to generalize Chetaev's Theorem I.6.31 on equi-asymptotic stability, while Exercise 3.10 is an extension of Theorem 3.4 to partial stability.

3.9. Exercise (N.P. Bhatia and V. Lakshmikantham [1965]). Suppose there exist two $\mathscr{C}^1$ functions $V: I \times \Omega \to \mathscr{R}$, $k: I \to \mathscr{R}$ and a continuous function $\omega: I \times \mathscr{R}^+ \to \mathscr{R}$ such that $\omega(t,0) = 0$, that uniqueness holds for the solutions of $\dot{u} = \omega(t,u)$, and at last, that for some function $a \in \mathscr{K}$ and every $(t,x) \in I \times \Omega$:

(i) $V(t,x) \geq a(||x||)$, $V(t,0) = 0$;

(ii) $k(t)\dot{V}(t,x) + \dot{k}(t)V(t,x) \leq \omega(t,k(t)V(t,x))$;

(iii) $k(t) > 0$;

if further

(iv) $k(t) \to \infty$ as $t \to \infty$;

then stability of $u = 0$ implies equi-asymptotic stability of $x = 0$.

The hypothesis that $k(t)$ be a $\mathscr{C}^1$ function is by no means essential. But to dispense with it, one should resort to a Dini derivative, a device to appear in subsequent chapters only. If $k(t)$ is continuous and $kV \in \mathscr{C}^1$, the statement above can be reduced to Theorem I.6.31 of Chetaev, by putting $\omega(t,u) \equiv 0$ and replacing $V(t,x)$ by $k(t)V(t,x)$.

3.10. Exercise (C. Corduneanu [1964]). The following statement pertains to the system of Equation I(4.1), described in detail in Section I.4.4.

Suppose there exist a function ω as described in Lemma 3.3, with $\omega(t,0) = 0$, and a $\mathscr{C}^1$ function $V: I \times \Omega \times \mathscr{R}^m \to \mathscr{R}$, such that for some function $a \in \mathscr{K}$

and every $(t,x,y) \in I \times \Omega \times \mathscr{R}^m$:

(i) $V(t,x,y) \geq a(||x||)$, $V(t,0,0) = 0$;

(ii) $\dot{V}(t,x,y) \leq \omega(t,V(t,x,y))$;

then

(a) stability of $u = 0$ implies stability with respect to x of $x = y = 0$;

(b) asymptotic stability of $u = 0$ implies equi-asymptotic stability with respect to x of $x = y = 0$, provided the solutions of I(4.1) do not approach ∞ in a finite time;

if moreover, for some function $b \in \mathscr{K}$ and every $(t,x,y) \in I \times \Omega \times \mathscr{R}^m$:

(iii) $V(t,x,y) \leq b(||x|| + ||y||)$;

then

(c) uniform stability of $u = 0$ implies uniform stability with respect to x of $x = y = 0$;

(d) uniform asymptotic stability of $u = 0$, along with the same assumptions as in (b) for the existence of solutions, implies uniform asymptotic stability with respect to x of $x = y = 0$.

4. Total Stability

4.1. All stability-like concepts considered up to this section pertain to variations of the initial conditions only. We introduce here a new type of stability, where variations of the second member of the equation are also taken into account. This is relevant to most practical problems, where significant perturbations occur not only at the initial time, but during the motion.

4.2. Our general hypotheses remain those of Section I.2.2. In particular, we still assume that the differential equation

$$\dot{x} = f(t,x) \tag{4.1}$$

is such that $f(t,0) = 0$ for all $t \in I$. Another differential equation will have to be considered along with (4.1), namely

$$\dot{y} = f(t,y) + g(t,y) \tag{4.2}$$

where $g\colon I \times \Omega \to \mathscr{R}^n$ satisfies the same regularity conditions as f, thus ensuring global existence and uniqueness for all solutions of (4.2). This function g will play the role of a perturbation term added to the second member of (4.1). Not much will be supposed about it besides these regularity conditions, and of course some kind of bound. Peculiarly, it will not be assumed that $g(t,0) = 0$, and therefore the origin will not be, in general, a solution of (4.2).

4.3. The solution $x = 0$ of (4.1) is called <u>totally stable</u> (or <u>stable under persistent disturbances</u>) whenever $(\forall\varepsilon > 0)(\exists\delta_1,\delta_2 > 0)(\forall t_0 \in I)(\forall y_0 \in B_{\delta_1})$ and for any g such that

$$(\forall t \geq t_0)(\forall x \in B_\varepsilon) \quad ||g(t,x)|| < \delta_2,$$

then

$$(\forall t \geq t_0) \qquad y(t;t_0,y_0) \in B_\varepsilon.$$

4.4. Theorem (I.G. Malkin [1944]). If there exist a $\mathscr{C}^1$ function $V: I \times \Omega \to \mathscr{R}$, three functions $a,b,c \in \mathscr{K}$ and a constant M such that, for every $(t,x) \in I \times \Omega$:

(i) $a(||x||) \leq V(t,x) \leq b(||x||)$;

(ii) $\dot{V}(t,x) \leq -c(||x||)$, $\dot{V}$ computed along the solutions of (4.1);

(iii) $||\frac{\partial V}{\partial x}(t,x)|| \leq M$;

then the origin is totally stable for (4.1).

Proof. Let us write $\dot{V}_P(t,y)$ for the time derivative of V along the solutions of the perturbed equation (4.2), i.e.

$$\dot{V}_P(t,y) = (\frac{\partial V}{\partial x}(t,y) \mid f(t,y) + g(t,y)) + \frac{\partial V}{\partial t}(t,y).$$

Choose δ_1, $0 < \delta_1 < \varepsilon$, such that $b(\delta_1) < a(\varepsilon)$, and choose $\delta_2 < kc(\delta_1)/M$ for some k, $0 < k < 1$. Observe that $V(t,x) \leq b(\delta_1) < a(\varepsilon)$ for $(t,x) \in I \times B_{\delta_1}$ and $V(t,x) \geq a(\varepsilon)$ for $(t,x) \in I \times \partial B_\varepsilon$. But if $(t_0,y_0) \in I \times B_{\delta_1}$ and if $||g(t,y)|| \leq \delta_2$, there can exist no $t_1 > t_0$ such that $y(t_1;t_0,y_0) \in \partial B_\varepsilon$. In fact, for $\delta_1 \leq ||y|| \leq \varepsilon$, one gets

$$\begin{aligned} \dot{V}_P(t,y) &\leq -c(||y||) + (\frac{\partial V}{\partial x}(t,y) \mid g(t,y)) \\ &\leq -c(\delta_1) + M\frac{kc(\delta_1)}{M} < 0. \end{aligned}$$

Q.E.D.

4.5. Theorem (I.G. Malkin [1944] and S. Gorsin [1948]). If f is lipschitzian in x uniformly with respect to t on $I \times \Omega$ and if the origin is uniformly asymptotically stable, then the origin is totally stable.

Proof. One has only to observe, using Theorem I.7.4 and its Corollary, that the hypotheses above imply those of Theorem 4.4. Q.E.D.

4.6. Obviously, the hypotheses of Theorem 4.4 do not imply that any solution $y(t;t_0,y_0)$ tends to 0 as $t \to \infty$: in fact, $g(t,y)$ does not vanish, nor does it fade down as $t \to \infty$. However, some kind of asymptotic property can be proved, as is apparent from the following theorem.

Theorem (cf. I.G. Malkin [1952]). In the hypotheses of Theorem 4.4, $(\forall\varepsilon > 0)(\exists\delta_1 > 0)(\forall\eta > 0)(\exists\delta_2' > 0)(\forall t_0 \in I)$ if

$$y_0 \in B_{\delta_1}$$

and if

$$(\forall t \geq t_0)(\forall x \in B_\varepsilon) \qquad ||g(t,y)|| < \delta_2' ,$$

then there is a $T > 0$ such that $(\forall t \geq T)\ y(t;t_0,y_0) \in B_\eta$.

Proof. For any $\varepsilon > 0$, let δ_1 and δ_2 be chosen as in the definition of total stability. Thus every solution starting from B_{δ_1} remains in B_ε, providing $||g(t,y)|| < \delta_2$ in the appropriate region. Consider an η, $0 < \eta < \delta_1$ and let δ_1' and $\delta_2' < \delta_2$ be chosen as in the definition of total stability, with ε replaced by η. Further let $\delta_2' < kc(\delta_1')/M$ for some k, $0 < k < 1$. Then, for $\delta_1' \leq ||y|| \leq \varepsilon$ and if $||g(t,y)|| < \delta_2'$ on $[t_0,\infty[\ \times B_\varepsilon$, one gets

$$\begin{aligned}\dot{V}_P(t,y) &\leq -c(||y||) + \left(\frac{\partial V}{\partial x}(t,y) \,\Big|\, g(t,y)\right) \\ &\leq -c(\delta_1') + M\,\frac{kc(\delta_1')}{M} < 0.\end{aligned}$$

Thus, for $y_0 \in B_{\delta_1}$, $y(t;t_0,y_0)$ enters at least once inside $B_{\delta_1'}$. But it remains thereafter in B_η, because of total stability. Q.E.D.

4.7. Exercise. Show that in the definition of total stability, δ_1 and δ_2 may be replaced by a single $\delta = \delta_1 = \delta_2$.

4.8. Exercise (W. Hahn [1967]). Prove that, in Theorem 4.4, $V(t,x) \leq b(||x||)$ can be replaced by: $V(t,0) = 0$.

4.9. Exercise (J.L. Massera [1956 and 1958]). If the origin is totally stable for a linear differential equation $\dot{x} = A(t)x$, (A a continuous matrix of order n), then it is uniformly asymptotically stable.

4.10. Exercise (J.L. Massera [1956 and 1958]). Prove that it is not true that, for an equation $\dot{x} = f(x)$, $f \in \mathscr{C}^1$, $f(0) = 0$, total stability implies uniform asymptotic stability. Hint: consider a function $f: \mathscr{R} \to \mathscr{R}$, vanishing at a sequence of points x_i approaching zero monotonically, and such that $xf(x) < 0$ for $x \neq 0$, $x \neq x_i$.

5. The Frequency Method for Stability of Control Systems

5.1. In this section, we shall illustrate, using the example of a nuclear reactor, one of the very few practical methods for constructing a Liapunov function, namely the so-called frequency method, due to V.M. Popov [1962].

5.2. The example of a nuclear reactor. Standard parameters to describe the state of a nuclear reactor are the fast neutron's density v, $v > 0$, and the temperatures $u \in \mathscr{R}^n$ of its various constituents (cf. H.B. Smets [1961]). The neutron's density satisfies an equation

$$\dot{v} = kv \tag{5.1}$$

where the reactivity k is a linear function of the state $k = k_0 - r^T u - \xi v$, $r \in \mathscr{R}^n$, r^T the transpose of r, $\xi \in \mathscr{R}$. Using Newton's law for heat transfer, one gets as equation for the temperatures

$$\dot{u} = Au - bv \tag{5.2}$$

with A a real matrix of order n and $b \in \mathscr{R}^n$. The system of equations (5.1), (5.2) has the equilibrium

$$u_0 = k_0 A^{-1} b/(\xi + r^T A^{-1} b),$$
$$v_0 = k_0/(\xi + r^T A^{-1} b).$$

Remembering that $v > 0$ and using the transformation of variables

$$x = u - u_0$$
$$\zeta = -\frac{v_0}{k_0}\left(\ln \frac{v}{v_0} + r^T A^{-1} x\right),$$

one obtains, after some computation, the system

$$\begin{aligned} \dot{x} &= Ax - b\phi(\sigma), \\ \dot{\zeta} &= \phi(\sigma), \\ \sigma &= c^T x - \gamma\zeta, \end{aligned} \tag{5.3}$$

where $\phi(\sigma) = v_0(e^\sigma - 1)$, $c^T = -r^T A^{-1}$ and $\gamma = k_0/v_0$. Notice by the way that $\sigma\phi(\sigma) > 0$ when $\sigma \neq 0$.

A system like (5.3) is equivalent to what is known as an indirect control system. On this notion, see e.g., S. Lefschetz [1965]. We shall not dwell any further here on the technical description of a nuclear reactor, nor on the derivation of its equation, but rather we shall prove some sufficient conditions for the global asymptotic stability of the critical point at the origin for the system (5.3) considered as defining the unknown functions x and σ. The following lemma, which we quote without proof, will be used to build an appropriate Liapunov function.

5.3. Lemma (V.A. Yacubovich [1962], R.E. Kalman [1963]). Let A be a matrix of order n, whose eigenvalues have strictly negative real parts, D a symmetric, positive definite matrix of order n, let $b \in \mathscr{R}^n$, $b \neq 0$, $k \in \mathscr{R}^n$ and let τ and ε be real scalars, $\tau \geq 0$, $\varepsilon > 0$. Then a necessary and sufficient condition for the existence of a symmetric matrix B of order n (necessarily positive definite) and a $q \in \mathscr{R}^n$ such that

(a) $A^T B + BA = -qq^T - \varepsilon D$;

(b) $Bb - k = \sqrt{\tau}\, q$,

is that ε be small enough and that the inequality

$$\tau + 2\mathrm{Re}(k^T(i\omega E - A)^{-1}b) > 0$$

be satisfied for all real ω.

5.4. The following theorem pertains to Equations (5.3). Remember that $\phi(\sigma)\sigma > 0$ whenever $\sigma \neq 0$.

Theorem (R.E. Kalman [1963]). Suppose all eigenvalues of A have strictly negative real parts, $\gamma > 0$ and $c^T b + \gamma > 0$. If further, there exist real constants α and β such that

(i) $\alpha \geq 0$, $\beta \geq 0$, $\alpha + \beta > 0$;

(ii) for all real ω:

$$\operatorname{Re}[(2\alpha\gamma+i\omega\beta)(c^T(i\omega E-A)^{-1}b + \tfrac{\gamma}{i\omega})] > 0, \tag{5.4}$$

then the origin $x = 0$, $\sigma = 0$ is globally asymptotically stable for the system (5.3).

Proof. Consider, for some matrix B to be determined later, the function

$$V(x,\sigma) = x^T Bx + \alpha(\sigma - c^T x)^2 + \beta\int_0^\sigma \phi(\xi)d\xi,$$

whose derivative along the solutions of (5.3) reads

$$\dot{V}(x,\sigma) = x^T(A^T B + BA)x - 2x^T(Bb-k)\phi(\sigma)$$
$$- \tau\phi^2(\sigma) - 2\alpha\gamma\phi(\sigma)\sigma,$$

where

$$k = \frac{1}{2}\beta A^T c + \alpha\gamma c,$$

$$\tau = \beta(c^T b + \gamma).$$

This quantity k should not be mistaken for the k of Section 5.2.

Now $V(x,\sigma)$ is positive definite and approaches ∞ with $||x|| + |\sigma|$. Using Lemma 5.3, we compute

$$\dot{V}(x,\sigma) = -\varepsilon x^T Dx - (x^T q + \sqrt{\tau}\phi(\sigma))^2 - 2\alpha\gamma\phi(\sigma)\sigma,$$

from which it follows readily that $\dot{V}(x,\sigma)$ is negative definite provided that, for $\omega \in \mathscr{R}$,

$$\tau + 2\operatorname{Re}[k^T(i\omega E-A)^{-1}b] > 0.$$

But this inequality is equivalent to (5.4) since

$$\begin{aligned} &\mathrm{Re}[(2\alpha\gamma+i\omega\beta)(c^T(i\omega E-A)^{-1}b + \tfrac{\gamma}{i\omega})] \\ &= \beta\gamma + \mathrm{Re}[2\alpha\gamma c^T(i\omega E-A)^{-1}b + \beta c^T(i\omega E-A+A)(i\omega E-A)^{-1}b] \\ &= \beta(c^Tb+\gamma) + \mathrm{Re}[(2\alpha\gamma c^T+\beta c^TA)(i\omega E-A)^{-1}b] \\ &= \tau + 2\mathrm{Re}[k^T(i\omega E-A)^{-1}b]. \end{aligned}$$

The thesis now follows from Corollary I.6.9. Q.E.D.

5.5. Remark. Of course, Theorem I.6.2 and its Corollary I.6.9 have been established for an equation in normal form $\dot{x} = f(t,x)$, which is not the form of (5.3) considered as defining the unknown functions x and σ. But it can be proved (e.g., S. Lefschetz [1965]) that (5.3) can be reduced to the normal form, and therefore, the difficulty mentioned is not essential.

5.6. Theorem 5.4 is typical of a class of propositions based on a frequency response diagram of the linear part of the control system. The results obtained by this method are, in a way, stronger and more effective than those considered in the preceding sections. Indeed, they apply to a whole class of systems, namely those corresponding to any function ϕ with $\phi(\sigma)\sigma > 0$ when $\sigma \neq 0$, and they yield the auxiliary function explicitly. Further, the so-called frequency criterion is the best possible in the sense that the proposed Liapunov function proves asymptotic stability if and only if this criterion is satisfied. However, the scope of the method is unquestionably narrower than the one of Liapunov's direct method as such.

6. Non-Differentiable Liapunov Functions

6.1. All auxiliary functions introduced up to here were $\mathscr{C}^1$ functions. But, as the example of a transistorized circuit will show, it may happen that the "natural" Liapunov function is not that regular. Hence, the interest of generalizing the theorems of Liapunov's second method to encompass the case of less smooth functions V. Using the results of Appendix I, it is possible to prove anew most theorems of Chapters I and II, while imposing on the auxiliary functions nothing more than a local Lipschitz condition with respect to x and continuity in (t,x). These considerations motivate the level of generality adopted in most succeeding chapters. As for the preceding theorems, it is a mere exercise to rewrite them in this new setting. For example, Theorems I.4.2 and I.6.2, and Corollary I.6.9 would read as follows, and their proofs use mainly Theorems 2.1, 3.3 and 4.3 of Appendix I.

6.2. Theorem. If there exists a continuous function $V: I \times \Omega \to \mathscr{R}$ which is locally lipschitzian in x and such that, for some $a \in \mathscr{K}$ and every $(t,x) \in I \times \Omega$:

(i) $V(t,x) \geq a(||x||)$; $V(t,0) = 0$;

(ii) $D^+V(t,x) \leq 0$;

then, the origin is stable.

6.3. Theorem. Suppose there exists a continuous function $V: I \times \Omega \to \mathscr{R}$ which is locally lipschitzian in x and such that, for some functions $a,b,c \in \mathscr{K}$ and every $(t,x) \in I \times \Omega$:

(i) $a(||x||) \leq V(t,x) \leq b(||x||)$;

(ii) $D^+V(t,x) \leq -c(||x||)$.

Choosing $\alpha > 0$ such that $\bar{B}_\alpha \in \Omega$, let us put for every $t \in I$

$$V_{t,\alpha}^{-1} = \{x \in \Omega: V(t,x) \leq a(\alpha)\}.$$

Then

(a) for any $t_0 \in I$ and any $x_0 \in V_{t_0,\alpha}^{-1}$: $x(t;t_0,x_0) \to 0$ uniformly in t_0,x_0 when $t \to \infty$;

(b) the origin is uniformly asymptotically stable.

6.4. Corollary. The origin is uniformly globally asymptotically stable if the assumptions of Theorem 6.3 are satisfied for $\Omega = \mathscr{R}^n$ and $a(r) \to \infty$ as $r \to \infty$.

6.5. Example of a transistorized network. In this section we assume that the reader has some familiarity with the terminology and symbolism of electrical engineering. Consider the network shown in Figure 2.2, where n transistors $T_1,\ldots,T_n$ are plugged into a linear resistive n-port with

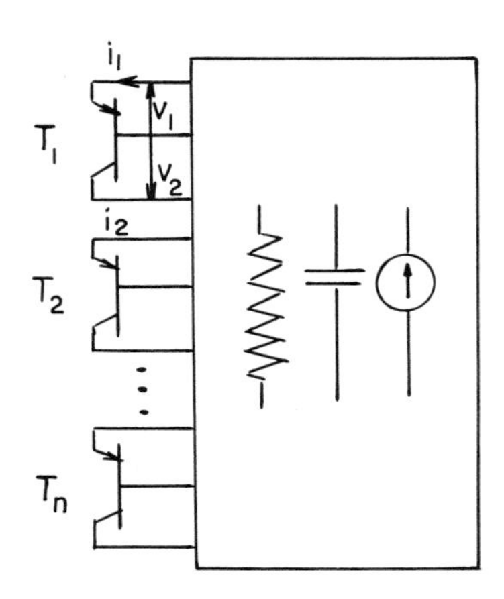

Figure 2.2. Transistor network

sources. Each transistor will be described by the circuit diagram of Figure 2.3, which embodies the models of

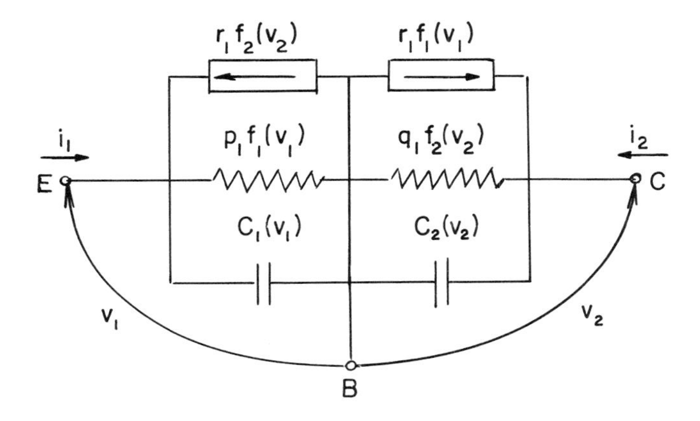

Figure 2.3. Model of the transistor

H.K. Gummel [1968], generalizing the classical Ebers and Moll's model. In Figure 2.3, the subscripts pertain to T_1. The circuit consists of nonlinear resistors $p_1 f_1(v_1)$ and $q_1 f_2(v_2)$, voltage controlled current sources $r_1 f_2(v_2)$, $r_1 f_1(v_1)$ and nonlinear junction capacitors $C_1(v_1), C_2(v_2)$. No bulk resistors for the base (B), emitter (E) or collector (C) have been considered, since these can be thought of as embedded in the linear resistive n-port. The corresponding equations are as follows:

$$i_1 = C_1(v_1)\dot{v}_1 + p_1 f_1(v_1) - r_1 f_2(v_2),$$

$$i_2 = C_2(v_2)\dot{v}_2 - r_1 f_1(v_1) + q_1 f_2(v_2).$$

For our purpose, we suppose the following:

(i) the constants p_i, q_i, r_i are > 0;

(ii) the functions $f_i\colon \mathscr{R} \to \mathscr{R}$ are strictly increasing continuous functions;

(iii) the $C_i\colon \mathscr{R} \to \mathscr{R}$ are continuous functions, such that for some $\varepsilon > 0$, $C_i(v) \geq \varepsilon > 0$ for any v.

Let $i = (i_1, i_2, \ldots, i_{2n})$, $v = (v_1, v_2, \ldots, v_{2n})$, $C(v) = \operatorname{diag}[C_1(v_1), C_2(v_2), \ldots, C_{2n}(v_{2n})]$,

$$T = \operatorname{diag}\left[\begin{pmatrix} p_1 & -r_1 \\ -r_1 & q_1 \end{pmatrix}, \ldots, \begin{pmatrix} p_n & -r_n \\ -r_n & q_n \end{pmatrix}\right],$$

$F(v) = [f_1(v_1), f_2(v_2), \ldots, f_{2n}(v_{2n})]$. The transistors equations then read

$$i = C(v)\dot{v} + TF(v) \tag{6.1}$$

In a similar way, we obtain the following admittance equation for the resistive n-port:

$$i = -Gv + b, \tag{6.2}$$

where G is a constant matrix of order $2n$ and b is a constant $2n$-vector. From (6.1) and (6.2), we get the network equation

$$C(v)\dot{v} = -TF(v) - Gv + b. \tag{6.3}$$

To investigate the stability of a critical point $v = v_0 = (v_{10}, \ldots, v_{2n\,0})$, i.e., a point such that

$$-TF(v_0) - Gv_0 + b = 0, \tag{6.4}$$

we shall use the auxiliary functions

$$V_i(v_i) = \left|\int_{v_{i0}}^{v_i} c_i(v_i)dv_i\right|, \qquad 1 \leq i \leq 2n.$$

If $v_1 \neq v_{10}$, we get using (6.4)

$$D^+V_1(v_1) = \operatorname{sign}(v_1-v_{10})[-p_1(f_1(v_1) - f_1(v_{10}))$$

$$+ r_1(f_2(v_2) - f_2(v_{20})) - \sum_{1\leq j\leq 2n} G_{1j}(v_j-v_{j0})].$$

Observing that at $v_1 = v_{10}$, the left derivative of V_1 equals minus the right derivative, we obtain

$$D^+V_1(v_{10}) = |r_1(f_2(v_2) - f_2(v_{20})) - \sum_{2\leq j\leq 2n} G_{1j}(v_j-v_{j0})|.$$

Therefore,

$$D^+V_1(v_1) \leq -p_1|f_1(v_1) - f_1(v_{10})| + r_1|f_2(v_2) - f_2(v_{20})|$$

$$- G_{11}|v_1-v_{10}| + \sum_{2\leq j\leq 2n} |G_{1j}|\cdot|v_j-v_{j0}|.$$

One obtains a similar inequality for $D^+V_2(v_2)$ and further for any $D^+V_i(v_i)$, i odd or even.

Consider next the auxiliary function

$$V(v) = \sum_{1\leq i\leq 2n} d_iV_i(v_i)$$

where $d_i > 0$. We obtain right away, for any v,

$$V(v) \geq \varepsilon \sum_{1\leq i\leq 2n} d_i|v_i-v_{i0}|$$

and

$$D^+V(v) \leq - \sum_{1\leq k\leq n} \{[d_{2k-1}p_k - d_{2k}r_k] |f_{2k-1}(v_{2k-1}) - f_{2k-1}(v_{2k-1\ 0})|$$

$$+ [-d_{2k-1}r_k + d_{2k}q_k] |f_{2k}(v_{2k}) - f_{2k}(v_{2k\ 0})|\}$$

$$- \sum_{1\leq i\leq 2n} (d_i G_{ii} - \sum_{\substack{1\leq j\leq 2n \\ j\neq i}} d_j |G_{ji}|) |v_i - v_{i0}|.$$

In order to state a condition for $D^+V(v)$ to be negative definite, the following definition will be useful: a matrix (a_{ij}) of order n will be said to be column sum dominant if

$$a_{ii} - \sum_{\substack{1\leq j\leq n \\ j\neq i}} |a_{ji}| > 0 \qquad 1 \leq i \leq n.$$

Then, of course, $D^+V(v)$ is negative definite if $D(T+G)$, where $D = \mathrm{diag}(d_i)$, is column sum dominant, and using Corollary 6.4, we obtain the following result.

6.6. Theorem (cf. I.W. Sandberg [1969]). The rest solution $v = v_0$ of the network equation (6.3) is globally asymptotically stable if there exists a positive diagonal matrix $\mathrm{diag}(d_i)$ which makes $D(T+G)$ column sum dominant.

6.7. Exercise. Consider a network without sources $(b = 0)$ and with transistors such that $f_{2i-1}(v) \equiv f_{2i}(v)$, $f_{2i}(0) = 0$. Suppose further that for any i: $p_i > r_i$ and $q_i > r_i$. Prove that the solution $v = 0$ is globally asymptotically stable if G is a positive definite matrix. Hint: use as Liapunov function

$$V(v) = \sum_{1\leq i\leq 2n} \int_0^{v_i} v_i C_i(v_i)\, dv_i.$$

7. Bibliographical Note

Theorem 1.3 and Corollary 1.5 prove asymptotic stability or similar properties by using the fact that the origin is the only complete positive semi-trajectory contained in the set of points where $\dot{V}(t,x) = 0$. This idea, applied to autonomous systems, appears in E.A. Barbashin and N.N. Krasovski [1952] for stability and in N.N. Krasovski [1956] for instability. The extension to periodic systems comes from N.N. Krasovski [1959]. More bibliography can be found in this last reference.

Theorem 2.5 is substantially due to V.M. Matrosov [1962], who gave also several variants of the same proposition, along with some instability results. Another theorem of the same type, but where W is a vector function (N. Rouche [1968]) will be introduced in Chapter VII, followed by several more recent results.

The comparison method, a detailed account of which will be the object of Chapter IX, has been used by R. Conti [1956] to prove a continuability result. C. Corduneanu applied it to stability problems [1960], as well as to partial stability [1964].

As for total stability, if the second member of the differential equation is viewed as belonging to some appropriate function space, the perturbations $g(t,x)$ in Definition 4.3 are estimated in the topology of uniform convergence. Various extensions of the concept of total stability have been proposed and studied. They correspond to variations of the topology of the space of second members. For instance, N.N. Krasovski [1959] studied stability under persistent disturbances that are bounded in the mean, and I. Vrkoč [1959], cited by W. Hahn [1967] introduced integral stability.

Concerning the frequency method, a subject matter which will not be considered any further here, the relevant references have been given in the text. For more details, see e.g. V.M. Popov [1966], or A. Halanay [1963].

Theorem 6.6 on the transistorized network is a slight generalization of the one given by I.W. Sandberg [1969], for whom the sufficient condition of global asymptotic stability was the existence of a positive diagonal matrix $\text{diag}(d_i)$ making $\text{diag}(d_i)G$ and $(\text{diag } d_i)T$ simultaneously column sum dominant. On an analogous problem of circuit theory, see D. Mitra and C. So. Hing [1972].

CHAPTER III

STABILITY OF A MECHANICAL EQUILIBRIUM

1. Introduction

This chapter is devoted to stability questions concerning mechanical equilibria, or in other words to stability of critical points for differential equations having the Lagrangian or Hamiltonian form, with or without friction forces. The central theorem in this context was stated by J.L. Lagrange [1788]: roughly speaking, it asserts that a mechanical equilibrium of a conservative system is stable at each point where the potential function is strictly minimum. Lagrange himself (as well as S.D. Poisson [1838]) failed to give a proof for a potential function more general than a quadratic form. G. Lejeune-Dirichlet [1846] gave an elegant general proof which yielded the model for the entire direct method of Liapunov.

In this chapter, we first study Lagrange-Dirichlet's theorem along with some of its variants: it concerns stability as such or some kind of partial stability. Next, we consider the inversion of this theorem, a classical incompletely solved problem of theoretical mechanics. Suppose that, at some equilibrium position, the potential function has no strict minimum. One knows of some cases where stability then still prevails (cf. Theorem 2.6). The question is: what additional conditions would enforce instability? Several answers will be proposed, first using an auxiliary function and then by considering the first approximation. But to find the mildest possible conditions remains an open problem, and will be left to the reader as a nice pastime for winter evenings. The last section of the chapter is devoted to asymptotic stability or instability of a mechanical equilibrium in the presence of complete dissipation.

2. The Lagrange-Dirichlet Theorem and Its Variants

2.1. We consider first a conservative Lagrangian system as described in Section 13 of Appendix II, but without assuming T to be positive definite with respect to $\dot{q}$. In this case, the Lagrangian equations of motion cannot be solved for the generalized acceleration vector $\ddot{q}$, and thus they are not reducible to the normal form $\dot{x} = f(t,x)$. That is why, rigorously speaking, no theorem of Chapters I or II can be applied to study the stability of the equilibrium $q = \dot{q} = 0$. However, it is possible to prove, by a direct line of reasoning from the integral of energy, that if the potential function $\Pi(q)$ has a strict minimum at $q = 0$, one gets what might be called partial stability of the equilibrium with respect to q. This is the content of Dirichlet's original contribution.

Modern versions of this theorem actually assume a bit more, namely positive definiteness of T, and prove a bit more, namely stability with respect to q and $\dot{q}$. In this last context it is substantially equivalent and more convenient to consider the Hamiltonian equations, which assume de facto the normal form. The equilibrium is then considered at the origin of phase space; i.e., the space of q and p. Concerning stability with respect to q as well as some further historical discussion, we refer the reader to N. Rouche and K. Peiffer [1967], where one will also find an extension of the Lagrange-Dirichlet theorem to mechanical systems with time-dependent constraints.

2.2. We now consider the Hamiltonian system described in Section 14 of Appendix II. Let us, for convenience, transcribe the equations of motion:

$$\dot{q} = \frac{\partial H}{\partial p}(q,p), \quad \dot{p} = -\frac{\partial H}{\partial q}(q,p) \tag{2.1}$$

where

$$H(q,p) = T(q,p) + \Pi(q).$$

The general hypotheses considered in the Appendix will remain in force up to Section 5 included. We recall them briefly.

2.3. General Hypotheses. The kinetic energy $T(q,p) = \frac{1}{2} p^T B(q) p$ where $B(q)$ is symmetric and $\mathscr{C}^1$, and $B(0)$ is positive definite. The potential energy $\Pi(q)$ is $\mathscr{C}^1$, and further $\Pi(0) = 0$ and $\frac{\partial \Pi}{\partial q}(0) = 0$.

2.4. Theorem. If $\Pi(q) > 0$ for $q \neq 0$, the origin $q = p = 0$ is stable.

This is an immediate consequence of Theorem I.4.2, when observing that H is an integral of the equations of motion and is positive definite on some neighborhood of the origin.

2.5. As observed first by P. Painlevé [1904] and then by A. Wintner [1941], Theorem 2.4 is not the best possible: the hypothesis can be weakened, as is shown by the following theorem. In a way, the improvement is academic, but it will throw some light on the inversion problem, to be dealt with in further sections.

2.6. Theorem. Suppose that for any $\eta > 0$ with $\overline{B}_\eta \subset \Omega$, there exists an open set $\Psi \subset \mathscr{R}^n$ such that $0 \in \Psi \subset B_\eta$ for every $q \in \partial\Psi$: $\Pi(q) > 0$; then the origin $q = p = 0$ is stable.

Proof. As was shown in Appendix II, if Ω has been chosen appropriately, there exists a function $a \in \mathscr{K}$ such that $T(q,p) \geq a(||p||)$. Let then $\varepsilon > 0$ be given arbitrarily and choose $\eta > 0$ such that $\overline{B}_\eta \subset \Omega$, and for every $q \in \overline{B}_\eta$: $|\Pi(q)| < a(\varepsilon)/2$. This is possible since Π is continuous and $\Pi(0) = 0$. Further, choose an open set $\Psi \subset B_\eta$ according to the hypothesis of the theorem, and select $\mu > 0$ such that

$$\mu < \inf\{\Pi(q): q \in \partial\Psi\}$$

and, therefore that

$$\mu < a(\varepsilon)/2.$$

Finally, choose $\delta > 0$ such that for every q_0, $||q_0|| < \delta$ and every p_0, $||p_0|| < \delta$: $H(q_0,p_0) < \mu$.

Then, (q_0, p_0) being the initial conditions, one gets firstly that $||q(t)|| < \eta < \varepsilon$ for every $t \geq t_0$, because

$$\Pi(q(t)) \leq H(q(t),p(t)) = H(q_0,p_0) < \mu,$$

and secondly that $||p(t)|| < \varepsilon$ for every $t \geq t_0$, because

$$a(||p(t)||) \leq T(q(t),p(t)) = H(q_0,p_0) - \Pi(q(t)) \leq a(\varepsilon).$$ Q.E.D.

2.7. Exercise. In the hypotheses of Theorem 2.6 and with δ chosen as in the proof, show that $||q(t)|| < \varepsilon$ and $||p(t)|| < \varepsilon$ for every $t \in \mathscr{R}$ (replacing "for every $t \geq t_0$"). Stability is thus proved for past as well as future.

2.8. Exercise (A. Wintner [1941]). Consider a system with one degree of freedom ($q \in \mathscr{R}$), such that

$$\Pi(q) = e^{-1/q^2}\cos\frac{1}{q}, \quad q \neq 0$$

and $\Pi(0) = 0$. Then the origin $q = p = 0$ is stable.

2.9. Let us now show that for a mechanical system with one degree of freedom ($n = 1$), the sufficient stability condition given by Theorem 2.6 is also necessary. Observe first that if $q \in \mathscr{R}$, negating the hypothesis of Theorem 2.6 amounts to asserting that, for some $\eta > 0$, with $[-\eta,\eta] \subset \Omega$, one has $\Pi(q) \leq 0$ either for any $q \in [0,\eta]$ or for any $q \in [-\eta,0]$. Both cases can be treated alike. Therefore, it will suffice to prove the following theorem.

2.10. Theorem. Suppose that

(i) $q \in \mathscr{R}$;

(ii) $B(0) > 0$;

(iii) $(\exists\eta > 0,\ [0,\eta] \subset \Omega)(\forall q \in [0,\eta])\ \Pi(q) \leq 0$;

then the origin $q = p = 0$ is unstable.

Proof. Suppose Ω has been chosen small enough in order that, for some constant $a > 0$ and every $q \in \Omega$: $a < B(q)$. Then choose initial conditions $t_0 = 0$, $q_0 = 0$ and $p_0 > 0$, p_0 arbitrarily small. The integral of energy reads

$$B(q(t))\frac{p(t)^2}{2} + \Pi(q(t)) = B(0)\frac{p_0^2}{2}. \tag{2.2}$$

Observe that $\dot{q}(0) = B(0)p_0 > 0$, and therefore $q(t) \geq 0$ for some time after $t = 0$. But one deduces from (2.2) and (iii) that as long as $0 \leq q(t) \leq \eta$:

$$p(t)^2 \geq \frac{B(0)}{B(q(t))} p_0^2,$$

and further, since $\dot{q}(t) = B(q(t))p(t)$, that $\dot{q}(t) \geq ap_0$. Therefore, $q(t)$ becomes never smaller than zero, and $q(t) \geq ap_0t$, which proves instability. Q.E.D.

2.11. Unfortunately, this line of reasoning cannot be extended to $\mathscr{R}^n$ for $n > 1$, except if one assumes the very strong condition that $\Pi(q) \leq 0$ on some line segment having one of its end points at $q = 0$ along with some further hypotheses. A particular case of this situation is proposed in the following exercise.

2.12. Exercise. For $q = (q_1,q_2) \in \mathscr{R}^2$, suppose $\Pi(q) \geq 0$ and $\Pi(q) = 0$ if and only if $q_2 = 0$. Find conditions on $T(q,p)$ to ensure the instability of the origin (cf. G. Hamel [1903], L. Silla [1908]).
Hint: try a solution $q(t) = (q_1(t),0)$.

2.13. Let us now show, by way of a counterexample (cf. M. Laloy [1975]), that the hypothesis of Theorem 2.6 is not a necessary condition of stability, as soon as $n \geq 2$. Indeed, consider the equations

$$\dot{q}_1 = p_1, \quad \dot{p}_1 = -\frac{\partial \Pi}{\partial q_1}(q), \tag{2.3}$$

$$\dot{q}_2 = p_2, \quad \dot{p}_2 = -\frac{\partial \Pi}{\partial q_2}(q), \tag{2.4}$$

where

$$\Pi(q) = e^{-1/q_1^2}\cos\frac{1}{q_1} - e^{-1/q_2^2}[\cos\frac{1}{q_2} + q_2^2]$$

for $q \neq 0$, and $\Pi(0) = 0$. One has $\Pi(q) < 0$ at every point where $q_1 = q_2 \neq 0$. However, the origin is stable, as is seen by applying Theorem 2.6 separately to (2.3) and to (2.4). Observe that Π is a $\mathscr{C}^\infty$ function.

2.14. So we know now that negating the hypothesis of Theorem 2.6 does not entail instability. It will be found interesting to discuss how far we can go, in these conditions, i.e., by negating the hypothesis in question, on the way towards proving instability. The present subsection utilizes some notions to be introduced in Chapter V only: thus, it will be wise not to include it in a first reading of the book.

The opposite of the hypothesis of Theorem 2.6 reads:

$$(\exists \eta > 0, \overline{B}_\eta \subset \Omega)(\forall \text{ open set } \Psi, 0 \in \Psi \subset B_\eta)(\exists q \in \partial\Psi) \quad \Pi(q) \leq 0. \tag{2.5}$$

The set $A = \{q \in \mathscr{R}^n : \Pi(q) \leq 0 \text{ or } ||q|| \geq \eta\}$ is closed and contains the origin. Let A_0 be the connected component of A which contains the origin, and suppose ab absurdo that $A_0 \subset B_\eta$. Then A_0 is compact and disjoint from $A \setminus A_0$. Let $\varepsilon > 0$ be such that $\Psi = \{q: d(A_0, q) < \varepsilon\} \subset C(A \setminus A_0) \subset B_\eta$. This Ψ is an open subset of B_η, and for every $q \in \partial\Psi$: $\Pi(q) > 0$, which contradicts (2.5).

Consider now the set

$$G = \{(t,q,p) \in \mathcal{R} \times \Omega \times \mathcal{R}^n: t \in \mathcal{R}, ||q|| + ||p|| < \eta,$$
$$T(q,p) + \Pi(q) \leq 0\}. \qquad (2.6)$$

Due to the existence of the energy integral, any solution starting in G cannot come out of it except by violating the condition $||q|| + ||p|| < \eta$. If (2.5) is verified, it is clear that the origin $(q,p) = (0,0)$ is, for any t, a cluster point for

$$G(t) = \{(q,p): (t,q,p) \in G\}.$$

Therefore, G is an absolute sector in the sense defined at V.2.3. We have proved even more: this sector possesses a connected component extending to the subset of $\mathcal{R} \times \Omega \times \mathcal{R}^n$ where $||q|| + ||p|| = \eta$.

Owing to the existence of the counterexample 2.13, the absolute sector G need not to be an expeller. This is illustrated by the following exercise.

2.15. <u>Exercise</u> (M. Laloy [1975]). Show that, for $q \in \mathcal{R}$ and $p \in \mathcal{R}$, if the equations of motion are

$$\dot{q} = p, \ \dot{p} = -\frac{\partial \Pi}{\partial q}(q),$$

where $\Pi(q) = -q^8 \sin^2 \frac{1}{q^2}$ for $q \neq 0$ and $\Pi(0) = 0$, the origin is unstable, whereas the set G defined in (2.6) is not an expeller. Notice that Π is a $\mathcal{C}^2$ function

2.16. <u>Conjecture</u>. The hypothesis of Theorem 2.6 becomes necessary if $\Pi(q)$ is analytic?

2.17. So, except for $n = 1$, we have no condition which would be necessary and sufficient for stability. In the next two sections, we shall try to find some sufficient conditions for instability. This question is improperly known as the problem of "inversion of the Lagrange-Dirichlet Theorem". Before proceeding to this part of our study, let us propose another exercise which yields a useful complement to Theorem 2.10.

2.18. Exercise. Consider the system

$$\left.\begin{aligned} \dot{q}_i &= p_i, \\ \dot{p}_i &= -\frac{\partial \Pi}{\partial q_i}(q) \end{aligned}\right\} \quad 1 \le i \le n$$

where Π is a $\mathscr{C}^2$ function such that,

(i) $\Pi(0) = 0$ and $\Pi(q) \le 0$ for some $\eta > 0$ with $\overline{B}_\eta \subset \Omega$ and all $q \in B_\eta$;

(ii) $\Pi(q) = -U(r)$, where $r = \frac{1}{2} \sum_{1 \le i \le n} q_i^2$.

Then the origin is unstable.

3. Inversion of the Lagrange-Dirichlet Theorem Using Auxiliary Functions

3.1. We come back now to a Hamiltonian system with n degrees of freedom. The following theorem is typical of a class of inverse Lagrange-Dirichlet theorems to be illustrated in the present section.

3.2. Theorem (N.G. Chetaev [1952]). If there exists $\varepsilon > 0$ (with $\overline{B}_\varepsilon \subset \Omega$) such that

(i) $\theta = \{q \in B_\varepsilon : \Pi(q) < 0\} \neq \phi$;

(ii) $0 \in \partial\theta$;

(iii) $(\frac{\partial \Pi}{\partial q} | \, q) < 0$ for every $q \in \theta$;

then, the origin $q = p = 0$ is unstable.

<u>Proof</u>. We first define the set

$$\Psi = \{(q,p): q \in \theta, \ ||p|| < \varepsilon, \ H(q,p) < 0, \ (q|p) > 0\}.$$

Clearly $(0,0) \in \partial\Psi$. The function $V = -(q|p)H$ is such that, for every $(q,p) \in \Psi$

(a) $0 < V(q,p)$;

(b) $\dot{V} = -[-(q| \frac{\partial H}{\partial q}) + (\frac{\partial H}{\partial p}| p)]H.$

Using Euler's theorem on homogeneous functions, we obtain that

$$\dot{V} = -[2T - (\frac{\partial T}{\partial q}| q) - (\frac{\partial \Pi}{\partial q}| q)]H.$$

As it changes nothing to lower the value of ε, we choose ε small enough to get

$$S(q,p) = 2T - (\frac{\partial T}{\partial q}| q) > 0 \quad \text{when} \quad ||q|| < \varepsilon \quad \text{and} \quad p \neq 0.$$

Such an ε exists since $S(q,p)$ is quadratic with respect to p and $S(0,p) = p^T B(0)p$ is positive definite. But then $\dot{V}(q,p) > 0$ for every $(q,p) \in \psi$. Instability is thus proved through Theorem I.5.1 and Remark I.5.5. Q.E.D.

3.3. <u>Exercise</u> (N.G. Chetaev [1952]). Prove the following generalization of Theorem 3.2.

Let assumptions (i) and (ii) of Theorem 3.2 be satisfied. Suppose further there exists a $\mathscr{C}^1$ function $f(q)$ on θ into $\mathscr{R}^n$ such that:

(i) $f(0) = 0$;

(ii) the matrix $[\frac{\partial f}{\partial q}(0)]^T B(0)^T + B(0) \frac{\partial f}{\partial q}(0)$ is positive definite;

(iii) $(\frac{\partial \Pi}{\partial q}| f) < 0$ for every $q \in \theta$;

then the origin $q = p = 0$ is unstable.

Hint: use the function $V = -(f,p)H$.

3.4. Theorems 3.2 and 3.3 are only partial inverses of the Lagrange-Dirichlet theorem. Indeed, they do not cover the cases of unstable systems with $\Pi(q) \geq 0$: compare with Theorem 2.10.

3.5. Exercise (L.N. Avdonin [1971]). Prove that assumption (iii) of Theorem 3.2 can be replaced by

(iii-a) $(\frac{\partial \Pi}{\partial q} | q) \neq 0$ for every $q \in \theta$.

Hint: (iii-a) implies (iii).

3.6. Corollary (N.G. Chetaev [1952]). If, in Theorem 3.2, assumption (iii) is replaced by

(iii-b) Π is analytic, $\Pi(q) = \sum_{2 \leq i < \infty} \Pi_i(q)$ where $\Pi_i(q)$ is a homogeneous form of degree i, and for some $k \geq 2$

$$\Pi_i(q) \geq 0 \text{ if } i < k, \ \Pi_i(q) \leq 0 \text{ if } i > k;$$

then the origin is unstable.

Proof. The proof follows from Theorem 3.2, since

$$(\frac{\partial \Pi}{\partial q} | q) = \sum_{2 \leq i < \infty} (\frac{\partial \Pi_i}{\partial q} | q) = k\Pi - \sum_{2 \leq i \leq k-1} (k-i)\Pi_i$$

$$+ \sum_{k+1 \leq i < \infty} (i-k)\Pi_i$$

and it is apparent that this quantity is < 0 for $q \in \theta$. Q.E.D.

3.7. Corollary (A.M. Liapunov [1897]). In Theorem 3.2, assumption (iii) can be replaced by:

(iii-c) $\Pi(q) = \Pi_m(q) + R(q)$ where $\Pi_m(q)$ is a negative definite, homogeneous form of degree m and $R(q)$ is such that

$$||\frac{\partial R}{\partial q}|| / ||q||^{m-1} \to 0 \quad \text{as} \quad q \to 0.$$

The proof is left as an exercise.

3.8. Exercise. Let Oxy be orthogonal coordinate axes in a plane. Suppose a material particle of mass m is connected to each of the four points $a = (1,0)$, $b = (0,1)$, $c = (-1,0)$, $d = (0,-1)$ by means of four identical springs of constant k and whose natural length (i.e., their length in the unstretched state) is equal to $l > 0$. Obviously, the origin 0 is an equilibrium. Show that it is stable or unstable according to whether $l < 2$ or $l > 2$. In other words, stability prevails as long as l is smaller than the diagonal of the square abcd, and instability when l is larger.

4. Inversion of the Lagrange-Dirichlet Theorem Using the First Approximation

4.1. The Hamiltonian system studied here is the same as in the preceding sections. Remember, in particular, that $B(0)$ is positive definite.

4.2. Theorem (A.M. Liapunov [1897]). Suppose $\Pi(q) = \Pi_2(q) + R(q)$, where $\Pi_2(q)$ is a quadratic form and $||\partial R/\partial q|| / ||q|| \to 0$ as $q \to 0$. Then, if $\Pi_2(q)$ takes strictly negative values, the origin is unstable.

Proof. Let $\Pi_2(q) = \frac{1}{2} q^T C q$ with C symmetric. Then Hamilton's equations (2.1) read

$$\dot{q} = B(0)p + g_1(q,p), \quad \dot{p} = -Cq + g_2(q,p),$$

where

$$g_1(q,p) = (B(q) - B(0))p$$

and

$$g_2(q,p) = -\frac{\partial}{\partial q}\left[R(q) + \frac{1}{2}p^T B(q)p\right]$$

are such that

$$\frac{||(g_1(q,p), g_2(q,p))||}{||(q,p)||} \to 0 \quad \text{as} \quad ||(q,p)|| \to 0.$$

From Theorem I.5.8, the origin is unstable whenever the matrix

$$\begin{pmatrix} 0 & -C \\ B(0) & 0 \end{pmatrix}$$

has at least one eigenvalue with strictly positive real part. The corresponding characteristic equation is

$$\det\begin{pmatrix} -\lambda E & -C \\ B(0) & -\lambda E \end{pmatrix} = 0.$$

For non-zero eigenvalues, it reads successively

$$0 = \det\begin{pmatrix} -\lambda E & -C \\ B(0) & -\lambda E \end{pmatrix}\begin{pmatrix} \lambda E & 0 \\ B(0) & \lambda^{-1}E \end{pmatrix}$$

$$= \det\begin{pmatrix} -\lambda^2 E - CB(0) & -\lambda^{-1}C \\ 0 & -E \end{pmatrix}$$

$$= \det(\lambda^2 E + CB(0)). \tag{4.1}$$

Since $B(0)$ is symmetric positive definite, there exists a regular matrix Q such that $B(0) = Q^TQ$. But then the equation may be transformed thus:

$$0 = \det(\lambda^2 E+CB(0)) = \det Q(\lambda^2 E+CQ^TQ)Q^{-1} = \det(\lambda^2 E+QCQ^T).$$

Putting $q = Q^T\overline{q}$, one gets that $\Pi_2 = \frac{1}{2}\,\overline{q}^TQCQ^T\overline{q}$. As Π_2 takes negative values, QCQ^T has a strictly negative eigenvalue. Therefore, (4.1) has a strictly positive solution. Q.E.D.

4.3. This theorem includes Corollary 3.7 when $m = 2$; in this case Π_2 is negative definite. But it does not include unstable systems such that $\Pi_2(q) = \frac{1}{2}\,q^TCq \geq 0$. In order to cope with such systems, we introduce the following lemma, that describes $\Pi(q)$ when C has a single zero eigenvalue.

4.4. Lemma. Suppose $\Pi(q)$ is analytic and $\frac{\partial^2\Pi}{\partial q^2}(0)$ is a positive semi-definite matrix with a single zero eigenvalue. Then, using appropriate coordinates, Π can be written as

$$\Pi(q) = \frac{1}{2}\,\overline{q}^TB(q_1,\overline{q})\overline{q} + \beta(q_1)q_1^m, \qquad (4.2)$$

where $\overline{q} = (q_2,\ldots,q_n)$, $B(q_1,\overline{q})$ and $\beta(q_1)$ are analytic, $B(0,0)$ is positive definite and $m \geq 3$.

Proof. It follows easily from the assumptions that, using appropriate coordinates

$$\Pi(q) = \frac{1}{2}\,\overline{q}^TB_0\overline{q} + f(q_1,\overline{q})$$

where B_0 is a positive definite constant matrix and

$$f(q_1,\overline{q}) = 0(||q||^3). \qquad (4.3)$$

For each q_1, Π is minimum for $\overline{q}$ such that

$$B_0\bar{q} + \frac{\partial f}{\partial \bar{q}}(q_1,\bar{q}) = 0. \tag{4.4}$$

Since

$$\frac{\partial}{\partial \bar{q}}\left(B_0\bar{q} + \frac{\partial f}{\partial \bar{q}}(q_1,\bar{q})\right)\Bigg|_{q_1 = 0,\ \bar{q} = 0} = B_0,$$

the implicit function theorem applies and there exists, in some neighborhood of the origin $q = 0$, a unique analytic solution $\bar{q} = y(q_1)$ of (4.4), such that $y(0) = 0$. Further

$$y'(0) = -B_0^{-1}\frac{\partial^2 f}{\partial q_1 \partial \bar{q}}(0,0) = 0$$

so that

$$y(q_1) = 0(q_1^2). \tag{4.5}$$

Next, consider the change of variables

$$r_1 = q_1,$$
$$\bar{r} = \bar{q} - y(q_1).$$

The potential function Π can be written

$$\Pi = \frac{1}{2}y^T(r_1)B_0y(r_1) + y^T(r_1)B_0\bar{r} + \frac{1}{2}\bar{r}^TB_0\bar{r} + f(r_1,y(r_1))$$
$$+ \frac{\partial f}{\partial \bar{q}}(r_1,y(r_1))\bar{r} + \frac{1}{2}\bar{r}^T\frac{\partial^2 f}{\partial \bar{q}^2}(r_1,y(r_1) + \theta\bar{r})\bar{r}, \qquad 0 \le \theta \le 1.$$

Using (4.4) with $\bar{q} = y(r_1)$, we obtain

$$\Pi = \frac{1}{2}\bar{r}^T\left(B_0 + \frac{\partial^2 f}{\partial \bar{q}^2}(r_1,y(r_1) + \theta\bar{r})\right)\bar{r}$$
$$+ \frac{1}{2}y^T(r_1)B_0y(r_1) + f(r_1,y(r_1)),$$

which is the desired form of Π if we set

$$B(r_1,\bar{r}) = B_0 + \frac{\partial^2 f}{\partial \bar{q}^2}(r_1, y(r_1) + \theta\bar{r}) = B_0 + o(1),$$

$$\beta(r_1)r_1^m = \frac{1}{2} y^T(r_1)B_0 y(r_1) + f(r_1, y(r_1)).$$

Using (4.3) and (4.5), one gets that

$$\beta(r_1)r_1^m = O(r_1^3), \text{ i.e. } m \geq 3. \qquad \text{Q.E.D.}$$

4.5. Theorem (W.T. Koiter [1965]). If there exists $\varepsilon > 0$ such that

(i) $\theta = \{q \in B_\varepsilon : \Pi(q) < 0\} \neq \phi$;

(ii) $0 \in \partial\theta$;

(iii) Π is analytic and $\frac{\partial^2 \Pi}{\partial q^2}(0)$ is a positive semi-definite matrix with a single zero eigenvalue;

then the origin $q = p = 0$ is unstable.

Proof. From Theorem 3.2 it is sufficient to prove that

$$(\frac{\partial \Pi}{\partial q} | \ q) < 0 \quad \text{for every} \quad q \in \theta.$$

On the other hand, from Lemma 4.4, we can choose $\Pi(q)$ as in (4.2). Since Π takes negative values arbitrarily near the origin and $B(q_1,\bar{q})$ is positive definite, we can choose m large enough to get $\beta(0) \neq 0$. Further

$$(\frac{\partial \Pi}{\partial q} | \ q) = \bar{q}^T[B(q_1,\bar{q}) + \frac{1}{2}\sum_{1\leq i\leq n}\frac{\partial B}{\partial q_i}(q_1,\bar{q})q_i]\bar{q}$$

$$+ q_1^m(m\beta(q_1) + \beta'(q_1)q_1)$$

$$= 2\Pi + \frac{m-2}{2}\beta(q_1)q_1^m + \frac{1}{2}\bar{q}^T[\sum_{1\leq i\leq n}\frac{\partial B}{\partial q_i}(q_1,\bar{q})q_i]\bar{q}$$

$$+ \frac{m-2}{2}\beta(q_1)q_1^m + \beta'(q_1)q_1^{m+1}.$$

But for $q \in \theta$ and ε small enough

$$\frac{1}{m-2} \bar{q}^T \Big(\sum_{1 \le i \le n} \frac{\partial B}{\partial q_i} (q_1, \bar{q}) q_i \Big) \bar{q} \le \frac{1}{2} \bar{q}^T B(q_1, \bar{q}) \bar{q} < -\beta(q_1) q_1^m,$$

$$\beta(q_1) q_1^m < 0,$$

and

$$\left| \frac{m-2}{2} \beta(q_1) q_1^m \right| > |\beta'(q_1) q_1^{m+1}|.$$

Hence

$$\left(\frac{\partial \Pi}{\partial q} \middle| q\right) < 2\Pi < 0 \quad \text{for} \quad q \in \theta. \quad \text{Q.E.D.}$$

5. Mechanical Equilibrium in the Presence of Dissipative Forces

5.1. Up to this point, we have considered conservative mechanical systems only. What happens when some dissipative forces are present? This question is examined below. Roughly speaking, complete dissipation ensures asymptotic stability of the origin of the $(q, \dot{q})$-space if Π is minimum there, and instability if Π has no minimum. However, the proof requires that the minimum be an isolated equilibrium. This is expressed in the following theorem, generalizing and making rigorous the considerations of Kelvin (see W. Thomson and P.G. Tait [1912]) on the subject. The mechanical system referred to here is the Lagrangian system described in Section 13 of Appendix II, but with some dissipative forces added to it: a vector $Q(q, \dot{q})$, defined and $\mathscr{C}^1$ on $\Omega \times \mathscr{R}^n$.

5.2. Theorem (L. Salvadori [1966]). If

(i) Π has a minimum at $q = 0$;

(ii) the equilibrium at $q = 0$ is isolated;

(iii) the dissipation is complete, i.e. for some function $a \in \mathcal{K}$: $(Q|\dot{q}) \leq -a(||\dot{q}||)$;

then the equilibrium $q = \dot{q} = 0$ is asymptotically stable.

If Hypothesis (i) is replaced by

(i-a) Π has no minimum at $q = 0$;

then the equilibrium $q = \dot{q} = 0$ is unstable.

Proof. We consider the total energy as the auxiliary function. It reads $V(q,\dot{q}) = T(q,\dot{q}) + \Pi(q)$. One knows that $\dot{V}(q,\dot{q}) = (Q|\dot{q})$ and due to (iii), $\dot{V} = 0$ if and only if $\dot{q} = 0$. With a view to applying Theorem II.1.3, one should first show that the set $M = \{(q,\dot{q}) \in B_\rho: \dot{q} = 0\}$ contains, besides the equilibrium $q = \dot{q} = 0$, no other positive semi-trajectory. But, by (ii), it contains no other static solution whereas any positive semi-trajectory contained in M is a static solution ($\dot{q} = 0$).

Let us now show that $V(q,\dot{q})$ is positive definite. One knows that T is positive definite with respect to $\dot{q}$. Hypotheses (i) and (ii) together imply that the minimum of Π is a strict minimum. Indeed, there is a neighborhood of $q = 0$ where $\Pi(q) \geq 0$. If the minimum were not strict, there would exist, in every neighborhood of $q = 0$, a point $q \neq 0$ where Π would equal zero. But such a point would be a minimum of Π, i.e. a static solution, and (ii) would be violated.

If, on the other hand, Π has no minimum at $q = 0$, there exists, in every neighborhood of $q = 0$, a point q_0 with $\Pi(q_0) < 0$. But then of course, $V(q_0,0) < 0$. Therefore, all the hypotheses of Theorem II.1.3 are verified for $\Omega = B_\rho$ and Theorem 5.2 is proved. Q.E.D.

On an extension of this theorem to some non-autonomous mechanical systems, cf. VIII 3.5.

5.3. Remark. Notice that we can state here a necessary and sufficient condition: if there is complete dissipation, the isolated equilibrium $q = \dot{q} = 0$ is stable if and only if Π has a minimum at $q = 0$. Further, the stability is always asymptotic in this case.

5.4. Exercise. Show that the condition $\Pi(q) \leq 0$ along with complete dissipation do not entail stability.

Hint: consider, for $q \in \mathscr{R}$, the equation $\ddot{q} + \dot{q}^3 = 0$.

5.5. Exercise. Hypothesis (ii) of Theorem 5.2 for instability can be weakened as follows: if, for some $\varepsilon > 0$, $\overline{B}_\varepsilon \subset \Omega$, there is no equilibrium in $B_\varepsilon \cap \{q: \Pi(q) < 0\}$, then instability still prevails. This is a significant extension of the theorem, as is shown by the simple example for $q \in \mathscr{R}^2$: $\Pi(q_1,q_2) = q_1^3 q_2^3$.

6. Mechanical Equilibrium in the Presence of Gyroscopic Forces

6.1. Statement of the problem. Consider a conservative system as described in Section 13 of Appendix II. The kinetic energy assumes the form $T = \frac{1}{2} \dot{q}^T A(q)\dot{q}$. We suppose that $A(q)$ is defined and $\mathscr{C}^2$ on some open neighborhood of the origin of q-space, and that $A(0)$ is positive definite. The potential function $\Pi(q)$ is defined and also $\mathscr{C}^2$ on the same set Ω

and is such that $\Pi(0) = \frac{\partial \Pi}{\partial q}(0) = 0$. The equations of motion read

$$\frac{d}{dt}\frac{\partial T}{\partial \dot{q}} - \frac{\partial T}{\partial q} = -\frac{\partial \Pi}{\partial q} . \tag{6.1}$$

One knows that, if Π has no minimum at $q = 0$, the equilibrium at $q = \dot{q} = 0$ may be unstable. The present section is devoted to studying the influence on this equilibrium of the adjunction of appropriate forces in the second member of (6.1). Except at the beginning of our study, these forces will be gyroscopic, or at least non-energic forces. The corresponding problem, which is practically important, is known under the name of "gyroscopic stabilization".

6.2. Suppose the forces $Q(q,\dot{q})$ to be added in the second member of (6.1) are of the form

$$Q(q,\dot{q}) = -G\dot{q} + P(q,\dot{q}) \tag{6.2}$$

where G is a constant matrix (not necessarily antisymmetric for the time being) and

$$\frac{P(q,\dot{q})}{||(q,\dot{q})||} \to 0 \quad \text{as} \quad ||(q,\dot{q})|| \to 0.$$

Suppose, moreover, that $\Pi(q)$ is a function of class $\mathscr{C}^3$ and therefore assumes the form

$$\Pi(q) = \frac{1}{2} q^T C q + R(q)$$

where $\frac{\partial R/\partial q}{||q||} \to 0$ as $q \to 0$, and C is a constant symmetric matrix. This last hypothesis will remain in force until Section 6.8. The equations of motion, which read

$$\frac{d}{dt}\frac{\partial T}{\partial \dot{q}} - \frac{\partial T}{\partial q} = -\frac{\partial \Pi}{\partial q} + Q(q,\dot{q})$$

can be reduced to the following form

$$\frac{d}{dt}\begin{pmatrix} q \\ \dot{q} \end{pmatrix} = \begin{pmatrix} 0 & E \\ -A(0)^{-1}C & -A(0)^{-1}G \end{pmatrix}\begin{pmatrix} q \\ \dot{q} \end{pmatrix} + \begin{pmatrix} 0 \\ F(q,\dot{q}) \end{pmatrix} \tag{6.3}$$

for some appropriate function F such that

$$\frac{F(q,\dot{q})}{||(q,\dot{q})||} \to 0 \quad \text{as} \quad ||(q,\dot{q})|| \to 0.$$

Equation (6.3) is the normal form for the differential equation of second order

$$A(0)\ddot{q} + G\dot{q} + Cq + F(q,\dot{q}) = 0.$$

The eigenvalues corresponding to its linear part are the roots of the polynomial

$$\Delta(\lambda) = \det(\lambda^2 A(0)+\lambda G+C). \tag{6.4}$$

A first instability result is immediate.

6.3. Theorem. In the general hypotheses above, if C has no vanishing eigenvalue and has an odd number of strictly negative eigenvalues, the origin is unstable for (6.3).

Proof. Since $\det C < 0$, one has $\Delta(0) < 0$. But $\Delta(\lambda)$ tends to ∞ as λ tends to ∞. Therefore, $\Delta(\lambda)$ has at least one strictly positive real root. Hence, the instability, owing to Theorem I.5.8. Q.E.D.

This result, already mentioned by Kelvin for the case

of an antisymmetric matrix G, appears also in N.G. Chetaev [1955].

6.4. Exercise. Let $A(0) = E$ and C be diagonal with an even number $2s$ of vanishing diagonal elements and an odd number of strictly negative diagonal elements. Suppose, without loss of generality, that the first $2s$ diagonal elements are equal to zero. Then, if the upper left principal diagonal minor of order $2s$ of G is > 0, the origin is unstable for (6.3). For an antisymmetric G, this result is due to Van Chzhao-Lin [1963].

6.5. The degree of instability. The characteristic polynomial of the linear part of (6.1) (i.e., the equations of motion considered before the adjunction of gyroscopic terms) reads $\det(\lambda^2 A(0) + C)$. But a classical result shows (see e.g. R. Bellman [1970]) that there exists a regular matrix P such that $P^T A(0) P = E$, while $P^T CP$ is diagonal. Of course,

$$\det(\lambda^2 A(0) + C) = (\det P)^{-2} \det(\lambda^2 E + P^T CP).$$

Therefore, the eigenvalues of the linear part of (6.1) are the roots of $\det(\lambda^2 E + P^T CP) = 0$. Now, according to Sylvester's theorem, the number of strictly negative diagonal elements of $P^T CP$ equals the number of strictly negative eigenvalues of C. The latter number equals, therefore, the number of eigenvalues with strictly positive real part for the linear part of (6.1). This number is called the degree of instability of the system. Theorem 6.3 may be reformulated thus: if C has no vanishing eigenvalue and if the degree of instability is odd, the equilibrium $q = \dot{q} = 0$ cannot be stabilized by the adjunction of forces of type (6.2).

6.6. Let us now prove an auxiliary result which will determine to a large extent the rest of our discussion, namely that _if G is antisymmetric_, $\Delta(\lambda) = \Delta(-\lambda)$. Indeed, one gets successively

$$\Delta(\lambda) = \det(\lambda^2 A(0) + \lambda G + C) = \det[(\lambda^2 A(0) + \lambda G + C)^T]$$
$$= \det(\lambda^2 A(0) - \lambda G + C) = \Delta(-\lambda).$$

It follows from this property that $\Delta(\lambda)$ has no root with strictly positive real part if and only if all the roots of $\Delta(\lambda)$ are pure imaginary.

The following theorem will give us a key to gyroscopic stabilization.

6.7. _Theorem_. Assume the matrix C possesses an even number $2s > 0$ of strictly negative eigenvalues, all the other eigenvalues being strictly positive. Then there exists an antisymmetric matrix G such that every root of $\Delta(\lambda)$ is pure imaginary and non-vanishing.

Proof. Let P be the matrix introduced in Section 6.5 and which is such that $P^T A(0) P = E$ and $C^* = P^T C P$ is diagonal. Putting $G^* = P^T G P$, we observe that the roots of $\Delta(\lambda)$ are also those of

$$\Delta^*(\lambda) = \det(\lambda^2 E + \lambda G^* + C^*).$$

Naturally, if G is antisymmetric, so is G^*. Further, the diagonal elements of C^* are the eigenvalues of C and there is no loss of generality to assume that the first $2s$ are the negative ones. Let us now construct G^* in the following way: choose $g^*_{2i,2i-1} = -g^*_{2i-1,2i} \neq 0$ for $1 \leq i \leq s$, and all the other elements of G^* equal to zero. Then

$$\Delta^*(\lambda) = \prod_{1 \leq k \leq s} [(\lambda^2 + c^*_{2k-1})(\lambda^2 + c^*_{2k}) + \lambda^2 g^{*2}_{2k,2k-1}] \prod_{2s+1 \leq i \leq n} (\lambda^2 + c^*_i).$$

One verifies readily that if $g^*_{2k,2k-1}$ is chosen large enough for every k, $1 \leq k \leq s$, then $\Delta^*(\lambda)$ has nothing but pure, imaginary roots. Q.E.D.

6.8. Suppose now the equations (6.1) are linear and the matrix C satisfies the hypotheses of Theorem 6.7. It follows from this theorem that the eigenvalues of the system can be transformed into pure imaginary eigenvalues by the adjunction of appropriate gyroscopic forces. Now a linear system of differential equations with constant coefficients, whose matrix can be diagonalized and whose eigenvalues have real parts smaller than or equal to zero has a <u>stable</u> equilibrium at the origin. (It may be an interesting exercise to prove this, and to prove that the matrix of (6.1) can be diagonalized.) Therefore, gyroscopic stabilization is possible for a system of this form. But it is a significant fact that all we can obtain in this way is stability, not asymptotic stability. The appearance of nonlinear terms in (6.1) can destroy the stabilizing effect of the gyroscopic forces. The example to follow is an illustration of this observation.

6.9. <u>A glance at the restricted problem of three bodies</u>. When the nonlinear terms F in (6.3) depend on q only, it may be very difficult to determine whether the equilibrium at the origin is or is not stable. This is the case for the equilibrium point known as L_4 in the restricted problem of three bodies (and also, by reason of symmetry, for the point L_5). The relevant equations, as given e.g. by H. Leipholz [1968], read

$$\ddot{q} + \begin{pmatrix} 0 & -2 \\ 2 & 0 \end{pmatrix} \dot{q} + Cq + F(q) = 0 \tag{6.5}$$

where $q = (q_1, q_2) \in \mathscr{R}^2$, C is the hessian matrix, computed at $q = 0$, of the potential function $\Pi(q)$, whereas $F(q) = \frac{\partial \Pi}{\partial q} - Cq$ contains but terms of order at least 2. The function $\Pi(q)$, through the change of variables $x = q_1 + \frac{1}{2} - p$, $y = q_2 + \frac{1}{2}\sqrt{3}$, where p is some strictly positive parameter, becomes

$$\Pi(x,y) = -\frac{1}{2}(x^2+y^2) - \frac{1-p}{[(x+p)^2 + y^2]^{1/2}} - \frac{p}{[(x+p-1)^2 + y^2]^{1/2}} .$$

The matrix C can be written explicitly as

$$C = \begin{pmatrix} -\frac{3}{4} & -\frac{3\sqrt{3}}{2}(\frac{1}{2} - p) \\ -\frac{3\sqrt{3}}{2}(\frac{1}{2} - p) & -\frac{9}{4} \end{pmatrix}.$$

It is readily verified that if

$$0 < p(1-p) < \frac{1}{27} , \tag{6.6}$$

the origin is stable for the system (6.5) where F has been replaced by zero. V.I. Arnold [1961] and A.M. Leontovich [1962] have shown that stability still prevails for (6.5) and p belonging to the interval (6.6), with the possible exception of a denumerable number of values. Later A. Deprit and A. Deprit-Bartholomé [1967] have shown that the critical values were only three in number, and they worked them out. And now, A.P. Markeev [1969] has proved that stability still prevails for one of these values, and one gets instability for the other two.

6.10. This example demonstrates the following proposition: let A and C be as above, C verifying the hypotheses of Theorem 6.7. Let G be an antisymmetric matrix chosen such that the origin be stable for the equation $A(0)\ddot{q} + G\dot{q} + Cq = 0$. It may be possible to choose a nonlinear function $R(q)$, with

$$\frac{\partial R/\partial q}{||q||} \to 0 \quad \text{as} \quad q \to 0,$$

such that the origin becomes unstable for

$$A\ddot{q} + G\dot{q} + Cq - \frac{\partial R}{\partial q} = 0. \tag{6.7}$$

Another problem, more important from the point of view of gyroscopic stabilization, is the following one: let A, C and R be given, C verifying the hypotheses of Theorem 6.7. Is it always possible to find an antisymmetric matrix G such that the origin be stable for (6.7)? Apparently, this problem remains open.

Notice finally that it is easy to exhibit some examples where a nonlinear function F depending on $\dot{q}$ destroys the stability of a gyroscopically stabilized system.

6.11. In this and the next few sections, we shall resort to Liapunov's direct method to dispose of the hypotheses according to which $\det C \neq 0$ and the gyroscopic forces do not depend on q. Let us consider the differential equations

$$\frac{d}{dt}\frac{\partial T}{\partial \dot{q}} - \frac{\partial T}{\partial q} = -\frac{\partial \Pi}{\partial q} - G(q)\dot{q} \tag{6.8}$$

where T verifies the hypotheses of 6.1, Π is a $\mathscr{C}^2$ function such that $\Pi(0) = 0$, $\frac{\partial \Pi}{\partial q}(0) = 0$ and G is $\mathscr{C}^1$ and antisymmetric.

Clearly, if Π has a strict minimum at $q = 0$, the origin $q = \dot{q} = 0$ is stable, since the function $H = T + \Pi$ verifies the hypotheses of Theorem I.4.2. On the other hand, even if $\Pi(q)$ admits at $q = 0$ a strict maximum with a negative definite hessian matrix, it may happen as follows from Theorem 6.7, that the equilibrium be stable, due to the presence of gyroscopic forces. The situation is, therefore, more intricate than it was in Sections 2 to 5.

6.12. Let us assume that the gyroscopic forces are Lagrangian, i.e. that for some $\mathscr{C}^2$ function $\mathscr{L}(q,\dot{q})$, one has $\mathscr{L}$

$$G(q)\dot{q} = \frac{d}{dt}\frac{\partial \mathscr{L}}{\partial \dot{q}} - \frac{\partial \mathscr{L}}{\partial q}.$$

By the way, this is always the case for a constant G, for it suffices to choose $\mathscr{L} = -\frac{1}{2}q^T G\dot{q}$. It is easy to show that $\mathscr{L}$ is of the form

$$\sum_{1\leq i\leq n} \alpha_i(q)\dot{q}_i,$$

where the α_i are $\mathscr{C}^2$ functions. At last, since $\mathscr{L}$ is defined up to a linear form in $\dot{q}$ with constant coefficients, one is free to assume that $\alpha_i(0) = 0$ for every i, in such a way that

$$\mathscr{R}(q,\dot{q}) = T + \mathscr{L} - \Pi$$

is stationary at $q = \dot{q} = 0$. The following theorem, yielding a sufficient instability condition of the origin for (6.8), is akin to Theorem 3.2.

6.13. <u>Theorem</u> (V.V. Rumiantsev [1966]). If there exists an $\varepsilon > 0$ (with $\overline{B}_\varepsilon \subset \Omega$) such that

(i) $\theta = \{q \in B_\varepsilon,\ \Pi(q) < 0\} \neq \phi$;

(ii) the set

$$\Psi = \{(q,\dot{q}) : q \in \theta,\ \dot{q} \in B_\varepsilon,\ H < 0 \text{ and } (q|\frac{\partial \mathscr{R}}{\partial \dot{q}}) > 0\}$$

is $\neq \phi$, and $(0,0) \in \partial\Psi$;

(iii) the function $2T + \mathscr{L} + (q|\frac{\partial \mathscr{R}}{\partial q})$ is strictly positive on Ψ;

then the origin is unstable.

<u>Proof</u>. Clearly $(0,0) \in \partial\Psi$. The function $V = -(q|\frac{\partial \mathscr{R}}{\partial \dot{q}})H$ is such that, for every $(q,\dot{q}) \in \Psi$:

(a) $0 < V(q,\dot{q})$

(b) $\dot{V} = -H(2T + \mathscr{L} + (q|\frac{\partial \mathscr{R}}{\partial q})) > 0.$

Instability then follows from Theorem I.5.1 and Remark I.5.5. Q.E.D.

6.14. <u>Corollary</u>. In Theorem 6.13, Hypothesis (iii) can be replaced by

(iv) $\mathscr{R}^{(2)}(q,\dot{q}) > 0$ on Ψ,

where $\mathcal{R}^{(2)}$ is the quadratic form associated with the hessian matrix of $\mathcal{R}$ at $q = \dot{q} = 0$.

Proof. Indeed, $\dot{V} = -H(\mathcal{R}^{(2)} + o(||(q,\dot{q})||^2))$.

6.15. Exercise. Suppose that F in equation (6.3) does not depend on $\dot{q}$. Suppose further that the hypotheses of Theorem 6.7 are satisfied. Then, whatever the form of F, an antisymmetric G can always be chosen such that Hypothesis (iii) of Theorem 6.13 is not verified.

Hint: by a suitable change of variables, bring the linear part of the equation to the form $E\ddot{r} + G^*\dot{r} + C^*r = 0$, with G^* and C^* as in the proof of 6.7.

6.16. Remark. There exist mechanical systems whose equation assumes the form (6.8) without the kinetic energy verifying the hypotheses on T in 6.11. Indeed, consider a system with potential function $\tilde{\Pi}$ and a kinetic energy $\tilde{T} = T_2 + T_1 + T_0$, with $T_2 = \frac{1}{2}\dot{q}^T A(q)\dot{q}$, $T_1 = b^T(q)\dot{q}$ and $T_0 = d(q)$, $b \in \mathcal{R}^n$ and $d \in \mathcal{R}$. The equation of motion is again (6.8) with $T = T_2$, $\Pi = \tilde{\Pi} - d$ and $G(q) = \left[\frac{\partial b^T}{\partial q} - \left(\frac{\partial b^T}{\partial q}\right)^T\right]$. As observed by P. Hagedorn [1972], if the differential forms $b^T(q)dq$ are exact, then $G \equiv 0$.

In much the same way, the equations of Routh with a fixed c (cf. Appendix II.14) assume the form (6.8) with T_2, T_1, T_0 replaced respectively by R_2, R_1 and R_0. That is why the instability theorems proved in this section are often known as "inverses of Routh's stability Theorem" (cf. IV.5 and IV.8).

7. Bibliographical Note

Good historical surveys of the inversions of Lagrange-Dirichlet's theorem will be found in L. Salvadori [1968] and P. Hagedorn [1971]. Both papers deal also with dissipative systems and stationary motions, and devote a subsection to systems with gyroscopic forces. We make an effort below to complete the bibliographies of Salvadori and Hagedorn, in such a way that, in regard to the inversion of the Lagrange-Dirichlet theorem, our list combined with theirs should not be too far from exhaustive.

Several authors claimed that there is always instability in the absence of a strict minimum at the equilibrium point (cf. e.g. P. Appel [1932] and B. Lanczos [1962]; see however P. Appel [1953]), which is not true, as we have shown. Besides those already cited, many partial inversions have been published, associated with the names of J. Hadamard [1897], P. Painlevé [1897], A. Kneser [1895-1897], G. Hamel [1903], L. Silla [1908].

P. Hagedorn, in the paper cited above [1971] gives an interesting partial inversion: in case T and Π are $\mathscr{C}^2$ functions, the equilibrium is unstable if Π has a strict maximum. We did not deal with this theorem here, because the type of proof, i.e. showing by variational methods the existence of a motion with appropriate properties, is too remote from the subject of this book. By the same method, P. Hagedorn [1975] proved recently an interesting sufficient condition of instability for the systems mentioned in Remark 6.16: if the function $\frac{1}{2}\, b^T A^{-1} b + (\tilde{\Pi}-d)$ has a relative strict maximum at $q = 0$ then the equilibrium $q = \dot{q} = 0$ is unstable.

As for dissipative systems, and here without any claim at completeness, let us mention, besides those already cited, the early contribution of P. Duhem [1902], as well as V.M. Matrosov $[1962]_1$ and W.T. Koiter [1965]. Matrosov considers time-dependent systems. As we shall see later (N. Rouche [1968]), the result embodied in Theorem 6.2 can be obtained straight-forwardly by using two Liapunov-like functions with a nice physical interpretation: first the total energy and second the vector of conjugate momenta: cf. Section VIII.3.

CHAPTER IV

STABILITY IN THE PRESENCE OF FIRST INTEGRALS

1. Introduction

The principal drawback of Liapunov's direct method is that no general procedure is known to construct auxiliary functions suiting specific theorems. That is why, in stability problems, one should a priori neglect no available information concerning the solutions. In particular, the first integrals will often be helpful, either to facilitate the search for auxiliary functions or to eliminate part of the variables and thus decrease the number of equations to examine. Both points of view will be developed later, in Sections 3 and 4 respectively. Section 5 deals with an important case where first integrals are known, namely the stationary motions of mechanical systems with ignorable coordinates. Section 6 studies a particular motion of this type: the orbiting particles in the betatron. And the last section gathers practical criteria concerning the various methods of constructing positive definite functions.

2. General Hypotheses

2.1. Starting with this chapter, our general hypotheses will be somewhat weakened. This will be done at practically no cost in the proofs. Further, as we have seen in Section II.6 and as we shall see again in Theorem 3.5, it may be quite helpful, if not necessary, to consider auxiliary functions which are not differentiable. Let us therefore describe the general setting of our future work.

For n an integer ≥ 2, $I =]\tau,\infty[$ for some $\tau \in \mathscr{R}$ and Ω an open connected subset of $\mathscr{R}^n$, consider a continuous function

$$f: I \times \Omega \to \mathscr{R}^n, \ (t,x) \to f(t,x),$$

and the corresponding Cauchy problem

$$\dot{x} = f(t,x) \tag{2.1}$$

$$x(t_0) = x_0 \tag{2.2}$$

for some point $(t_0,x_0) \in I \times \Omega$. There passes through (t_0,x_0) at least one non-continuable solution. Such solutions will be written

$$x: J \to \mathscr{R}^n, \ t \to x(t)$$

where J is, of course, an open interval. We shall often write $J =]\alpha,\omega[$ and $J^+ = [t_0,\omega[$. Without mention to the contrary, uniqueness of the solutions will not be required. Further, we assume that the origin 0 of $\mathscr{R}^n$ belongs to Ω and is a critical point of (2.1): i.e. for any $t \in I$: $f(t,0) = 0$.

2.2. A function $W: I \times \Omega \to \mathscr{R}$, $(t,x) \to W(t,x)$, locally lipschitzian with respect to x and continuous, is called a <u>first integral</u> of Equation (2.1) if $D^+W(t,x) = 0$ for every $(t,x) \in I \times \Omega$. This condition implies (see Appendix I) that for any solution $x(t)$ of (2.1), (2.2), $W(t,x(t))$ is simultaneously increasing and decreasing: therefore it is constant.

3. How to Construct Liapunov Functions

3.1. A trivial case is when one knows a first integral which is also positive definite. Then, of course, stability of the origin follows immediately from Theorem I.4.2. Suppose now that there are m first integrals $W_i(t,x)$, $1 \leq i \leq m < n$, and let us write $W(t,x)$ for the m-vector $(W_1(t,x),\ldots,W_m(t,x))$. Without loss of generality, we assume that $W(t,0) = 0$ for $t \in I$. A natural approach is to try combining the W_i into a single positive definite function. The following theorem proves helpful in this direction.

3.2. <u>Theorem</u> (G.K. Pozharitskii [1958]). There exists a continuous function $\phi: \mathscr{R}^m \to \mathscr{R}$ such that $\phi(W(t,x))$ is positive definite* if and only if $||W(t,x)||$ is positive definite, where $||\cdot||$ is any norm in $\mathscr{R}^m$.

<u>Proof</u>. The condition is obviously sufficient. To prove its necessity, notice first that since ϕ is continuous and $\phi(0) = 0$, there exists a function $b \in \mathscr{K}$ with $\phi(W) \leq b(||W||)$. Moreover, $\phi(W(t,x))$ being positive definite, there is a

---

*A $\mathscr{C}^1$ positive definite function was defined in I.3.3. What is used here is a trivial extension of this definition to continuous functions.

function $a \in \mathcal{K}$ such that $\phi(W(t,x)) \geq a(||x||)$. Thus, $a(||x||) \leq \phi(W(t,x)) \leq b(||W(t,x)||)$ and $||W(t,x)|| \geq b^{-1}(a(||x||))$, where $b^{-1}\circ a \in \mathcal{K}$. Q.E.D.

3.3. Practically, one will choose the norm, $||W(t,x)|| = \sqrt{\sum_i W_i(t,x)^2}$, in such a way that if, as has been assumed, the W_i are locally lipschitzian with respect to x, the same will be true of $||W(t,x)||$. And, therefore, $||W(t,x)||$ is a first integral and an auxiliary function suiting Theorem I.4.2.

3.4. If this test of Pozharitskii fails, one may try to find a supplementary function $V(t,x)$ and some continuous combination $\phi(V(t,x),W(t,x))$ such that this last function be positive definite, and decreasing along the solutions. A reasonable choice for ϕ would be $\max(V(t,x),\psi(W(t,x)))$, with ψ a continuous function on $\mathcal{R}^m$ into $\mathcal{R}$, or equivalently (see Exercise 3.6), $\max(V(t,x),||W(t,x)||)$. Hence, the following theorem concerning two real valued functions.

3.5. <u>Theorem</u>. Let $V(t,x),W(t,x)$ be two functions on $I \times \Omega$ into $\mathcal{R}$, both locally lipschitzian in x and continuous, W a first integral, and let $a \in \mathcal{K}$ be such that

(i) $V(t,0) = 0$, $W(t,0) = 0$ for $t \in I$;

(ii) $\max(V(t,x),W(t,x)) \geq a(||x||)$ for $(t,x) \in I \times \Omega$;

(iii) $D^+V(t,x) \leq 0$ for every $(t,x) \in I \times \Omega$ such that $V(t,x) \geq W(t,x)$;

then the origin is stable. If there exists further a function $b \in \mathcal{K}$ such that

(iv) $\max(V(t,x),W(t,x)) \leq b(||x||)$ for $(t,x) \in I \times \Omega$,

the stability is uniform.

Proof. As appears from Theorem 2.7 of Appendix I, $D^+[\max(V(t,x),W(t,x))] \le 0$ for $(t,x) \in I \times \Omega$. Therefore, Theorem I.4.2 applies. Q.E.D.

3.6. Exercise. Let $V(t,x)$ and $W(t,x)$ be two continuous functions on $I \times \Omega$ into $\mathscr{R}$ and $\mathscr{R}^m$ respectively, $1 \le m \le n$, with $V(t,0) = 0$ and $W(t,0) = 0$. There exists a continuous function $\psi: \mathscr{R}^m \to \mathscr{R}$ such that $\max(V(t,x),\psi(W(t,x)))$ is positive definite, if and only if $\max(V(t,x),||W(t,x)||)$ is positive definite, where $||\cdot||$ is any norm in $\mathscr{R}^m$.

3.7. Exercise. In the notations of Exercise 3.6, suppose V and W do not depend on t. We write their values $V(x)$ and $W(x)$. Then the following statements are equivalent (and this equivalence has some practical value):

(a) there exists an open neighborhood N of the origin, $N \subset \Omega$, such that for $x \in N\backslash\{0\}$: $\max(V(x),||W(x)||) > 0$;

(b) the origin 0 is for $V(x)$ a strict minimum constrained by the equation $W(x) = 0$.

3.8. Remark. It may happen that hypothesis (ii) of Theorem 3.5 is difficult to check. The situation becomes simpler if one knows of a continuous function $W^*(x)$ on Ω into $\mathscr{R}$ such that $W(t,x) \ge W^*(x) \ge 0$ for every $(t,x) \in I \times \Omega$. Let us put $G = \{x \in \Omega: W^*(x) = 0\}$ and let $N \subset \Omega\backslash\{0\}$ be an open neighborhood of $G\backslash\{0\}$. Hypothesis (ii) will be verified in each of the two following circumstances:

(a) there exists a function $a^* \in \mathscr{K}$ such that $V(t,x) \ge a^*(||x||)$ on $I \times N$;

(b) there exists a continuous function $V^*(x)$ on Ω into $\mathscr{R}$ such that $V(t,x) \geq V^*(x)$ on $I \times N$ and $V^*(x) > 0$ on $G\backslash\{0\}$.

In the former case, noticing that Ω can be restricted to a ball B_ρ for some $\rho > 0$, there exists a function $a_* \in \mathscr{K}$ such that, for every $(t,x) \in I \times (B_\rho\backslash N)$

$$W(t,x) \geq \inf\{W^*(y): ||x|| \leq ||y|| \leq \rho,\ y \notin N\} \geq a_*(||x||).$$

One may choose $a(||x||) = \min(a^*(||x||), a_*(||x||))$.

In the latter case, it is clear that, for some open neighborhood M of $G\backslash\{0\}$, with $\overline{M}\backslash\{0\} \subset N$, one has $\forall x \in \overline{M}\backslash\{0\}$: $V^*(x) > 0$.

Then there exists a function $a^* \in \mathscr{K}$ such that, for every $(t,x) \in I \times M$,

$$V(t,x) \geq \inf\{V^*(y): ||x|| \leq ||y|| \leq \rho,\ y \in \overline{M}\} \geq a^*(||x||),$$

and condition 3.8 a) is satisfied with M substituted to N.

The hypothesis on $D^+V(t,x)$ in Theorem 3.5 can be weakened at the expense of an extra requirement on $V(t,x)$. This is the object of the following theorem, where W^* and G are defined as above.

3.9. <u>Theorem</u>. Let $V(t,x), W(t,x)$ be two functions on $I \times \Omega$ into $\mathscr{R}$, both locally lipschitzian in x and continuous, W a first integral with $W(t,0) = 0$, $W(t,x) \geq W^*(x) \geq 0$, where W^* is continuous. If

(i) $b(||x||) \geq V(t,x) \geq a(||x||)$ for every $(t,x) \in I \times N$ and some functions $a \in \mathscr{K}$ and $b \in \mathscr{K}$, and $N \subset \Omega\backslash\{0\}$ is an open neighborhood of $G\backslash\{0\}$;

(ii) $D^+V(t,x) \leq 0$ for every $(t,x) \in I \times N$ such that $V(t,x) \geq W(t,x)$;

then the origin is stable. If there exists further a function $c \in \mathscr{K}$ such that

(iii) $c(||x||) \geq W(t,x)$ for every $(t,x) \in I \times \Omega$;

the stability is uniform.

3.10. Exercise (A.M. Liapunov [1893]). In the general hypotheses of this section, suppose $W(t,x)$ on $I \times \Omega$ into $\mathscr{R}^m$ is a first integral. Suppose $V(t,x)$ on $I \times \Omega$ into $\mathscr{R}$ is a "Liapunov function" subject to the constraint $W(t,x) = 0$, i.e., for some $a \in \mathscr{K}$ and every $(t,x) \in I \times \Omega$ such that $W(t,x) = 0$:

(i) $V(t,x) \geq a(||x||)$, $V(t,0) = 0$;

(ii) $\dot{V}(t,x) \leq 0$.

Prove that the origin is stable with respect to the perturbations satisfying the equation $W(t,x) = 0$ i.e. $(\forall t_0 \in I)(\forall \varepsilon > 0)(\exists \delta > 0)(\forall x_0 \in B_\delta\colon W(t_0,x_0) = 0)(\forall t \in J^+)$ $||x(t;t_0,x_0)|| < \varepsilon$ (conditional stability).

3.11. Exercise. Use the following differential equation

$$\dot{x} = \sin(ty) - x,$$
$$\dot{y} = 0$$

with x and y in $\mathscr{R}$, to prove that the hypotheses in Exercise 3.10 do not entail stability (unconditional stability). Hence, the interest of Theorem 3.5.

4. Eliminating Part of the Variables

4.1. Let us now see what kind of results can be obtained by using the first integrals to eliminate part of the variables and thus decrease the number of equations to deal with. Suppose we know m, $1 \leq m < n$, first integrals, or in other words a vectorial first integral $W(t,x) = (W_1(t,x),\ldots,W_m(t,x))$. Let us separate Equation (2.1) into two subsystems as

$$\begin{aligned} \dot{y} &= g(t,y,z), \\ \dot{z} &= h(t,y,z), \end{aligned} \tag{4.1}$$

where y, $g \in \mathscr{R}^{n-m}$ and z, $h \in \mathscr{R}^m$. We shall write $x = (y,z)$. Suppose at last that the equation $W(t,y,z) = \beta$ where $\beta \in \mathscr{R}^m$ can be uniquely solved in some neighborhood of the origin of x-space and yields the continuous solution $z = z(t,y,\beta)$, such that $z(t,0,0) = 0$. The Equations (4.1) can be replaced by the system

$$\begin{aligned} \dot{y} &= g_0(t,y) + r(t,y,\beta), \\ \dot{\beta} &= 0, \end{aligned} \tag{4.2}$$

where $g_0(t,y) = g(t,y,z(t,y,0))$ and $r(t,y,\beta) = g(t,y,z(t,y,\beta)) - g_0(t,y)$. For $\beta = 0$, the first of these equations reduces to

$$\dot{y} = g_0(t,y). \tag{4.3}$$

Total stability of the origin for this reduced equation is a key hypothesis in the following theorem. Sufficient conditions insuring total stability have been given in Section II.4.

4.2. <u>Theorem</u>. Let $\Omega = B_\rho$ for some $\rho > 0$ and suppose there exists a function $a \in \mathscr{K}$ such that for every $(t,y,\beta) \in I \times B_\rho$

(i) $||z(t,y,\beta)|| \leq a(||(y,\beta)||)$,

(ii) $||r(t,y,\beta)|| \leq a(||\beta||)$,

(iii) the origin is totally stable for Equation (4.3);

then the origin $y = 0$, $z = 0$ is stable for Equation (4.1).

<u>Proof</u>. Hypotheses (i), (iii) and (ii) imply respectively that:

$$(\forall\varepsilon > 0)(\exists\eta > 0)\ [||y|| < \eta \text{ and } ||\beta|| < \eta] \Rightarrow [||z|| < \varepsilon];$$

$$(\forall\eta > 0)(\forall t_0 \in I)(\exists\delta_1 > 0)(\exists\delta_2 > 0)\ [||y_0|| < \delta_1,\ ||r(t,y,\beta)|| < \delta_2 \text{ and } t \in J^+] \Rightarrow [||y(t)|| < \eta];$$

$$(\forall\delta_2 > 0)(\exists\delta_3,\ 0 < \delta_3 < \eta)\ [||\beta|| < \delta_3] \Rightarrow [||r(t,y,\beta)|| < \delta_2].$$

But $W(t,x)$ is continuous and $W(t,0) = 0$. Therefore,

$$(\forall t_0 \in I)(\forall\delta_3 > 0)(\exists\delta_4 > 0)\ [||x_0|| < \delta_4] \Rightarrow [||\beta|| < \delta_3].$$

We conclude that

$$(\forall t_0 \in I)(\forall\varepsilon > 0)(\exists\delta > 0)\ [||x_0|| < \delta \text{ and } t \in J^+] \Rightarrow ||x(t)|| < \varepsilon.$$

Indeed, it will be sufficient to choose $\delta = \min(\delta_1,\delta_4)$. Q.E.D.

4.3. In the next theorem, Theorem 3.5 is applied to the reduced Equation (4.3). In a way, it combines the advantages of both methods: eliminating part of the variables and using the first integrals to construct an appropriate auxiliary function.

4.4. <u>Theorem</u>. Suppose there exist a continuous function $V(t,y)$ on $I \times \Omega'$ into $\mathscr{R}$ for some neighborhood Ω' of the origin of y-space, a function $p(t)$ on I into $\mathscr{R}$ and two functions $a, c \in \mathscr{K}$ such that $(\forall(t,y) \in I \times \Omega')(\forall(t,y') \in I \times \Omega')$

(i) $|V(t,y) - V(t,y')| \leq p(t)||y - y'||$,

(ii) $V(t,y) \geq a(||y||)$; $V(t,0) = 0$,

(iii) $D^+_{(4.3)}V(t,y) = -c(V(t,y))$, where the subscript (4.3) indicates a derivative computed along the solutions of Equation (4.3).

Suppose further, that there exists a function $g(t)$ on I into $\mathscr{R}$, a function $\phi \in \mathscr{K}$ and a constant L such that, at each point where r is defined,

(iv) $||r(t,y,\beta)|| \leq q(t)\phi(||\beta||)$,

(v) $p(t)q(t) \leq L$.

Then the origin $(y,\beta) = 0$ is stable for Equation (4.2). Moreover, if there exists a function $b \in \mathscr{K}$ such that, for every $(t,y) \in I \times \Omega'$

(vi) $V(t,y) \leq b(||y||)$,

the stability is uniform.

Proof. Consider System (4.2) and its first integral $W(\beta) = c^{-1}(L\phi(||\beta||))$. This integral is not locally lipschitzian, but it does not matter! Because of (ii), V and W verify the hypotheses stated in Remark 3.8, and therefore Hypothesis (ii) of Theorem 3.5 is satisfied. Hypothesis (i) of the same theorem is trivially satisfied. As for Hypothesis (iii), notice that the derivative of $V(t,y)$ along the solutions of (4.2), namely

$$D^+V(t,y,\beta) = \lim_{h\to 0^+} \sup \frac{V(t + h,y + hg_0(t,y) + hr(t,y,\beta)) - V(t,y)}{h}$$

may be estimated thus, using (i) above

$$D^+V(t,y,\beta) \leq D^+_{(4.3)}V(t,y) + p(t)||r(t,y,\beta)||.$$

But using (iii), (iv) and (v):

$$D^+V(t,y,\beta) \leq -c(V(t,y)) + L\phi(||\beta||).$$

The second member of this inequality is negative if $V(t,y) \geq c^{-1}(L\phi(||\beta||)) = W(\beta)$. Hypothesis (iii) of Theorem 3.5 is thus verified. Q.E.D.

4.5. As is well known (cf. e.g. L. Cesari [1959]), stability may be destroyed by a change of coordinates, even if this change is continuous. Therefore, stability of the origin $(y,z) = 0$ for (4.1) is not necessarily equivalent to stability of $(y,\beta) = 0$ for (4.2). The following exercise gives a clue to settle this question.

4.6. Exercise. Suppose the origin of the x-space, $x \in \mathscr{R}^n$, is stable for some differential equation, and let $x = \phi(t,y)$ be a continuously differentiable change of coordinates such that $\phi(t,0) = 0$ for every t and the jacobian matrix $\frac{\partial\phi}{\partial y}(t,y)$ be everywhere regular. Show that the origin of y-space is stable for the corresponding differential equation in y, if ϕ^{-1} is continuous in x at $x = 0$, uniformly with respect to $t \in I$.

5. Stability of Stationary Motions

5.1. The above theorems were largely inspired by the classical stability problem of stationary motions of Lagrangian systems. For such motions, E.J. Routh [1877] (see also [1975]) gave simple conditions insuring stability, but only with respect to perturbations leaving unchanged the values of the conjugate momenta. Such a drastic restriction is seldom if ever realized. Hence, the interest of the following generalizations.

We consider the Routh's equations (Appendix II(11.4)) and assume that the stationary motion to be studied corresponds to the critical point $q = \bar{q} = 0$, $\dot{q} = 0$, $c = \bar{c}$ of these equations. That $\bar{q}$ has been chosen to be 0 does not reduce the generality of our study. Introducing the new variables $q = x$, $\dot{q} = y$, $\beta = c - \bar{c}$, we bring the Routh's equations to the general form

$$\begin{aligned} \dot{x} &= y, \\ \dot{y} &= Y(t,x,y,\beta), \\ \dot{\beta} &= 0. \end{aligned} \tag{5.1}$$

The critical point is now $(x,y,\beta) = (0,0,0)$. It will be convenient to write Routh's function (Appendix II(11.3)) in terms of the new variables in the obvious form

$$\mathscr{R}(t,x,y,\beta) = \mathscr{R}_2(t,x,y) + \mathscr{R}_1(t,x,y,\beta) + \mathscr{R}_0(t,x,\beta) - \Pi(t,x)$$

where $\mathscr{R}_2$ and $\mathscr{R}_1$ are respectively quadratic and linear in y. The stability of the critical point will be studied using the Hamiltonian function

$$H(t,x,y,\beta) = (\frac{\partial\mathscr{R}}{\partial y}|y) - (\mathscr{R} - \mathscr{R}(t,0,0,0)) = \mathscr{R}_2 + W$$

where $W(t,x,\beta) = \Pi - \mathscr{R}_0 + \mathscr{R}(t,0,0,0)$ does not depend on y. Notice by the way that $W(t,0,0) = 0$ for all t.

The time derivative of H along the motions reads

$$\dot{H}(t,x,y,\beta) = -\frac{\partial}{\partial t}(\mathscr{R} - \mathscr{R}(t,0,0,0)) + (Q|y)$$

and does not depend on β if $\mathscr{R}$ is independent of t. The following theorem is a consequence of Theorem 3.9, if one identifies V with H and W with $||\beta||$.

5.2. <u>Theorem</u>. If there exists an open neighborhood $N \subset \Omega\backslash\{0\}$ of $\{(x,y,\beta) \in \Omega: \beta = 0, (x,y) \neq 0\}$ such that for every $(t,x,y,\beta) \in I \times N$ and some functions a,b,a', $b' \in \mathscr{K}$:

(i) $a(||y||) \leq \mathscr{R}_2(t,x,y) \leq b(||y||)$;

(ii) $a'(||x||) \leq W(t,x,\beta) \leq b'(||x|| + ||\beta||)$;

(iii) $-\frac{\partial}{\partial t}(\mathscr{R} - \mathscr{R}(t,0,0,0)) + (Q|y) \leq 0$ wherever $H(t,x,y,\beta) \geq ||\beta||$;

then

(a) the equilibrium $(x,y,\beta) = (0,0,0)$ of (5.1) is uniformly stable and therefore the corresponding generalized steady motion is stable with respect to $(q,\dot{q})$;

(b) if the integral of momentum $\frac{\partial T}{\partial \dot{r}}(t,q,\dot{q},\dot{r})$ is continuous in $(q,\dot{q},\dot{r})$ uniformly with respect to t, this partial stability is uniform;

(c) if $\dot{r}(t,q,\dot{q},\overline{c}+\beta)$ is continuous in $q,\dot{q},\beta$ at $(q,\dot{q},\beta) = (0,0,0)$ uniformly with respect to t, stability (or, in case b), uniform stability) obtains with respect to all variables $q,\dot{q},\dot{r}$.

5.3. We consider next the case of a system with time-independent constraints and let the forces Q be dissipative: $(Q|y) \leq 0$. We allow Π and Q to be time-dependent. The constraints being time-independent, one gets, using obvious notations, that $\mathscr{R}(t,x,y,\beta) = \mathscr{R}_2(x,y) + \mathscr{R}_0(x,\beta) - \Pi(t,x)$.

Further $\mathscr{R}_2$ is positive definite in y and one observes that $\dot{H}$ does not depend on β. The following theorem is then a simple consequence of Theorem 3.5 and Remark 3.8 b)

5.4. Theorem. If there exists a continuous function $W_*(x,\beta)$ and an open neighborhood $N \subset \Omega\setminus\{0\}$ of the set $\{(x,y,\beta) \in \Omega\colon \beta = 0,\ (x,y) \neq 0\}$ such that for $(t,x,y,\beta) \in I \times N$ and some $a \in \mathscr{K}$:

(i) $W(t,x,\beta) \geq W_*(x,\beta)$;

(ii) $W_*(x,0) \geq a(||x||)$;

and for $(t,x,y,\beta) \in I \times \Omega$

(iii) $\frac{\partial}{\partial t}[\Pi(t,x) - \Pi(t,0)] \leq 0$; $(Q|y) \leq 0$;

then, the steady motion is stable with respect to $q,\dot{q},\dot{r}$. If moreover, for some $b \in \mathscr{K}$:

(iv) $W(t,x,\beta) \leq b(||x|| + ||\beta||)$,

the stability is uniform.

5.5. Example: regular precessions of a symmetrical top (V.V. Rumiantsev [1971]). Consider a rigid body of mass m with an axis of symmetry and a fixed point 0 on this axis. We assume 0 to be the origin of some inertial frame of reference $0XYZ$ with $0Z$ vertical in the field of gravity. $0xyz$ will be a system of orthogonal axes, fixed in the body, with $0z$ along the axis of symmetry. In this system, the coordinates of the center of mass will be written $(0,0,z_0)$, $z_0 > 0$. The Eulerian angles specifying the position of the body with respect to $0XYZ$ will be written ψ,θ,ϕ: we use for these angles the definitions of H. Goldstein [1950]. If A is the common value of the two equal principal moments of inertia and C is the value of the third one, the Lagrangian function reads

$$L = T - V = \frac{1}{2}(A\dot{\theta}^2 + A\dot{\phi}^2\sin^2\theta) + \frac{1}{2}C(\dot{\psi} + \dot{\phi}\cos\theta)^2 - mgz_0\cos\theta.$$

The variables ϕ and ψ are ignorable and yield the first integrals

$$C(\dot{\psi} + \dot{\phi}\cos\theta) = c_1,$$

$$A\dot{\phi}\sin^2\theta + C\cos\theta(\dot{\psi} + \dot{\phi}\cos\theta) = c_2.$$

The Routh's function reads

$$\mathcal{R}(\theta,\dot{\theta},c_1,c_2) = \frac{1}{2}A\dot{\theta}^2 - \frac{(c_2 - c_1\cos\theta)^2}{2A\sin^2\theta} - mgz_0\cos\theta$$

and the Routh's equations are

$$A\ddot{\theta} + \frac{c_1}{A}\frac{(c_2 - c_1\cos\theta)}{\sin\theta} - \frac{(c_2 - c_1\cos\theta)^2\cos\theta}{A\sin^3\theta}$$

$$- mgz_0\sin\theta = 0,$$

$$\dot{c}_1 = \dot{c}_2 = 0.$$

They admit the static solution

$$\theta = \theta_0 \in]0, \tfrac{\pi}{2}[, \dot{\theta} = 0,$$

$$c_1 = \bar{c}_1, \bar{c}_1^2 \geq 4Amgz_0\cos\theta_0,$$

$$c_2 = \bar{c}_2 = \bar{c}_1\cos\theta_0 + \frac{\sin^2\theta_0}{2\cos\theta_0}(\bar{c}_1 \pm \sqrt{\bar{c}_1^2 - 4Amgz_0\cos\theta_0})$$

corresponding to the stationary motions

$$\theta = \theta_0, \ \dot{\theta} = 0, \ \dot{\phi} = \frac{\bar{c}_1 \pm \sqrt{\bar{c}_1^2 - 4Amgz_0 \cos\theta_0}}{2A\cos\theta_0}$$
$$\dot{\psi} = \bar{c}_1 \cos\theta_0 + A\dot{\phi}\sin^2\theta_0 \qquad (5.2)$$

Noticing that

$$W(\theta,c_1,c_2) = \frac{(c_2 - c_1\cos\theta)^2}{2A\sin^2\theta} + mgz_0\cos\theta$$

possesses a relative strict minimum at $\theta = \theta_0$ for $c_1 = \bar{c}_1$ and $c_2 = \bar{c}_2$ (Exercise: prove this!), we deduce from Theorem 5.4 that the stationary motion (5.2) is stable in θ, $\dot{\theta}$, $\dot{\phi}$ and $\dot{\psi}$.

6. Stability of the Betatron

6.1. Description of the system. The betatron is an axially symmetric accelerator where a particle with some electric charge e describes a circular trajectory in a transverse time-varying magnetic field. The hatched parts in Figure 4.1

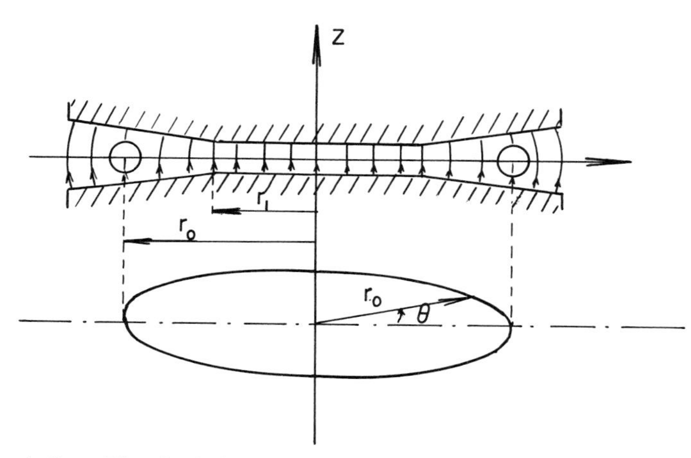

Figure 4.1. The betatron: cross-section and trajectory.

represent a section through the pole pieces of the electromagnet. We choose the axis of symmetry of the device as $0z$ axis in a system of cylindrical coordinates (r,θ,z), and we write (h_r,h_θ,h_z) for the three unit-vectors of the associated orthonormal base. For symmetry reasons, it may be assumed that the magnetic induction $\vec{B}$ does not depend on θ and has a vanishing component along h_θ. Further, it seems to be a fair approximation to assume that in the plane of symmetry $z = 0$, one has $\operatorname{rot} \vec{B} = 0$. Indeed, let us start from Maxwell's equations

$$\begin{aligned} \operatorname{rot} \vec{E} &= -\frac{1}{c}\frac{\partial \vec{B}}{\partial t} \\ \operatorname{rot} \vec{B} &= \frac{1}{c}\frac{\partial \vec{E}}{\partial t} + \frac{4\pi}{c}\vec{i}. \end{aligned} \tag{6.1}$$

where $\vec{E}$ is the electric field, $\vec{i}$ the current density and c the velocity of light. A first approximation will be to consider that the current density $\vec{i}$ is negligible. Strictly speaking, it does not vanish, because of the presence of the particles being accelerated. But it is very small anyhow. Therefore, the second equation becomes

$$\operatorname{rot} \vec{B} = \frac{1}{c}\frac{\partial \vec{E}}{\partial t} . \tag{6.2}$$

Using the symmetry hypotheses, one deduces from (6.1) and (6.2) that

$$\frac{\partial B_r}{\partial z}(t,r,0) - \frac{\partial B}{\partial r}(t,r,0) = -\frac{1}{c^2}\frac{1}{r}\int_0^r s\,\frac{\partial^2 B}{\partial t^2}(t,s,0)\,ds \tag{6.3}$$

where B has been written for B_z. But, for the values of r to be considered,

$$\frac{1}{r}\int_0^r s\,\frac{\partial^2 B}{\partial t^2}(t,s,0)\,ds$$

is practically negligible as compared to $1/c^2$. It is easily verified that, equating the second member of (6.3) to 0 amounts to obtain $\text{rot}\,\vec{B} = 0$ for $z = 0$, as was announced above.

Let us now assume that the dependence of $\vec{B}$ on t and the spatial coordinates can be made explicit as in the following formula:

$$B(t,r,z) = g(t)[B_r(r,z)h_r + B(r,z)h_z],$$

where

(i) $B_r(r,0) = 0$, and $g(t)$ is proportional, for every t, to the current intensity through the coils of the electro-magnet.

The following realistic hypotheses will be used:

(ii) for some quantities $\alpha, \beta > 0$ and every t

$$0 < \alpha < g(t) < \beta;\ 0 \leq \dot{g}(t),\ \dot{g}(t) \not\equiv 0;$$

(iii) $\vec{B}$ is a $\mathscr{C}^2$ function, and $B(r,z) > 0$.

6.2. <u>Equations of motion</u>. As $\text{div}\,\vec{B} = 0$, one obtains that, for every $r \geq 0$,

$$rB_r(r,z) = -\int_0^r s\,\frac{\partial B}{\partial z}(s,z)\,ds \tag{6.4}$$

and, using (i), that

$$\frac{\partial B}{\partial z}(r,0) = 0. \tag{6.5}$$

The relativistic Lagrangian function for a particle with mass at rest m_R and charge e (positive or negative) reads

$$L = -m_R c^2 \left(1 - \frac{\dot{r}^2 + r^2\dot{\theta}^2 + \dot{z}^2}{c^2}\right)^{1/2} + \frac{e}{c}\, g(t)\dot{\theta} \int_0^r s\, B(s,z)ds,$$

where c is the velocity of light, whereas the mass of the moving particle is $m = m_R\left(1 - \frac{v^2}{c^2}\right)^{-1/2}$ with $v^2 = \dot{r}^2 + r^2\dot{\theta}^2 + \dot{z}^2$. For the derivation of this function L through the use of a vector potential, we refer to H. Goldstein [1950] and M. Laloy [1973]$_3$. The equations of motion read

$$\frac{dr}{dt} = \dot{r}, \quad \frac{dz}{dt} = \dot{z},$$

$$\frac{d}{dt}(m\dot{r}) = mr\dot{\theta}^2 + \frac{e}{c}\,\dot{\theta} r g(t) B(r,z),$$

$$w(t,r,z,\dot{\theta}) = mr^2\dot{\theta} + \frac{e}{c}\, g(t) \int_0^r s\, B(s,z)ds = k,$$

$$\frac{d}{dt}(m\dot{z}) = \frac{e}{c}\,\dot{\theta} g(t) \int_0^r s\, \frac{\partial B}{\partial z}(s,z)ds,$$

where $w(t,r,z,\dot{\theta})$ is a first integral and k a constant depending on the initial conditions. We observe, for future reference that if $H = \frac{1}{2}\, m(\dot{r}^2 + r^2\dot{\theta}^2 + \dot{z}^2)$, one gets

$$\frac{d}{dt}(mH) = -m\, \frac{e}{c}\, \dot{\theta}\dot{g}(t) \int_0^r s\, B(s,z)ds \tag{6.6}$$

the derivative being computed along the motions.

6.3. <u>Existence of a circular orbit</u>. Let us now investigate the possibility of generating a circular orbit of the type

$$r(t) = r_0, \quad \dot{r}(t) = 0, \quad z(t) = 0, \quad \dot{z}(t) = 0, \quad \dot{\theta} = \dot{\theta}_0(t). \tag{6.7}$$

It corresponds to a generalized stationary motion (see Appendix II.12), since $\dot{\theta}_0$ is not necessarily constant. Substituting this solution in the equations of motion and setting

$$m_0 = m_R\left(1 - \frac{r_0^2\dot{\theta}_0^2(t)}{c^2}\right)^{-1/2},$$

we get

$$m_0 r_0 \dot{\theta}_0^2(t) = -\frac{e}{c}\,\dot{\theta}_0(t) r_0 g(t) B(r_0,0), \tag{6.8}$$

$$m_0 r_0^2 \dot{\theta}_0(t) + \frac{e}{c}\, g(t) \int_0^{r_0} s\, B(s,0)\,ds = k,$$

$$\dot{\theta}_0(t) g(t) \int_0^{r_0} s\, \frac{\partial B}{\partial z}(s,0)\,ds = 0.$$

Due to (6.5), the latter condition is always verified. Assuming $\dot{\theta}_0(t) \neq 0$, and since $\dot{g}(t) \not\equiv 0$, one deduces from the two former ones the equality

$$r_0^2 B(r_0,0) = \int_0^{r_0} s\, B(s,0)\,ds, \tag{6.9}$$

which is necessary and sufficient for the existence of a stationary motion of type (6.7). This is a condition on the pole-pieces shape. It implies that for some value of r, $0 < r < r_0$, one has $B(r,0) > B(r_0,0)$, which can be realized for instance if $B(r,0)$ is constant, for some r_1, on $0 < r < r_1 < r_0$, and strictly decreasing on $[r_1,r_0]$. The <u>index of the magnetic field</u> is the number n defined by

$$\frac{\partial B}{\partial r}(r_0,0) = -\frac{n}{r_0} B(r_0,0). \tag{6.10}$$

One often chooses $B(r,0) = B(r_0,0)\left(\frac{r_0}{r}\right)^n$ for r near r_0. If (6.9) is satisfied, $\dot{\theta}_0$ is given by (6.8):

$$m_0\dot{\theta}_0(t) = -\frac{e}{c}\, g(t)B(r_0,0). \tag{6.11}$$

6.4. The stability question. Let us now investigate the stability of the stationary motion (6.7). The transformation of variables

$$r = r,\ z = z,\ \dot{r} = \frac{g(t)}{m}\,\dot{\bar{r}},\ \dot{\theta} = \frac{g(t)}{m}\,\dot{\bar{\theta}},\ \dot{z} = \frac{g(t)}{m}\,\dot{\bar{z}}$$

does not affect the stability to be studied (cf. 4.6): remember the existence, assumed in Section 6.1, of the bounds on $g(t)$. The new equations of motion admit the first integral $g(t)\bar{w}(r,z,\dot{\bar{\theta}})$ where

$$\bar{w}(r,z,\dot{\bar{\theta}}) = r^2\dot{\bar{\theta}} + \frac{e}{c}\int_0^r s\ B(s,z)ds. \tag{6.12}$$

But (6.9) and (6.11) imply that

$$\bar{w}(r_0,0,\dot{\bar{\theta}}_0) = r_0^2\dot{\bar{\theta}}_0 + \frac{e}{c}\int_0^{r_0} s\ B(s,0)ds = 0, \tag{6.13}$$

and therefore $\dot{\bar{\theta}}_0$ is a constant which, by (iii), is such that $e\dot{\bar{\theta}}_0 < 0$. By the transformation of variables, the stationary motion has been changed into

$$r = r_0,\ z = 0,\ \dot{\bar{r}} = 0,\ \dot{\bar{z}} = 0,\ \dot{\bar{\theta}} = \dot{\bar{\theta}}_0. \tag{6.14}$$

Its stability will be studied using the auxiliary function $\bar{H}_1$ defined by

$$\bar{H}_1 = \bar{H} \quad \text{where} \quad e\bar{w} \le 0$$

and

$$\bar{H}_1 = \bar{H} - 2\dot{\bar{\theta}}_0\bar{w} \quad \text{where} \quad e\bar{w} > 0,$$

with

$$\bar{H} = \frac{1}{2}\,(\dot{\bar{r}}^2 + r^2\dot{\bar{\theta}}^2 + \dot{\bar{z}}^2) - \frac{1}{2}\,r_0^2\dot{\bar{\theta}}_0^2\ .$$

6.5. <u>Lemma</u>. $\dot{\bar{H}}_1 \leq 0$ along the motions of the system, in a sphere of radius $|\dot{\bar{\theta}}_0|$ and center at point (6.14).

<u>Proof</u>. Taking into account (6.6), (6.11) and the fact that $g(t)\bar{w}$ is a constant, we get

$$\text{if } e\bar{w} \leq 0: \qquad \dot{\bar{H}}_1 = -\frac{\dot{g}(t)}{g(t)} [\dot{\bar{r}}^2 + \dot{\bar{z}}^2 + \dot{\bar{\theta}}\,\bar{w}],$$

$$\text{if } e\bar{w} > 0: \qquad \dot{\bar{H}}_1 = -\frac{\dot{g}(t)}{g(t)} [\dot{\bar{r}}^2 + \dot{\bar{z}}^2 + \bar{w}(\dot{\bar{\theta}} - 2\dot{\bar{\theta}}_0)].$$

The lemma is proved, because $(\dot{\bar{\theta}} - 2\dot{\bar{\theta}}_0)\dot{\bar{\theta}}_0 < 0$ in the sphere cited above. Q.E.D.

6.6. <u>Lemma</u>. If $0 < n < 1$, $\bar{H}_1$ has an isolated minimum at point (6.14) under the constraint $\bar{w} = 0$.

<u>Proof</u>. Using (6.12) and (6.13), we get for $\bar{w} = 0$:

$$\bar{H}_1 = \frac{1}{2}(\dot{\bar{r}}^2 + \dot{\bar{z}}^2) + \frac{1}{2}\frac{e^2}{c^2}[\psi^2(r,z) - \psi^2(r_0,0)]$$

where $\psi(r,z) = \frac{1}{r}\int_0^r s\,B(s,z)ds$. By (6.5) and (6.9) we obtain

$$\frac{\partial\psi}{\partial r}(r_0,0) = \frac{\partial\psi}{\partial z}(r_0,0) = \frac{\partial^2\psi}{\partial r\partial z}(r_0,0) = 0,$$

$$\frac{\partial^2\psi}{\partial r^2}(r_0,0) = \frac{\partial B}{\partial r}(r_0,0) + \frac{1}{r_0}B(r_0,0).$$

The latter equation together with (6.10) shows that $\frac{\partial^2\phi}{\partial r^2}(r_0,0) > 0$ if and only if $n < 1$. On the other hand, given (6.4)

$$\frac{\partial^2\psi}{\partial z^2}(r_0,0) = -\frac{\partial B_r}{\partial z}(r_0,0).$$

Remember now that $\text{rot}\,\vec{B} = 0$ in the plane $z = 0$ and therefore, as was shown above:

$$\frac{\partial B_r}{\partial z}(r,0) - \frac{\partial B}{\partial r}(r,0) = 0.$$

Using (6.10), we obtain

$$\frac{\partial^2 \psi}{\partial z^2}(r_0,0) = \frac{n}{r_0} B(r_0,0).$$

Thus, if $n > 0$, both members of this equality are strictly greater than 0 and the lemma is proved. Q.E.D.

6.7. Theorem. If the magnetic index n is such that $0 < n < 1$, the generalized stationary motion (6.7) is stable.

Proof. Indeed, referring to Theorem 3.5 and Remark 3.8(b), $|g(t)\overline{w}|$ plays the role of the first integral $W(t,x)$ and $\overline{H}_1$ the role of the function $V(t,x)$. Thus (6.14) is stable for the system in the new variables, and therefore (6.7) is also stable for the original equation of motion. Q.E.D.

6.8. Remark. The model adopted here for the betatron includes the following features, which are of course conditions of validity for our conclusion: the pole pieces admit the Oz axis as an axis of symmetry and the plane $z = 0$ as a plane of symmetry; the electromagnetic field due to the particle itself, or to other particles in the beam (space charge effect) can be neglected.

7. Construction of Positive Definite Functions: Practical Criteria

7.1. It was proved in Theorem 3.2 that, $W(t,x)$ being an m-dimensional first integral for Equation (2.1), there is a continuous function ϕ such that $\phi(W(t,x))$ is positive definite if and only if $||W(t,x)||$ is positive definite.

Observe that the character of $W(t,x)$ being a first integral is immaterial here: the only requirement is that $W(t,x)$ be continuous and $W(t,0) = 0$. Theorem 3.2 is typical of a class of useful propositions to be studied in the present section.

7.2. First of all, several helpful comments can be made on the positive definiteness of $||W||$, in case $W = W(x)$ is time independent.

(a) To start with, $||W(x)||$ is positive definite if and only if equation $W(x) = 0$ has no root besides 0 in some neighborhood of the origin.

(b) Therefore, by the implicit function theorem and if W is $\mathscr{C}^1$, a necessary condition for $||W(x)||$ to be positive definite is that the jacobian matrix $\frac{\partial W}{\partial x}(0)$ be of rank strictly smaller than m.

(c) Coming back now to a continuous W, one sees that $||W(x)||$ is positive definite whenever one of the components of W is different from zero on the set of points $x \neq 0$ where the other components vanish simultaneously.

(d) Of course, if $x = (y,z)$ with $y \in \mathscr{R}^p$ and $z \in \mathscr{R}^{n-p}$ for some p, $0 < p < m$, and if the equations $W_1(x) = \ldots = W_p(x) = 0$ can be solved in some neighborhood of the origin, yielding a unique solution $y = Y(z)$ then $||W(x)||$ is positive definite if and only if $W_{p+1}^2(Y(z),z) + \ldots$ $\ldots + W_m^2(Y(z),z)$ is positive definite, of course in z-space.

7.3. <u>Exercise</u>. Consider condition 7.2(c) above, restated in the quasi-equivalent form:

(i) W_1 is different from zero on the set of points $x \neq 0$ where $W_2,\ldots,W_m$ vanish simultaneously.

One natural way to satisfy this condition is to choose W_1 such that:

(ii) W_1 possesses a strict minimum constrained by the equations $W_2(x) = \ldots = W_m(x) = 0$.

Exhibit an example where (i) is satisfied, but not (ii), and check, on the example, the positive-definiteness of $||W(x)||$.

7.4. It would be practically useful, in case W is twice continuously differentiable, to get criteria for the existence of a function ϕ such that $\phi(W(x))$ be positive definite and the positive-definiteness be recognizable at the matrix of its second derivatives. This means that this matrix, which by the way is usually called the <u>hessian matrix</u>, should be positive definite at the origin. The remainder of the present section, where W is always assumed to be independent of t and $\mathscr{C}^2$, is devoted to this problem. In this respect, the function

$$\sum_{1\leq i\leq m} W_i^2(x)$$

should be discarded at once. In fact, by putting

$$v_i(x) = \sum_{1\leq j\leq n} \left(\frac{\partial W_i}{\partial x_j}\right)_0 x_j$$

one gets

$$\sum_{1\leq i\leq m} W_i^2(x) = \sum_{1\leq i\leq m} v_i^2(x) + o(||x||^2)$$

where $\frac{o(||x||^2)}{||x||^2} \to 0$ as $x \to 0$. But, as $m < n$, the equations $v_i(x) = 0$, $i = 1,\ldots,m$, define a subspace of $\mathscr{R}^n$ of dimension higher than 0. The following theorem is the basis of the so-called <u>Chetaev's method</u> to construct positive definite functions.

7.5. <u>Theorem</u>. There exists a $\mathscr{C}^2$ function $\phi: \mathscr{R}^m \to \mathscr{R}$, with $\phi(W(x))$ positive definite and the hessian matrix of $\phi(W(x))$, computed at the origin, also positive definite, if and only if there exist parameters $\lambda_1,\ldots,\lambda_m$ such that

(i) $\sum_{1\le i\le m} \lambda_i v_i(x) \equiv 0$;

(ii) the hessian matrix of

$$\sum_{1\le i\le m} [\lambda_i W_i(x) + W_i^2(x)] \tag{7.1}$$

is positive definite at the origin.

<u>Proof</u>. The conditions are clearly sufficient since (i) implies that the function (7.1) contains no terms linear in x. Let us prove that they are necessary. Considering the function $\phi(\xi_1,\ldots,\xi_m)$, we put

$$\alpha_i = \left(\frac{\partial\phi}{\partial\xi_i}\right)_0, \quad \beta_{ik} = \frac{1}{2}\left(\frac{\partial^2\phi}{\partial\xi_i\partial\xi_k}\right)_0$$

and

$$\mathscr{U}_i(x) = \frac{1}{2}\sum_{1\le p\le n}\sum_{1\le q\le n}\left(\frac{\partial^2 W_i}{\partial x_p\partial x_q}\right)_0 x_p x_q .$$

Then

$$\phi(W(x)) = \sum_{1 \leq i \leq m} \alpha_i [v_i(x) + \mathscr{U}_i(x)] + \sum_{1 \leq i,k \leq m} \beta_{ik} v_i(x) v_k(x) + o(||x||^2).$$

where $\dfrac{o(||x||^2)}{||x||^2} \to 0$ as $x \to 0$. As $\phi(W(x))$ is positive definite, $\sum_i \alpha_i v_i(x) \equiv 0$. Further, in $\mathscr{R}^n$

$$\sum_i \alpha_i \mathscr{U}_i(x) + \sum_{i,k} \beta_{ik} v_i(x) v_k(x) \geq k||x||^2$$

for some $k > 0$ and $||x||^2 = \sum_i x_i^2$. But $\sum_{i,k} \beta_{ik} v_i v_k \leq K\left(\sum_i v_i^2\right)$ for some $K > 0$. Therefore,

$$\sum_{1 \leq i \leq m} [\frac{\alpha_i}{K} \mathscr{U}_i(x) + v_i^2(x)] \geq \frac{k}{K} ||x||^2 \qquad (7.2)$$

and this proves the thesis for $\lambda_i = \alpha_i/K$. Q.E.D.

Finding appropriate parameters λ_i is made easier by the following corollary.

7.6. <u>Corollary</u>. In Theorem 7.5, (ii) may be replaced by

(ii)' $\sum_{1 \leq i \leq m} \lambda_i \mathscr{U}_i(x) > 0$ for every $x \neq 0$ such that $v_i(x) = 0$ for $i = 1,\ldots,m$.

<u>Proof</u>. The new condition is necessary, as is shown by (7.2) and using the choice above: $\lambda_i = \alpha_i/K$. Sufficiency is also easily proved for the new theorem. Indeed, let $\sigma > 0$ be such that $B_\sigma \subset \Omega$ and let $H = \{x \in \Omega: v_i(x) = 0, i = 1,\ldots,m\}$. There exists an open neighborhood N of $H \cap \partial B_\sigma$ such that $\sum_i \lambda_i \mathscr{U}_i(x) > 0$ over N. Further, there exist constants $\alpha > 0$ and $\beta > 0$ such that $\sum_i \lambda_i \mathscr{U}_i(x) > -\alpha$ and $\sum_i v_i^2(x) > \beta$

over the compact set $\partial B_\sigma \backslash N$. Therefore, one has, for $x \in \partial B_\sigma$

$$\frac{\beta}{\alpha} \sum_i \lambda_i \mathscr{U}_i(x) + \sum_i v_i^2(x) > 0,$$

and, since this function is a quadratic form, the inequality is true for all $x \neq 0$. Therefore, the function

$$\phi(W(x)) = \frac{\beta}{\alpha} \sum_i \lambda_i W_i(x) + \sum_i W_i^2(x)$$

is positive definite with a hessian matrix positive definite at the origin. Q.E.D.

Corollary 7.6 can be exploited as follows: solving the equations for some variables (say $p < m$), one substitutes into $\sum_i \lambda_i \mathscr{U}_i(x)$ and then uses any available criterion (e.g. Sylvester's criterion) to check the positive definiteness of the quadratic form thus obtained, which possesses $n - p$ variables only.

7.7. As observed in Exercise 7.3, $||W(x)||$ is positive definite if one of the components of W, say W_1, admits a strict minimum constrained by the equations $W_2(x) = \ldots = W_m(x) = 0$. The next two theorems result in a criterion to recognize this fact, using derivatives of W of order not higher than the second.

7.8. <u>Theorem</u>. Suppose there exist real numbers $\lambda_2, \ldots, \lambda_m$ such that

$$v_1(x) + \sum_{2 \leq i \leq m} \lambda_i v_i(x) \equiv 0.$$

Then

(a) there exists a $\mu \in \mathscr{R}$ such that

$$\phi(W(x)) = W_1(x) + \sum_{2\le i\le m} [\lambda_i W_i(x) + \mu W_i^2(x)]$$

be positive definite, with the hessian matrix of $\phi(W(x))$, computed at the origin, also positive definite, if and only if

(b) $\mathscr{U}_1(x) + \sum_{2\le i\le m} \lambda_i \mathscr{U}_i(x) > 0$ for every $x \neq 0$ such that $v_i(x) = 0$, $i = 2,\ldots,m$.

Proof. The condition is necessary because

$$\phi(W(x)) = \mathscr{U}_1(x) + \sum_{2\le i\le m} [\lambda_i \mathscr{U}_i(x) + \mu v_i^2(x)] + o(||x||^2)$$

whereas, by hypothesis

$$\mathscr{U}_1(x) + \sum_{2\le i\le m} [\lambda_i \mathscr{U}_i(x) + \mu v_i^2(x)] > 0 \quad \text{for} \quad x \neq 0.$$

It is also sufficient, for it can be proved as in Corollary 7.6, that if

$$\mathscr{U}_1(x) + \sum_{2\le i\le m} \lambda_i \mathscr{U}_i(x) > 0 \quad \text{for} \quad x \neq 0,\ v_i(x) = 0,\ i = 2,\ldots,m,$$

then

$$\mathscr{U}_1(x) + \sum_{2\le i\le m} [\lambda_i \mathscr{U}_i(x) + \mu v_i^2(x)] > 0 \quad \text{for} \quad x \neq 0. \qquad \text{Q.E.D.}$$

7.9. Corollary. Thesis (a) in Theorem 7.8 implies

(a)' $W_1(x)$ admits at $x = 0$ a strict minimum constrained by the equations $W_2(x) = \ldots = W_m(x) = 0$.

7.10. Let us introduce a last theorem concerning the case, already touched upon in Section 7.2(d), where some of the equations $W_i(x) = 0$ can be solved with respect to some of their arguments. More precisely, suppose $\left(\frac{\partial W}{\partial x}\right)_0$ is of rank p for some p, $1 \leq p < m$, and

$$\left[\frac{\partial(W_1,\ldots,W_p)}{\partial(x_1,\ldots,x_p)}\right]_0 \tag{7.3}$$

is regular. Then, if we write $y = (x_1,\ldots,x_p)$, $z = (x_{p+1},\ldots,x_n)$, the equations $W_1(x) = \ldots = W_p(x) = 0$ can be solved in some appropriate neighborhood of the origin, under the form $y = Y(z)$.

7.11. <u>Theorem</u>. Suppose the above conditions concerning $(\partial W/\partial x)_0$ and the jacobian matrix (7.3) are satisfied. Then there exists a $\mathscr{C}^2$ function $\psi: \mathscr{R}^{m-p} \to \mathscr{R}$, such that $\psi(0) = 0$ and $\psi(W_{p+1}(Y(z),z),\ldots,W_m(Y(z),z))$ is positive definite in z, with a hessian matrix, computed at the origin, also positive definite, if and only if there exist real numbers $\lambda_{p+1},\ldots,\lambda_m$ such that the function $\sum_{p+1\leq k\leq m} \lambda_k W_k(Y(z),z)$ be positive definite with a hessian matrix, computed at the origin, also positive definite.

<u>Proof</u>. Let us show first that the linear terms are lacking in the expansion of the function $\overline{W}_k(z) \equiv W_k(Y(z),z)$, $k = p+1,\ldots,m$ around $z = 0$. This happens because, for every x such that $W_i(x) = 0$, $i = 1,\ldots,p$, one has

$$W_k(x) = q_k(x) + o(||x||^2) \qquad k = p+1,\ldots,m, \tag{7.4}$$

where $q_k(x)$ is quadratic in x and $\dfrac{o(||x||^2)}{||x||^2} \to 0$ as $x \to 0$. Indeed, if $W_i(x) = 0$, one has

$$v_i(x) = -\mathscr{U}_i(x) + o_i(||x||^2) \tag{7.5}$$

where $o_i(||x||^2)/||x||^2 \to 0$ as $x \to 0$. The hypothesis concerning (7.3) implies that

$$v_k(x) \equiv \sum_{1\le i\le p} \gamma_{ki} v_i(x) \qquad k = p+1,\ldots,m$$

for some appropriate quantities γ_{ki}. Therefore,

$$W_k(x) = \sum_{1\le i\le p} \gamma_{ki} v_i(x) + \mathscr{U}_k(x) + o_k(||x||^2).$$

and the expected result (7.4) is obtained through replacing $v_i(x)$ by its value (7.5).

The condition in the statement of the theorem is obviously sufficient. Let us show that it is also necessary. Suppose $\psi(\overline{W}_{p+1}(z),\ldots,\overline{W}_m(z))$ is positive definite with a positive definite hessian matrix, where $\psi(\xi_{p+1},\ldots,\xi_m)$ is a $\mathscr{C}^2$ function. Let $\alpha_k = (\partial\psi/\partial\xi_k)_0$, $k = p+1,\ldots,m$. Remembering (7.4), one gets that

$$\psi(\overline{W}_{p+1}(z),\ldots,\overline{W}_m(z)) = \sum_{p+1\le k\le m} \alpha_k \tilde{\mathscr{U}}_k(z) + o(||z||^2)$$

where $o(||z||^2)/||z||^2 \to 0$ as $z \to 0$, and $\tilde{\mathscr{U}}_k(z)$ is the quadratic part (in z) of $q_k(Y(z),z)$. By hypothesis,

$$\sum_{p+1\le k\le m} \alpha_k \tilde{\mathscr{U}}_k(z) \ge h||z||^2$$

for some $h > 0$. Hence, the function

$$\sum_{p+1\le k\le m} \alpha_k \overline{W}_k(z)$$

is positive definite with a positive definite hessian matrix. Q.E.D.

7.12. Exercise. Find two quadratic forms $W_1(x_1,x_2)$, $W_2(x_1,x_2)$, $x_1, x_2 \in \mathscr{R}$, such that $W_1^2(x_1,x_2) + W_2^2(x_1,x_2)$ is positive definite, whereas no linear combination $\lambda_1 W_1 + \lambda_2 W_2$ is positive definite.

Hint: observe, in connection with Theorem 7.11, that the hessian matrix of W_1 and W_2, computed at the origin, cannot be positive definite.

7.13. Vertical rotations of a top. The example of the top illustrates several possibilities and drawbacks of the various theorems presented in this section. Let us formulate the problem as in N.G. Chetaev [1955]. We consider a rigid body with a fixed point 0. Let Oxyz be a system of orthogonal axes chosen along the principal axes of inertia of the body at 0, and let A, $B = A$ and C be the moments of inertia with regard to these axes. Let $x = y = 0$ and $z > 0$ be the coordinates of the center of inertia of the body. Let p,q,r be the components, in Oxyz, of the instantaneous angular velocity of the body with respect to some inertial reference frame. At last, let γ,γ',γ'' be the directional cosines of the upward vertical with respect to the moving axes Oxyz.

We are interested in the stability of the vertical rotations

$$p = 0,\ q = 0,\ r = r_0,\ \gamma = 0,\ \gamma' = 0,\ \gamma'' = 1,$$

this stability being considered with regard to the variables $p,q,r,\gamma,\gamma',\gamma''$. In fact, there will be at our disposal a sufficient number of first integrals to enable us to solve the stability question without resorting to any other equations of motion. The variations of the variables with respect to the critical point will be designated by $\xi,\eta,\zeta,\alpha,\beta,\gamma$, with

$$p = \xi,\ q = \eta,\ r = r_0 + \zeta,\ \gamma = \alpha,\ \gamma' = \beta,\ \gamma'' = 1 + \delta.$$

Well-known first integrals for this problem read

$$W_1 = A(\xi^2 + \eta^2) + C(\zeta^2 + 2r_0\zeta) + 2mgz\delta,$$

$$W_2 = A(\xi\alpha + \eta\beta) + C(\delta\zeta + r_0\delta + \zeta),$$

$$W_3 = \alpha^2 + \beta^2 + \delta^2 + 2\delta.$$

$$W_4 = \zeta.$$

When trying to apply Pozharitskii's Theorem 3.2 by first computing $||W||^2 = \sum_{1\leq i\leq 4} W_i^2$, one observes that in this expression, the terms of second degree do not constitute a positive definite quadratic form, and therefore deciding of the positive-definiteness of $||W||$ will not be easy. Theorem 7.5 probably yields a better clue, for here only the quadratic terms would have to be examined. But finding the λ_i and computing the expression (7.1) remains cumbersome, all four squares W_i^2 appearing in (7.1). The method used by N.G. Chetaev [1955] is somewhat simpler. He looks for a set of real quantities $\lambda_1,\ldots,\lambda_4$ and μ such that

$$\sum_{1\leq i\leq 4} \lambda_i W_i + \mu W_4^2$$

be positive definite. For details of computation, we refer to Chetaev's book. It appears, in fact, that the easiest procedure is suggested by 7.2(d) and Theorem 7.11. The jacobian matrix of the W_i is of rank 2 at the origin. The equations $W_3 = 0$ and $W_4 = 0$ can be solved for ξ and δ, yielding, up to terms of order 2,

$$\xi = 0, \ \delta = -\frac{\alpha^2 + \beta^2}{2} + \ldots .$$

Substituting in W_1 and W_2, we get

$$\overline{W}_1 = A(\xi^2 + \eta^2) - mgz(\alpha^2 + \beta^2) + \cdots$$

$$\overline{W}_2 = A(\xi\alpha + \eta\beta) - Cr_0 \frac{\alpha^2 + \beta^2}{2} + \cdots .$$

At last, looking for a linear combination

$$\overline{W}_1 + \lambda\overline{W}_2 = A\xi^2 + \lambda A\xi\alpha - (mgz + \lambda \frac{Cr_0}{2})\alpha^2$$

$$+ A\eta^2 + \lambda A\eta\beta - (mgz + \lambda \frac{Cr_0}{2})\beta^2 + \cdots$$

we see that both quadratic forms, in (ξ,α) and (η,β) respectively, are positive definite if and only if $C^2r_0^2 > 4Amgz$. This becomes apparent by applying Sylvester's criterion

$$\begin{vmatrix} A & \frac{A\lambda}{2} \\ \frac{A\lambda}{2} & -(mgz + \frac{Cr_0\lambda}{2}) \end{vmatrix} > 0$$

or

$$\frac{A}{4}\lambda^2 + \frac{Cr_0}{2}\lambda + mgz < 0.$$

This inequality will be verified for some λ if and only if

$$C^2r_0^2 > 4Amgz$$

which is a sufficient condition for stability.

7.14. <u>Regular precessions of a satellite.</u> Consider a rigid satellite attracted by newtonian forces towards the earth center O and let $Oxyz$ be an inertial reference frame. Let G be the center of mass of the satellite and $G\xi\eta\zeta$ a system of orthogonal axes chosen along the central axes of inertia. Let A, $B = A$ and C be the central moments of inertia. The satellite will be assumed to be sufficiently remote from the earth to justify the hypothesis of the classical restricted problem: i.e., the motion of the body around G has no effect upon the motion of G, which then of course obeys Kepler's laws. We suppose the orbit to be located in the plane $z = 0$, with radius R and angular velocity $\vec{\omega}_0$, anticlockwise around $0z$. The problem is thus reduced to studying the motion of the satellite around G under the action of the gravity torque, which reads

$$\vec{M}_G = 3\omega_0^2(C-A)(\gamma\beta\vec{e}_1-\gamma\alpha\vec{e}_2).$$

Here $\vec{e}_1,\vec{e}_2,\vec{e}_3$ are the unit vectors along the axes $G\xi\eta\zeta$, and α,β,γ are the directional cosines of $\overrightarrow{0G}$ with respect to $G\xi\eta\zeta$ (see e.g. E. Leimanis [1965] or V.V. Beletskii [1966]).

Let us now introduce the orbital system of axes $Gx'y'z'$, as shown in Figure 4.2, with Gx' directed as $\overrightarrow{OG}$, Gy' tangent to the orbit and Gz' parallel to Oz. Let $\vec{\omega} = p\vec{e}_1 + q\vec{e}_2 + r\vec{e}_3$ be the instantaneous angular velocity of the satellite with respect to $Gx'y'z'$. Following

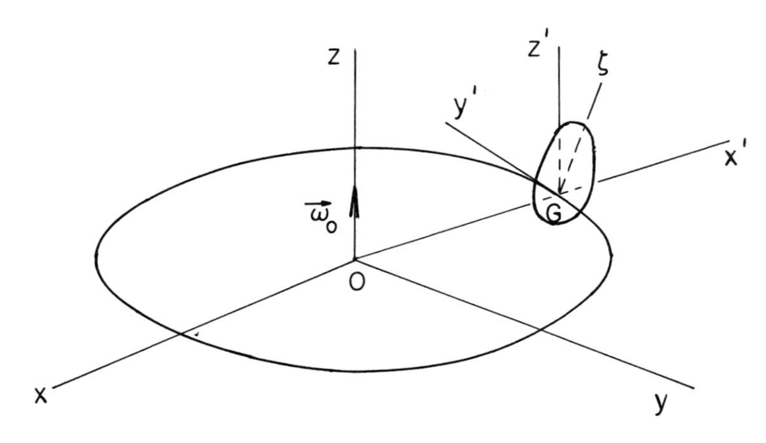

Figure 4.2. Orbiting satellite

F.L. Chernous'ko [1964], there is a class of motions called regular precessions, characterized by a fixed position of the axis $G\zeta$ in the orbital system. Every motion of this type is described by specifying the directional cosines γ, γ', γ'' of $G\zeta$ with respect to $Gx'y'z'$ and, since $p = q = 0$, the value of r, which is a constant. There are three subfamilies of motions:

$$r = -\omega_0 \frac{C-A}{C} \bar{\gamma}'', \quad \gamma = 0, \quad \gamma' = \pm\sqrt{1 - \bar{\gamma}''^2} \qquad \text{(I)}$$
$$\gamma'' = \bar{\gamma}'' \quad \text{with} \quad -1 < \bar{\gamma}'' < 1,$$

$$r = -4\omega_0 \frac{C-A}{A} \bar{\gamma}'', \gamma = \pm\sqrt{1 - \bar{\gamma}''^2}, \gamma' = 0, \qquad \text{(II)}$$
$$\gamma'' = \bar{\gamma}'' \quad \text{with} \quad -1 < \bar{\gamma}'' < 1,$$

$$r = \bar{r} \quad \text{with} \quad -\infty < \bar{r} < \infty, \quad \gamma = \gamma' = 0, \; \gamma'' = \pm 1, \qquad \text{(III)}$$

whose stability will be studied using the following first integrals of the equations of the relative motion:

$$W_1 = r + \omega_0 \gamma'', \tag{7.6}$$

$$W_2 = A(p^2+q^2) + Cr^2 - \omega_0^2(C-A)\gamma''^2 + 3\omega_0^2(C-A)\gamma^2, \tag{7.7}$$

$$W_3 = \gamma^2 + \gamma'^2 + \gamma''^2 = 1. \tag{7.8}$$

For the detailed calculation, we refer to Chernous'ko's paper.

Let us consider the variations p, q, $\xi = r + \omega_0 \frac{C-A}{C} \bar{\gamma}''$, γ, $\delta = \gamma'' - \bar{\gamma}''$ of the variables appearing in (7.6) and (7.7) with respect to the unperturbed motion (I). The integral W_1 written in the new variables and adjusted in such a way that $W_1(0) = 0$ reads $W_1 = \xi + \omega_0\delta$. When solving the equation

$W_1 = 0$ for ξ and substituting in W_2, one gets

$$W_2 = A(p^2+q^2) + A\omega_0^2\delta^2 + 3\omega_0^2(C-A)\gamma^2$$

a function which is positive definite for $C > A$. This entails stability with respect to p,q,r,γ and γ'', whereas stability with respect to γ' is a direct consequence of Equation (7.8).

For the unperturbed motion (II), the new variables will be $\xi = r + 4\omega_0 \frac{C-A}{C} \overline{\gamma}''$, $\delta = \gamma - \overline{\gamma}$, $z = \gamma'' - \overline{\gamma}''$, where $\overline{\gamma} = \pm\sqrt{1 - \overline{\gamma}''^2}$, and the first integrals W_1 and W_3 become

$$W_1 = \xi + \omega_0 z,$$
$$W_3 = 2\overline{\gamma}\delta + 2\overline{\gamma}''z + \delta^2 + z^2 + \gamma'^2.$$

Solving $W_1 = W_3 = 0$ for δ and ξ and substituting in W_2, one gets, up to terms of second order,

$$W_2 = A(p^2+q^2) + [A\omega_0^2 - 3\omega_0^2(C-A)]z^2 - 3\omega_0^2(C-A)\gamma'^2 + \ldots$$

and this function is positive definite if $A > C$.

We reason alike for the unperturbed motion (III), using as new variables $\xi = r - \overline{r}$ and $\delta = \gamma'' \mp 1$. After computation, W_2 becomes, up to terms of second order,

$$W_2 = A(p^2+q^2) \pm C\overline{r}\omega_0(\gamma^2+\gamma'^2) + \omega_0^2(C-A)(\gamma^2+\gamma'^2) + 3\omega_0^2(C-A)\gamma^2.$$

This function is positive definite, if $\gamma'' = +1$, when $\overline{r} > -\omega_0 \frac{C-A}{C}$ or $\overline{r} > +4\omega_0 \frac{A-C}{C}$ accordingly as $C > A$ or $A > C$, and if $\gamma'' = -1$, when $\overline{r} < \omega_0 \frac{C-A}{C}$ or $\overline{r} < 4\omega_0 \frac{C-A}{C}$ accordingly as $C > A$ or $A > C$.

The general conclusion is that the stationary motion (I) is stable with respect to p,q,r,γ,γ' and γ'', if the ellipsoid of inertia of the satellite is of the flat type, the motion (II) if it is of the oblong type, whereas the motion (III) can be stable in both cases, but only when the angular velocity $\bar{r}$ satisfies a suitable inequality.

7.15. Exercise. What kind of difficulty would one encounter if, to study the stability of motion (I) hereabove, one tries, as for (II) and (III), first to solve $W_1 = W_3 = 0$, and then to substitute in W_2?

8. Bibliographical Note

The earliest result concerning stability in the presence of first integrals concerns stationary motions of autonomous lagrangian systems and is due to E.J. Routh [1877]. With the notations of Section 5 adapted, in some obvious way, to the autonomous case, let $\mathscr{R}_2 = \mathscr{R}_2(x,y)$, $\mathscr{R}_0 = \mathscr{R}_0(x,\beta)$, $\Pi = \Pi(x)$ and $Q = 0$. Routh observed that if, for $\beta = \bar{\beta}$, $\Pi(x) - \mathscr{R}_0(x,\beta)$ admits a strict minimum at $x = 0$, then the stationary motion defined by $x = y = 0$ and $\bar{\beta}$ is stable for all perturbations satisfying the equation $\beta = \bar{\beta}$. This proposition of Routh was extended by A.M. Liapunov [1893] to non-autonomous equations $\dot{x} = f(t,x)$ possessing a first integral $W(t,x)$: see Exercise 3.10.

The awkward restriction concerning the perturbations was partially removed by A.M. Liapunov: on this point see also V.V. Rumiantsev [1968]. It was completely removed, in the setting of autonomous mechanical systems, by L. Salvadori [1953]. Indeed, using a suitable Liapunov function in connection with Theorem I.4.2, he proved stability of the stationary motion for arbitrary perturbations. On this extension, see also G.K. Pozharitskii [1958] and V.V. Rumiantsev [1968].

On the other hand, L. Salvadori [1966] extended his result of 1953 to a class of dissipative systems, again by constructing an appropriate auxiliary function and using Theorem I.4.2. Of course, the dissipation is limited to non-ignorable coordinates. This result is still generalized in L. Salvadori [1969] to the case of a function $\Pi(x) - \mathscr{R}_0(x,\beta)$ considered along with a more general dissipation, reduced however to non-ignorable coordinates. The type of proof is here completely different: it makes use of families of Liapunov functions with one parameter, an ingenious technical trick to be dealt with at length in Chapter VIII. In the mean time, C. Risito [1967] proved Theorem 4.2 and used it to get a similar extension to dissipative systems, with the restriction, however, that the dissipation, although again limited to non-ignorable coordinates, has to be complete in this setting.

As becomes apparent from these historical considerations, the methods used to tackle this problem of extending Routh's initial observation have been varied ones. At the end, it became clear that the utmost generalization, namely the one presented in P. Habets and C. Risito [1973] and in Section 5 above, which by the way includes non-autonomous systems, is but a simple consequence of Theorem 3.5. This theorem appears in the same paper of P. Habets and C. Risito. Its forerunners can be found in P. Habets and K. Peiffer [1973] and M. Laloy $[1973]_3$.

A generalized version of Theorem 4.2 appears in C. Risito [1974] where the bound on r can be time-dependent and where, moreover, one may find conditions for uniform stability. A first version of Theorem 4.4 appears in C. Risito [1971]. It has been generalized in P. Habets and C. Risito [1973] and further in C. Risito [1974].

A survey of stability results for stationary motions was given by S. Pluchino [1971], along with an extension of Salvadori's result to non-holonomic dissipative systems.

The study of the betatron is due to M. Laloy $[1973]_4$.

Theorem 3.2 has already been attributed to G.K. Pozharitskii [1958]. The method of constructing weighted combinations of first integrals and of their squares goes back to N.G. Chetaev [1961]. The theorems of Section 7 emphasize the usefulness of those criteria where positive-definiteness can be recognized at the derivatives of second order, thus enabling one to apply Sylvester's criterion. They come essentially from C. Risito [1975]. Interesting comparisons between the various methods of constructing Liapunov functions can also be found in V.V. Rubanovskii and S. Ia. Stepanov [1969].

CHAPTER V

INSTABILITY

1. Introduction

1.1. Inasmuch as stability is a desired property in many circumstances, it is important to have at one's disposal some effective means of recognizing instability. This is the object of the present chapter. However, before studying instability as such, we shall deal at some length with new concepts such as sectors, expellers, etc., and this deserves some preliminary comments.

Let us go back to Chetaev's Theorem I.5.1, where the role played by the function $V(t,x)$ is twofold: it is used to prove first that no solution starting from inside the open set Ψ crosses $\partial\Psi \cap B_\varepsilon$ and then that no such solution can remain in Ψ as t approaches infinity. Roughly speaking, the first of these properties will be referred to hereafter as Ψ being a sector, and the second as Ψ being an expeller. When combined, they imply instability. One should here emphasize the following fact which, as we shall see, has important theoretical and practical consequences: it is possible to prove separately, i.e. by using two distinct auxiliary functions, that Ψ is a sector and that it is an expeller. The following example makes it clear.

In the general hypotheses used for Chetaev's Theorem I.5.1, suppose for simplicity that the differential equation on hand is autonomous. Suppose further that for some $\varepsilon > 0$, with $\overline{B_\varepsilon} \subset \Omega$, there exists an open set $\Psi \subset B_\varepsilon$, a $\mathscr{C}^1$ function $V: B_\varepsilon \to \mathscr{R}$ and a function $b \in \mathscr{K}$ such that, for every x on Ψ:

(i) $V(x) > 0$;

(ii) $\dot{V}(x) \geq b(V(x))$;

assume further that there exists a neighborhood N of $\partial\Psi$ and a second auxiliary function $W: N \to \mathscr{R}$ such that

(iii) $W(x) = 0$ on $\partial\Psi \cap B_\varepsilon$;

(iv) $\dot{W}(x) \geq 0$ and $W(x) > 0$ on $\Psi \cap N \cap B_\varepsilon$;

if, at last, $0 \in \partial\Psi$, then the origin is unstable, as can be shown by reasoning almost as in Chetaev's Theorem, the function W being such that no solution issued from some point of Ψ can approach $\partial\Psi$.

1.2. <u>Exercise</u>. Give a detailed proof of the statement above.

1.3. This Chapter will show how the ideas of sector and expeller can be used to decompose the concept of instability into two simpler ones, each of them becoming the object of a separate study. This systematic approach appears as a natural continuation of Chetaev's results and, even more directly, of K. P. Persidski's theory of sectors. It is worth mentioning finally that our treatment takes closed as well as open sectors into account, and that the topological principle of Wazewski is used to get a useful generalization of the basic theory. From a practical point of view, closed sectors will often coincide with hypersurfaces defined by first integrals.

2. Definitions and General Hypotheses

2.1. Our general hypotheses remain here those of Section IV.2.1, to which we add however the requirement that, for any $(t_0, x_0) \in I \times \Omega$, the Cauchy problem

$$\dot{x} = f(t,x) \tag{2.1}$$

$$x(t_0) = x_0 \tag{2.2}$$

has a unique solution. This supplementary hypothesis is essential in the present context, because we shall often resort to the argument of continuity of the solutions with respect to the initial conditions. The solution of (2.1) and (2.2) will be written $x(t;t_0,x_0)$.

Instability is of course the contrary of stability. It will prove helpful, for reference purposes, to recall here the explicit definition of instability: the critical point at the origin is said to be unstable if

$$(\exists\varepsilon > 0)(\exists t_0 \in I)(\forall\delta > 0)(\exists x_0 \in B_\delta)(\exists t \in J^+)\quad x(t;t_0,x_0) \notin B_\varepsilon.$$

All the following theorems have the form of sufficient conditions and they should, in a complete version, begin as Chetaev's Theorem, by: "If there exist an $\varepsilon > 0$ with

$\overline{B_\varepsilon} \subset \Omega$ and a $t_0 \in I$ ...". By reason of simplicity, we assume that ε and t_0 are chosen here once and for all.

Let $C_\varepsilon = I \times B_\varepsilon$. For any $G \subset C_\varepsilon$ and $t \in I$, we define

$$G(t) = \{x: (t,x) \in G\},$$
$$G^* = \{(t,x) \in G: x \neq 0\},$$
$$L = \{(t,x) \in \partial G \cap C_\varepsilon: x \neq 0\}.$$

We call L the side-boundary of G. Another general hypothesis is that all sets G to be mentioned below are such that the origin of $\mathscr{R}^n$ is a cluster point for $G(t_0)$.

2.2. A set G is called a sector if, for every $\delta > 0$, one at least of the two following conditions is satisfied:

(i) $(\exists x_0 \in G^*(t_0) \cap \overline{B}_\delta)(\forall t \in J^+)$ $(t,x(t;t_0,x_0)) \in G$;

(ii) $(\exists x_0 \in \overline{B}_\delta)(\exists t \in J^+)$ $x(t;t_0,x_0) \notin B_\varepsilon$.

2.3. A set G is called an absolute sector if, for every $x_0 \in G^*(t_0)$, one at least of the two following conditions is satisfied:

(i) $(\forall t \in J^+)$ $(t,x(t;t_0,x_0)) \in G$;

(ii) $(\exists t \in J^+)$ $x(t;t_0,x_0) \notin B_\varepsilon$.

Clearly C_ε is an absolute sector, and every absolute sector is a sector. It may be somewhat surprising to observe that, in the definition of a sector G, some points are mentioned which do not belong to G. This peculiarity will be justified a posteriori by the role it will play in several proofs below.

2.4. A set G is called an expeller if

$$(\forall\delta > 0)(\exists x_0 \in G^*(t_0) \cap \overline{B}_\delta)(\exists t \in J^+) \quad (t, x(t;t_0,x_0)) \notin G.$$

2.5. It will be called an <u>absolute expeller</u> if

$$(\forall\delta > 0)(\forall x_0 \in G^*(t_0) \cap \overline{B}_\delta)(\exists t \in J^+) \quad (t, x(t;t_0,x_0)) \notin G.$$

3. Fundamental Proposition

<u>The following statements are equivalent</u>:

(a) <u>the origin is unstable</u>;

(b) <u>there exists a sector which is an absolute expeller</u>;

(c) <u>there exists an absolute sector which is an expeller</u>;

(d) <u>there exists an absolute sector which is an absolute expeller</u>.

<u>Proof</u>. Obviously (b) $\Rightarrow$ (a), (c) $\Rightarrow$ (a) and (d) $\Rightarrow$ (b) and (c). Therefore, it will be sufficient to prove that (a) $\Rightarrow$ (d). But if the origin is unstable, there exists a sequence of points $x_{0i} \in B_\varepsilon$ and a sequence of time-values $t_i > t_0$ such that $x(t_i;t_0,x_{0i}) \notin B_\varepsilon$ and $x_{0i} \to 0$ for $i = 1,2,\ldots$. The set G defined as the intersection with C_ε of the trajectories of all solutions $x(t;t_0,x_{0i})$ is an absolute sector and an absolute expeller. Q.E.D.

Conditions ensuring instability will therefore be obtained by combining, in various obvious ways, sufficient conditions for the existence of a sector, an absolute sector, an expeller or an absolute expeller. By the way, the set Ψ mentioned in Chetaev's Theorem I.5.1 is an absolute sector and an absolute expeller. The next two sections will be devoted to a fairly detailed study of sectors and expellers respectively.

4. Sectors

4.1. Sectors and absolute sectors will often be characterized by the way the solutions of the differential equation cross their side boundary. Roughly speaking, they can cross it all from outside to inside (see Theorem 4.2, (i) to (iii)) and this type of behavior leads to absolute sectors, or all from inside to outside (see Theorem 4.5 and 4.6) and this yields sectors, or in a more complicated way, some from outside to inside and the others from inside to outside (see Theorem 4.7), and this yields again a sector. A few more definitions are needed to describe precisely such behaviors of the solutions.

A point $(s,a) \in L$ will be called

an <u>ingress point</u> of G if $(\exists T > 0,\ s + T \in J^+(s,a))$ $(\forall t \in \,]s,s + T])$ $(t,x(t;s,a)) \in \overset{\circ}{G}$;

an <u>egress point</u> of G if $(\exists T > 0,\ s + T \in J^+(s,a))$ $(\forall t \in \,]s,s + T])$ $(t,x(t;s,a)) \notin \overline{G}$;

a <u>consequent point</u> of G if $(\exists T > 0,\ s - T \in J(s,a))$ $(\forall t \in [s - T,s[)$ $(t,x(t;s,a)) \in \overset{\circ}{G}$.

To recognize the fact that a point of the side-boundary is, or is not, an ingress, an egress or a consequent point is often an easy matter and can, in many instances, be deduced from a careful examination of the second member of the differential equation. One may also resort to some simple criteria making use of auxiliary functions. Several of them will be given below: see e.g. 4.3 and 4.4 and some lemmas in Section 6. For the time being, assume that we can in general recognize such points and let us prove sufficient conditions for a set G to be a sector or an absolute sector. We deal first with absolute sectors, because they are simpler.

4.2. <u>Theorem</u>. Each of the following conditions is sufficient for a set G to be an absolute sector:

(i) G is closed in C_ε and no point of L is an egress point;

(ii) G is open and no point of L is a consequent point;

(iii) G is open, $\partial G = \partial \overline{G}$ and no point of L is an egress point;

(iv) $G = H \cap C_\varepsilon$ for some H positively invariant.

<u>Proof</u>. (i) and (ii) are obvious; (iii) follows from (ii) and from Lemma 4.3 below; (iv) is also obvious. Q.E.D.

4.3. <u>Lemma</u>. If G is open, $\partial G = \partial \overline{G}$ and no point of L is an egress point, then no point of L is a consequent point.

<u>Proof</u>. Suppose a point $P = (s,a) \in L$ is a consequent point, i.e. $(\exists T > 0,\ s - T \in J(s,a))(\forall t \in [s - T, s[)$ $(t, x(t;s,a)) \in G$. Let N be a neighborhood of P which, because $\partial G = \partial \overline{G}$, contains points outside $\overline{G}$. The continuity with respect to the initial conditions implies that there exist solutions starting from some neighborhood N' of $(s - T,\ x(s - T;\ s,a))$, $N' \subset G$ and reaching these points outside $\overline{G}$. Thus there would exist an egress point, which is excluded. Q.E.D.

4.4. <u>Exercise</u>. If G is open and no point of L is a consequent point, then no point of L is an egress point.

As appears from the theorems to follow, sufficient conditions to get a sector are more involved.

4.5. <u>Theorem</u>. Assume that, for some set G closed in C_ε and for every $t \geq t_0$:

(i) $\overset{\circ}{G}(t)$ is connected;

(ii) the origin is a cluster point of $\overset{\circ}{G}(t)$;

(iii) ∂B_ε contains at least one cluster point of $\overset{\circ}{G}(t)$;

(iv) no point of L is an ingress point;

then G is a sector.

<u>Proof</u>. Assume, on the contrary, that G is not a sector. Then, for some δ, $0 < \delta < \varepsilon$, we may write the opposites of propositions (i) and (ii) in the definition of a sector.

(a) The opposite of (ii) reads:

$$(\forall x_0 \in \overline{B}_\delta)(\forall t \in J^+) \qquad x(t;t_0,x_0) \in B_\varepsilon. \tag{4.1}$$

Let us put

$$H(t) = \{x\colon \ x = x(t;t_0,x_0),\ x_0 \in B_\delta\}.$$

This set is defined for every $t \geq t_0$, and is an open neighborhood of the origin of $\mathscr{R}^n$. Therefore, due to (ii) hereabove, $\overset{\circ}{G}(t) \cap H(t) \neq \phi$. On the other hand, (4.1) implies that $\overline{H}(t) \subset B_\varepsilon$ and then, because of (iii): $\overset{\circ}{G}(t) \cap \complement\, \overline{H}(t) \neq \phi$, where $\complement$ means "the complementary of". Thus, for every $t \geq t_0$,

$$\overset{\circ}{G}(t) \cap \partial H(t) \neq \phi \tag{4.2}$$

for, otherwise, $\overset{\circ}{G}(t)$ would be the union of the two open disjoint and non empty sets $\overset{\circ}{G}(t) \cap H(t)$ and $\overset{\circ}{G}(t) \cap \complement \overline{H}(t)$, and $\overset{\circ}{G}(t)$ would not be connected.

(b) From the opposite of (i) in the definition of a sector, one deduces that

$$(\forall x_0 \in G(t_0) \cap \partial B_\delta)(\exists t' \in J^+) \quad (t', x(t'; t_0, x_0)) \notin G.$$

Therefore, for every $x_0 \in G(t_0) \cap \partial B_\delta$, there exist a $\overline{t}(x_0)$ and a $t' > \overline{t}$ such that $(\overline{t}, x(\overline{t})) \in L$ and $(t, x(t)) \notin G$ for $t \in [\overline{t}, t']$. For every x_0, let $T(x_0)$ be the infinimum of these $\overline{t}(x_0)$. Let us show that $T(x_0)$ is upper semicontinuous (on this notion, see e.g. E. J. McShane [1944]). If it were not so, there would exist in $G(t_0) \cap \partial B_\delta$ a point x_0^0 and an infinite sequence $\{x_0^i\}$, $x_0^i \neq x_0^0$, such that x_0^i approaches x_0^0 as $i \to \infty$ and $\lim T(x_0^i) = T > T(x_0^0)$. Then, for every small enough $\eta < T - T(x_0^0)$ and every i large enough, one would get

$$(T(x_0^0) + \eta, x(T(x_0^0) + \eta; t_0, x_0^i)) \in G$$

and

$$(T(x_0^0) + \eta, x(T(x_0^0) + \eta; t_0, x_0^0)) \notin G.$$

But this is impossible because x is continuous with respect to the initial conditions and G is closed in C_ε.

Therefore, there exists a $\tilde{T}$ such that, for every $x_0 \in G(t_0) \cap \partial B_\delta$: $T(x_0) \leq \tilde{T}$. But due to (4.1), none of the solutions on hand comes out of B_ε (and therefore ceases to exist) and further, no point of L is an ingress point. Thus, on the one hand

$$(\forall x_0 \in G(t_0) \cap \partial B_\delta)(\forall t \geq \tilde{T}) \quad x(t; t_0, x_0) \notin \overset{\circ}{G}(t),$$

and on the other, due to (iv),

$$(\forall x_0 \in [G(t_0) \cap \partial B_\delta)\ (\forall t \geq t_0)\quad x(t;t_0,x_0) \notin \overset{\circ}{G}(t).$$

One concludes that $(\forall t \geq \tilde{T})\ \overset{\circ}{G}(t) \cap \partial H(t) = \phi$, but this assertion contradicts (4.2). Q.E.D.

This theorem leads naturally to the following one, concerning the case of a set G, no more necessarily closed. Notice however that if G is closed, Theorem 4.6 is weaker than Theorem 4.5.

4.6. <u>Theorem</u>. Assume that, for some open set G and for every $t \geq t_0$:

(i) $\overset{\circ}{G}(t)$ is connected;

(ii) the origin is a cluster point of $\overset{\circ}{G}(t)$;

(iii) ∂B_ε contains at least one cluster point of $\overset{\circ}{G}(t)$;

(iv) every point of L is an egress point;

then G is a sector.

<u>Proof</u>. All the hypotheses of Theorem 4.5 are satisfied if one substitutes to G the set $H = \overline{G} \cap C_\varepsilon$. Then for any given δ, either proposition (ii) in the definition of a sector is satisfied, or it is not. In the latter case

$$(\exists x_0 \in H^*(t_0) \cap \overline{B}_\delta)(\forall t \in J^+)\quad (t,x(t;t_0,x_0)) \in H = \overline{G} \cap C_\varepsilon.$$

But as every point of L is an egress point, there will be no value of t such that $(t,x(t;t_0,x_0)) \in L$, hence the theorem. Q.E.D.

In Theorems 4.5 and 4.6, the points of L are constrained to some kind of uniform behavior: in one case, no one of them can be an ingress point, whereas in the other, they all have to be egress points. Such a restriction is somewhat relaxed in the following theorem, which is adapted from Wazewski's topological principle (see e.g. P. Hartman [1964]). Before we can state it, two more definitions are needed. If X is a topological space and if $A \subset B \subset X$, A is called a <u>retract</u> of B if there exists a continuous mapping on B into A, which is the identity on A. This kind of mapping is called a <u>retraction</u> (of B into A).

4.7. <u>Theorem</u>. For some open set G, assume that

(i) all consequent points of G are egress points, and let S be the set of these consequent points.

Suppose there exists for every $\delta \in]0,\varepsilon[$ a set $Z_\delta \subset [G^*(t_0) \cup S(t_0)] \cap \overline{B}_\delta$ such that

(ii) $Z_\delta \cap G^*(t_0) \neq \phi$, $Z_\delta \cap S(t_0) \neq \phi$;

(iii) writing $Z_\delta' = \{(t_0,x_0): x_0 \in Z_\delta\}$, $Z_\delta' \cap S$ is a retract of S but not of Z_δ';

then G is a sector and even, which is a little more, for every $\delta \in]0,\varepsilon[$, there exists an $x_0 \in Z_\delta \cap G^*(t_0)$ such that $(t,x(t;t_0,x_0)) \in G$ for every $t \in J^+$, or the positive semi-trajectory starting from (t_0,x_0) meets ∂G for the first time at a point of ∂C_ε.

<u>Proof</u>. If the thesis is wrong $(\exists\delta > 0)(\forall x_0 \in Z_\delta)(\exists T \geq t_0)$ and

$$(\forall t \in [t_0,T[) \quad (t,x(t;t_0,x_0)) \in G \tag{4.3}$$
$$(T,x(T;t_0,x_0)) \in S.$$

Notice that, if $x_0 \in S(t_0)$: $T = t_0$ and $[t_0,T[\; = \phi$. The points of S being egress points, one shows as in Theorem 4.5, that $T(x_0)$ is upper semi-continuous. Owing to (4.3), one may show alike that it is lower semi-continuous. Thus, the function $\pi_1: Z'_\delta \to S$, $(t_0,x_0) \to (T(x_0),x(T(x_0);t_0,x_0))$ is continuous. If π is a retraction mapping S on $Z'_\delta \cap S$, then $\pi \circ \pi_1$ is a retraction of Z'_δ on $Z'_\delta \cap S$, and this cannot exist. Q.E.D.

4.8. Some geometrical insight is necessary to realize what happens in Theorem 4.7. The picture presented in Fig. 5.1 serves this purpose. It has been drawn for $n = 3$ and the t-axis

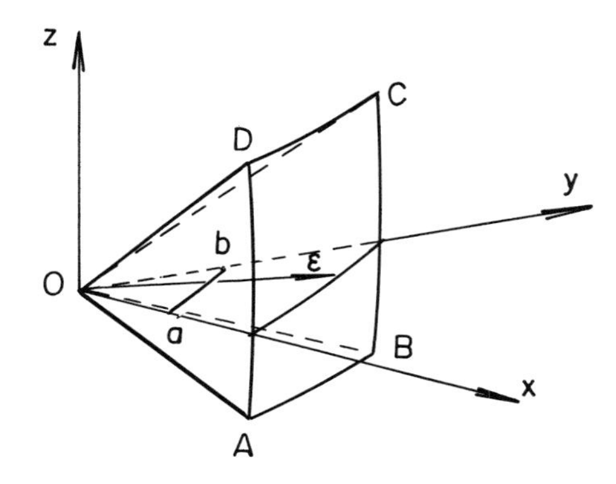

Fig. 5.1. The geometrical situation of Theorem 4.7.

is not represented. The set $G(t_0)$ is the "pyramid" OABCD. One assumes that $G(t) = G(t_0)$ for $t \geq t_0$ and, for simplicity, that the differential equation is autonomous. Then of course $S(t) = S(t_0)$. Assume that $S(t_0)$ is the union of the faces OAD and OBC (remember that the origin is excluded). For some $\delta < \varepsilon$, let the arc $\widehat{ab}$ represent the intersection of $\overline{G(t_0)} \cap \partial B_\delta$ with the Oxy plane. If one identifies Z_δ with the arc $\widehat{ab}$, then $Z_\delta \cap S(t_0) = \{a\} \cup \{b\}$.

4.9. Remark. It is interesting to compare Theorems 4.6 and 4.7. Suppose G is some open set satisfying Hypotheses (i) to (iii) of 4.6. It can be shown that if every point of L is an egress point (last hypothesis of 4.6), then every point of L is a consequent point, and therefore, in the notations of 4.7: $L = S$. In this setting, 4.6 is not a particular case of 4.7, for (iii) of 4.7 is an additional assumption. However, it is easy to prove that if $G(t) = G(t_0)$ for $t \geq t_0$ and if $G(t_0)$ is a cone (i.e. if $G(t_0)$ is the intersection with B_ε of a set which is a union of rays issuing from the origin), then, again in the setting described above, the hypotheses of 4.6 imply those of 4.7.

4.10. Remark. As it is usually more difficult to prove the existence of a retraction than to prove its non existence, the following generalization of Theorem 4.7 will prove helpful: Hypothesis (iii) of Theorem 4.7 may be replaced by: S is not a retract of $Z'_\delta \cup S$. This statement, as well as those contained in Remark 4.9 may be considered as exercises. The proofs as well as some further discussion of the relations between Theorems 4.6 and 4.7 appear in M. Laloy $[1974]_2$.

5. Expellers

5.1. There is a wide variety of theorems giving sufficient conditions for a set to be an expeller or an absolute expeller. Some of them are given below, selected from the most simple and effective ones. The general idea behind them is that a function $V(t,x(t))$ should be bounded from above in some way, and simultaneously increasing at a sufficient pace for the bound to be attained after some finite time if the solution $x(t)$ were to remain in the set. The upper bound on V can of course vary with t, as in the following theorem.

5.2. Theorem. Let there exist a function $V(t,x)$ on C_ε into $\mathscr{R}$, locally lipschitzian in x and continuous, two real continuous functions $a(t)$, $c(t)$ defined on $[t_0,\infty[$ and a function $b \in K$ such that, for some set $G \subset C_\varepsilon$:

(i) $(\forall x_0 \in G^*(t_0))\ \ V(t_0,x_0) > 0$;

(ii) $c(t) \geq 0$ and $\int_{t_0}^{t} c(s)ds \to \infty$ as $t \to \infty$;

(iii) $(\forall (t,x) \in G^*$ such that $V(t,x) > 0)$

$$V(t,x) \leq a(t), \tag{5.1}$$

$$D^+V(t,x) \geq 0, \tag{5.2}$$

$$D^+V(t,x) \geq c(t)b(V(t,x)) + D^+a(t); \tag{5.3}$$

then G is an absolute expeller.

Proof. If this were wrong, there would exist an $x_0 \in G^*(t_0)$ such that $(t,x(t;t_0,x_0)) \in G$ for every $t \in J^+$. Therefore $J^+ = [t_0,\infty[$, and, due to (5.2), $V(t,x(t)) \geq V(t_0,x_0) > 0$ for every $t \geq t_0$. It follows then from (5.3) that for every $t \geq t_0$

$$V(t,x(t)) - a(t) \geq V(t_0,x_0) - a(t_0) + b(V(t_0,x_0))\int_{t_0}^{t} c(s)ds,$$

and finally, by (5.1),

$$b(V(t_0,x_0))\int_{t_0}^{t} c(s)ds \leq a(t_0) - V(t_0,x_0),$$

an inequality becoming wrong for large enough t. Q.E.D.

5.3. Exercise. Prove in detail the following proposition which has been used implicitly in the proof of Theorem 5.2.

Let $h(t)$ be a continuous function on $[t_0,\infty[$ into $\mathscr{R}$, and assume that $h(t_0) > 0$ and further that $D^+h(t) \geq 0$ for every $t \geq t_0$ such that $h(t) > 0$. Then $h(t) > 0$ for every $t \geq t_0$.

5.4. Corollary. If Hypothesis (i) in Theorem 5.2 is replaced by (i)' the origin of $\mathscr{R}^n$ is a cluster point of the set $G(t_0) \cap \{x: V(t_0,x) > 0\}$, then G is an expeller.

Theorem 5.2 leads to a particularly simple corollary when neither V nor $\dot{V}$ depend on t (which happens of course mainly in the case of autonomous equations).

5.5. Corollary. Suppose $G(t) = G(t_0)$ for every $t \geq t_0$. Suppose further that for some $\varepsilon' > \varepsilon$ with $B_{\varepsilon'} \subset \Omega$, there exists a $\mathscr{C}^1$ function $V(x)$ on $B_{\varepsilon'}$ into $\mathscr{R}$, such that $\dot{V}$ does not vary with t and that:

(i) $(\forall x \in G^*(t_0))$ $V(x) > 0$;

(ii) $(\forall x \in \overline{G}^*(t_0),\ V(x) > 0)$ $\dot{V}(x) > 0$;

then G is an absolute expeller.

Proof. Let $\eta' = \max\{V(x): x \in \overline{G}(t_0)\}$. For every $\eta \in]0,\eta']$, the set $\{x: x \in \overline{G}(t_0),\ V(x) \geq \eta\}$ is non empty and compact. If $\lambda(\eta)$ is the infimum of $\dot{V}(x)$ on this set, $\lambda(\eta)$ is an increasing function of η, with $\lambda(\eta) > 0$. One knows then (see e.g. N. Rouche and J. Mawhin [1973]) that there exists a function $b \in \mathscr{K}$ with $b(\eta) \leq \lambda(\eta)$. Of course $\dot{V}(x) \geq b(V(x))$. By choosing $a(t) = \eta'$ and $c(t) = 1$, one verifies all the hypotheses of Theorem 5.2. Q.E.D.

In the proof of this corollary, the inequality $\dot{V}(x) > 0$ has been used in an essential way. It can be weakened however and be replaced by $\dot{V}(x) \geq 0$, at the expense of another hypothesis making sure that the solution will not spend too much time in the region where $\dot{V}(x) = 0$. In this sense, the following theorem is a natural extension of Corollary 5.5.

5.6. Theorem. Let Equation (2.1) be autonomous and $G = [t_0,\infty[\times \Psi$ for some Ψ in B_ε. Suppose there exists a real function $V(x)$ defined and locally lipschitzian on $B_{\varepsilon'}$ for some $\varepsilon' > \varepsilon$, $B_{\varepsilon'} \subset \Omega$, and such that

(i) $(\forall x \in \Psi^*)$ $V(x) > 0$;

(ii) $(\forall x \in \overline{\Psi}: V(x) > 0)$ $D^+V \geq 0$;

(iii) $(\forall \eta > 0)$ the set $F_\eta = \{x: x \in \overline{\Psi}, V(x) = \eta, D^+V(x) = 0\}$ contains no compact, invariant, non-empty subset;

then G is an absolute expeller.

Proof. If the thesis is wrong, there is an $x_0 \in \Psi^*$ such that for every $t \geq t_0$: $x(t;x_0) \in \Psi$. But then the positive limit set(*) Λ^+ of this solution is non-empty and $\Lambda^+ \subset \overline{\Psi}$. Due to (i), $V(x_0) > 0$, and therefore, by (ii), $V(x(t;x_0))$ is increasing. But $V(x)$ is bounded on $\overline{\Psi}$ and thus $V(x(t;x_0)) \to V_0$ as $t \to \infty$, for some V_0. Of course $V(x) = V_0$ on Λ^+. But as Λ^+ is invariant, $D^+V(x) = 0$ on Λ^+, and (iii) is violated. Q.E.D.

(*) On this notion, cf. Appendix III.

As is apparent, this theorem is akin to LaSalle's invariance principle VII.3 as well as to Theorem II.1.3 of N. N. Krasovski. Some other technical means can be used to ensure that the solution leaves the set $D^+V(x) = 0$ and, in particular, one can resort to a second auxiliary function possessing suitable properties on an appropriate neighborhood of this set. This kind of idea, which led to Matrosov's Theorem II.2, will not be illustrated here, for the sake of conciseness. On the other hand, and going back now to Theorem 5.2, let us demonstrate another way of weakening the main assumption on D^+V, this time by introducing a hypothesis on the second derivative of V.

5.7. Theorem. Let there exist a real $\mathscr{C}^1$ function $V(t,x)$ defined on C_ε with $\dot{V}(t,x)$ locally lipschitzian in x and continuous, such that for some set $G \subset C_\varepsilon$:

(i) $(\forall x_0 \in G^*(t_0))\quad \dot{V}(t_0,x_0) > 0$;

(ii) $(\forall (t,x) \in G^*)$ if $\dot{V}(t,x) \geq 0$, then $V(t,x) \leq 0$ and $D^+\dot{V}(t,x) \geq 0$;

then G is an absolute expeller.

The proof is similar to that of Theorem 5.2 and is left to the reader.

5.8. Remark. A corollary similar to 5.4 can be appended to 5.5, 5.6 and 5.7, thus yielding three sufficient conditions for a set G to be an expeller.

6. Example of an Equation of Nth Order

6.1. Let f be a real continuous function defined on some

interval $]\alpha,\beta[$, where $\alpha < 0 < \beta$, and such that $f(0) = 0$. Suppose further that f is regular enough to ensure uniqueness of the solutions of the nth order equation

$$\frac{d^n z}{dt^n} = f(z)$$

or of the equivalent system

$$\begin{aligned} \dot{x}_1 &= x_2, \\ \dot{x}_2 &= x_3, \\ &\cdots\cdots \\ \dot{x}_n &= f(x_1). \end{aligned} \tag{6.1}$$

It will be shown under these conditions that <u>if</u> $n \geq 3$ <u>and</u> <u>if</u> 0 <u>is an isolated root of</u> $f(z)$, <u>the origin is unstable</u> <u>for</u> (6.1).

Of course, if 0 is an isolated root, there exists an open interval $]\alpha',\beta'[$, with $\alpha' < 0 < \beta'$, such that $f(z)$ does not vanish on $]\alpha',\beta'[$ except at the origin. Without loss of generality, we assume hereafter that $]\alpha',\beta'[=]\alpha,\beta[$. The instability will be proved successively for the various following cases:

a) $zf(z) > 0$ on $]\alpha,0[$ or on $]0,\beta[$;

b) $zf(z) < 0$ on $]\alpha,0[$ and on $]0,\beta[$, and

b_1) n is odd;

b_2) $n = 2m$ with m even;

b_3) $n = 2m$ with m odd.

The following simple lemma on consequent points will be found useful.

6.2. <u>Lemma</u>. For any open set $G \subset C_\varepsilon$ and any point $P \in L$, if there exists an open neighborhood N of P and a real function $W(t,x)$ defined on $N \cap G$, locally lipschitzian in x and continuous and such that $W(t,x) > 0$, $D^+W(t,x) \geq 0$ on $N \cap G$, and $W(t,x) = 0$ at P, then P is not a consequent point of G.

6.3. Exercise. The following variant of Lemma 6.2 will be used in Section 8: the lemma remains valid if the conditions on W are replaced by the requirement that $D^+W(t,x) \geq 0$ and $W(t,x) \to -\infty$ when $(t,x) \to P$.

6.4. Let us now prove step by step the expected property.

a) We consider only the case $zf(z) > 0$ on $]0,\beta[$. The other one would be treated alike. Choose an arbitrary ε, $0 < \varepsilon < \min(|\alpha|,\beta)$, and consider the open set $G = \{(t,x)\colon t \in \mathscr{R}, x \in \mathscr{R}^n, 0 < x_i < \varepsilon \text{ for } i = 1,\ldots,n\}$. Any point P of the side boundary L of G is such that $x_i \geq 0$ for $i = 1,\ldots,n$, and at least one of the x_i, say x_j, vanishes. Using the function $W = x_j$ and Lemma 6.2, we see that P is not a consequent point. Then, by Theorem 4.2 (ii), G is an absolute sector. But it is also an absolute expeller, as is readily seen by using Theorem 5.7 and the function $V = x_1 - \varepsilon$. By the way, this proof holds also for $n = 2$.

b_1) Let $\varepsilon < \min(|\alpha|,\beta)$ and $G = \mathscr{R} \times \Psi$, $\Psi = \{x \in B_\varepsilon, V_1(x) > 0\}$, where $V_1(x) = -x_1x_n + x_2x_{n-1} - x_3x_{n-2} + \cdots + (-1)^m x_m x_{m+2} + (-1)^{m+1} \frac{1}{2} x_{m+1}^2$, for $m = (n-1)/2$. One calculates that

$$\dot{V}_1(x) = -x_1 f(x_1),$$

an expression which is larger than 0 for $x_1 \neq 0$. Then G is an absolute sector by Lemma 6.2 and Theorem 4.2 (ii). But it is also an absolute expeller, for V_1 satisfies the hypotheses of Theorem 5.6. In particular, $\dot{V}_1(x) = 0$ if and only if $x_1 = 0$, and the hyperplane $x_1 = 0$ contains no invariant set except for the origin. Indeed $x_1(t) \equiv 0$ implies $\dot{x}_1(t) \equiv 0$, and thus $x_2(t) \equiv 0$, etc.

b_2) Let ε be as above and $G = \mathscr{R} \times \Psi$, $\Psi = \{x \in B_\varepsilon, V_2(x) > 0\}$ for $V_2(x) = -x_1x_n + x_2x_{n-1} - x_3x_{n-2} + \cdots + x_mx_{m+1}$. One computes

$$\dot{V}_2(x) = x_{m+1}^2 - x_1f(x_1)$$

an expression which is larger than 0 for $x_1 \neq 0$. Then, as above, G is an absolute sector by Lemma 6.2 and Theorem 4.2 (ii). But it is also an absolute expeller by Theorem 5.6, because $\dot{V}_2 = 0$ only if $x_1 = 0$, and the hyperplane $x_1 = 0$ contains no invariant set, except for the origin.

b_3) One verifies easily that

$$W = \sum_{2\leq i\leq m} (-1)^i x_i x_{n-i+2} + \frac{1}{2} x_{m+1}^2 - \int_0^{x_1} f(z)dz$$

is a first integral for (6.1). Let $G = [t_0,\infty[\times \Psi$, with $\Psi = \{x \in B_\varepsilon, W(x) = 0\}$, and ε is defined as above. By Theorem 4.2 (iv), G is an absolute sector. Let us show that it is an expeller. In every neighborhood of the origin, there are points x where $W(x) = 0$ and $V_3(x) > 0$, with

$$V_3(x) = \sum_{1\leq i\leq \frac{m-1}{2}} \left(-\frac{2i-1}{2} x_{2i}x_{n-2i+1} + ix_{2i+1}x_{n-2i}\right).$$

Choose e.g. $x_2 \neq 0$, $x_{n-1} \neq 0$ and all the other x_i's equal to 0. One calculates that for $x \in \Psi$

$$\dot{V}_3(x) = \frac{2m-1}{4} x_{m+1}^2 - \frac{1}{2}\int_0^{x_1} f(z)dz \geq 0.$$

But the hyperplane $x_{m+1} = 0$ contains, except for the origin, no nonempty invariant subset. Therefore, the hypotheses of Theorem 5.6 as modified in Remark 5.8 are satisfied.

7. Instability of the Betatron

7.1. The stability of the circular stationary motion in the betatron has been studied in Section IV.6, to which we constantly refer hereafter. The result to be proved now is that this motion, as described by equations IV.(6.7) is unstable if the index n of the magnetic field is strictly larger than 1. The proof will be facilitated by first establishing the following two lemmas.

7.2. Lemma. If $n > 1$, the function $\psi(u)$ defined on $]-r_0,\infty[$ as

$$\psi(u) = \frac{\int_0^{u+r_0} sB(s,0)\,ds}{u + r_0}$$

admits a strict maximum at $u = 0$.

Proof. It results from IV.(6.9) that $\frac{d\psi}{du}(0) = 0$. Further, and using IV.(6.9) again,

$$\frac{d^2\psi}{du^2}(0) = \frac{1}{r_0} B(r_0,0) + \frac{\partial B}{\partial r}(r_0,0).$$

And by the definition IV.(6.10) of the index of the magnetic field,

$$\frac{d^2\psi}{du^2}(0) = \frac{B(r_0,0)}{r_0}(1 - n),$$

hence the result. Q.E.D.

7.3. Lemma. If $\phi(u)$ is a real function defined on $]-r_0,\infty[$ as

$$\phi(u) = \frac{1}{2}[\psi^2(u) - \psi^2(0)],$$

there exists an $\varepsilon_0 > 0$ such that $u\frac{d\phi}{du} < 0$ on $[-\varepsilon_0,+\varepsilon_0] \setminus \{0\}$.

7.4. Let us now introduce the new variables, $\bar{r}$, $\dot{\bar{r}}$, $\dot{\bar{\theta}}$ and $\dot{\bar{z}}$ by the equations

$$\bar{r} = r - r_0,$$

$$\dot{\bar{r}} = \frac{m}{g(t)}\dot{r},$$

$$\dot{\bar{\theta}} = \frac{m}{g(t)}\dot{\theta} - \dot{\bar{\theta}}_0 \quad \text{where} \quad \dot{\bar{\theta}}_0 = \frac{m_0}{g(t)}\dot{\theta}_0, \tag{7.1}$$

$$\dot{\bar{z}} = \frac{m}{g(t)}\dot{z},$$

the variable z remaining unchanged. The origin in the space $(\bar{r},\dot{\bar{r}},\dot{\bar{\theta}},z,\dot{\bar{z}})$ of the new coordinates corresponds to the stationary motion on hand. Considering, in the space of $(r,\dot{r},\dot{\theta},z,\dot{z})$, some compact neighborhood K of the orbit of the solution IV.(6.7), we see that the image of K in the space $(\bar{r},\dot{\bar{r}},\dot{\bar{\theta}},z,\dot{\bar{z}})$ contains a ball with center at the origin. Let ε' be its radius, and write $\varepsilon = \min(\varepsilon',\varepsilon_0)$. Further, let k be the supremum of m over K.

It is readily verified that the jacobian of the transformation (7.1) vanishes nowhere. Let $(\bar{S})$ be the system of differential equations deduced from the original equations of motion (which we shall call (S) here) by the transformation of coordinates, and let $\bar{m}(t,\bar{r},\dot{\bar{r}},\dot{\bar{\theta}},\dot{\bar{z}})$ be the new expression of the mass $m(r,\dot{r},\dot{\theta},\dot{z})$. It follows from the hypotheses on g (cf. (ii) in Section IV.6.1) that the instability of the origin for $(\bar{S})$ is equivalent to the instability of the stationary motion for (S). Indeed, one has

$$\frac{\alpha}{k} |\dot{\bar{r}}| \leq |\dot{r}| \leq \frac{\beta}{m_R} |\dot{\bar{r}}|,$$

and similar inequalities for $\dot{\bar{z}}$ and $\dot{z}$ on the one hand, and $\dot{\bar{\theta}}$ and $\dot{\theta} - \dot{\theta}_0$ on the other. It will therefore be enough to prove that the origin is unstable for $(\bar{S})$.

Of course, $(\bar{S})$ admits the first integral

$$\bar{W}(t,\bar{r},\dot{\bar{\theta}},z) = g(t)[(\bar{r} + r_0)^2(\dot{\bar{\theta}} + \dot{\bar{\theta}}_0) + \frac{e}{c}\int_0^{\bar{r}+r_0} sB(s,z)ds],$$

and, using IV.(6.11) and IV.(6.9), one verifies that $\bar{W}(t,0,0,0) = 0$. From IV.(6.5) and the last equation of motion, which reads

$$\frac{d}{dt}(m\dot{z}) = \frac{e}{c}\dot{\theta}\, g(t)\int_s^r s \frac{\partial B}{\partial z}(s,z)ds,$$

and from the fact that $\dot{\bar{W}} = 0$, one deduces that the set

$$H = \{(t,\bar{r},\dot{\bar{r}},\dot{\bar{\theta}},z,\dot{\bar{z}}): t \in I,\ \bar{W} = 0,\ z = \dot{\bar{z}} = 0\}$$

is positively invariant. Hence, by Theorem 4.2.(iv), $G = H \cap C_\varepsilon$ is an absolute sector.

Let us now prove, using the function $V = g(t)\bar{r}\,\dot{\bar{r}}$ in connection with Corollary 5.4, that it is an expeller. In every neighborhood N of the origin, there are points of $N \cap G(t_0)$ where $V > 0$. Further, it requires but a few calculations to verify, in particular by using the third equation of motion, i.e.

$$\frac{d}{dt} m\dot{r} = mr\dot{\theta}^2 + \frac{e}{c}\dot{\theta}rg(t)B(r,z),$$

that on G

$$\dot{V} = \frac{g^2(t)}{\bar{m}}[-\frac{e^2}{c^2}\bar{r}\frac{d\phi}{du}(\bar{r}) + \dot{\bar{r}}^2].$$

From Lemma 7.3 and from Hypothesis (ii) of IV.6.1, it follows that on G:

$$\dot{V} \geq \frac{\alpha^2}{k} \left[\frac{e^2}{c^2} \left| \bar{r} \frac{d\phi}{du} (\bar{r}) \right| + \dot{\bar{r}}^2\right].$$

Finally, as $\dot{\bar{r}} = V/g(t)\bar{r}$ for $V \neq 0$, one obtains, writing x for $(\bar{r},\dot{\bar{r}},\dot{\bar{\theta}},z,\dot{\bar{z}})$, that for every $(t,x) \in G^*$ such that $V > 0$:

$$\dot{V} \geq \frac{\alpha^2}{\beta^2 k \varepsilon_0} V^2.$$

All the hypotheses of Corollary 5.4 are therefore verified.

8. Example of an Equation of Third Order

8.1. Let Ω be some open neighborhood of the origin of $\mathscr{R}^3$ and consider a continuous function $g(t,x)$ on $I \times \Omega$ into $\mathscr{R}$, where $I =]\tau,\infty[$ for some $\tau \in \mathscr{R}$. Assume that g is regular enough to ensure uniqueness of the solutions of the differential equation

$$\frac{d^3y}{dt^3} = g(t,y,\frac{dy}{dt}, \frac{d^2y}{dt^2}),$$

which is equivalent to the system

$$\begin{aligned} \dot{x}_1 &= x_2, \\ \dot{x}_2 &= x_3 \\ \dot{x}_3 &= g(t,x_1,x_2,x_3). \end{aligned} \tag{8.1}$$

Assume further that $g(t,0) = 0$ for every $t \in I$. Suppose at last that for some real quantities α, β, $0 < \alpha < \beta$, and for any $x_1 < 0$, one has

$$g(t,x_1,\beta x_1,\beta^2 x_1) < \beta^3 x_1$$

and

$$g(t,x_1,\alpha x_1,\alpha^2 x_1) > \alpha^3 x_1.$$

Under these conditions, the origin of $\mathscr{R}^3$ is, as we shall see, unstable for (8.1). The following two lemmas will be used in proving this proposition. For their own interest, they are stated with some more generality than is strictly needed.

8.2. <u>Lemma</u>. For any open set $G \subset C_\varepsilon$ and any point $P \in L$, if there exists an open neighborhood N of P and a function $W(t,x)$ on N into $\mathscr{R}$, locally lipschitzian in x and continuous, such that

(i) $(\forall(t,x) \in N \cap G)$ $W(t,x) > 0$ and $D^+W(t,x) \leq 0$;

(ii) $(\forall(t,x) \in N \cap L)$ $W(t,x) = 0$ and $D^+W(t,x) \neq 0$;

then P is an egress point and a consequent point of G.

<u>Proof</u>. Let $P = (s,a)$ and suppose P is not an egress point. Then, in every interval $]s,s + T[$, there is a t_1 such that $(t_1,x(t_1;s,a)) \in \overline{G}$. Because of (i) and (ii), there is no T such that, for every $t \in]s, s + T[$: $(t,x(t)) \in \overline{G}$. Therefore, there is a $t_2 \in [s,t_1[$ such that $(t_2,x(t_2;s,a)) \notin \overline{G}$, and the continuity of the solutions with respect to the initial conditions implies the existence of an ingress point.

Suppose on the other hand that P is not a consequent point. By arguing as above, one shows that in this case also, there is an ingress point. The conclusion is that, in both cases, the following obvious proposition would be violated: for any open set $G \subset C_\varepsilon$ and any point $P \in L$, if there exists an open neighborhood N of P and a function $W(t,x)$ on N into $\mathscr{R}$, locally lipschitzian in x and continuous, such that:

(i) $(\forall(t,x) \in N \cap G)$ $W(t,x) > 0$ and $D^+W(t,x) \leq 0$;

(ii) $W(t,x) = 0$ at P;

then P is not an ingress point of G. Q.E.D.

8.3. Lemma. For any open set $G \subset C_\varepsilon$ and any point $P = (s,a) \in L$, if there exists an open neighborhood N of P, an even integer $m \geq 1$ and a $\mathscr{C}^m$ function $W(t,x)$ on N into $\mathscr{R}$ such that

(i) $(\forall (t,x) \in N \cap \bar{G})$ $W(t,x) \geq 0$;

(ii) $W^{(i)}(s,a) = 0$ for $0 \leq i \leq m$;

(iii) $W^{(m)}(s,a) < 0$;

then P is an egress point and is not a consequent point.

8.4. Let us now prove the instability of the origin for (8.1). First of all, Theorem 4.7 will be used to prove that, whatever $t_0 \in I$ and $\varepsilon > 0$ with $\bar{B}_\varepsilon \subset \Omega$, the set

$$G_1 = \{(t,x) : t \in I,\ \beta x_1 < x_2 < \alpha x_1\} \cap C_\varepsilon$$

is a sector. Notice that $x_1 < 0$ for any point of G_1. To start with, let us show that every consequent point of G_1 is an egress point. For every point $P \equiv (t,x_1,x_2,x_3)$ of L, one has

either $\quad x_1 < 0$ and $x_2 = \alpha x_1,$ (8.2)

or $\quad x_1 < 0$ and $x_2 = \beta x_1,$ (8.3)

or $\quad x_1 = x_2 = 0.$ (8.4)

Every point P verifying (8.2) and $\alpha^2 x_1 - x_3 < 0$ is an egress point and a consequent point, as is shown by Lemma 8.2, along with the auxiliary function $W_1 = \alpha x_1 - x_2$. In the same way, every point P verifying (8.3) and $x_3 - \beta^2 x_1 < 0$ is a consequent and egress point: use again Lemma 8.2 and $W_2 = x_2 - \beta x_1$.

Moreover, no other point of L is a consequent point. Indeed, if $P \in L$ verifies (8.4) with $x_3 > 0$, then $W_3 = 1/(x_2 - \alpha x_1)$ for which $\dot{W}_3 = (\alpha x_2 - x_3)/(\alpha x_1 - x_2)^2$, verifies the hypotheses of Lemma 6.2 as modified in Exercise 6.3. If P verifies (8.4) and $x_3 < 0$, the same conclusion is obtained using $W_4 = 1/(\beta x_1 - x_2)$. Let now P verify (8.2): if $\alpha^2 x_1 - x_3 > 0$, then P is not a consequent point by Lemma 6.2 considered along with W_1; if $\alpha^2 x_1 - x_3 = 0$, P is not a consequent point either, for $W_1(P) = \dot{W}_1(P) = 0$ whereas $\ddot{W}_1(P) = \alpha^3 x_1 - g(t,x_1,\alpha x_1,\alpha^2 x_1) < 0$ and the hypotheses of Lemma 8.3 are satisfied. The reasoning is the same if P satisfies (8.3) and $x_3 - \beta^2 x_1 \geq 0$.

Let us now check the other hypotheses of Theorem 4.7. For any $\delta \in]0,\varepsilon[$, let us choose two consequent points $P_1 = (t_0,x_{11},x_{21},x_{31})$ and $P_2 = (t_0,x_{12},x_{22},x_{32})$ belonging to C_δ and verifying (8.2) and (8.3) respectively. Consider the line segment $Z'_\delta = \{(t,x): t = t_0,\ x_i = (1 - \lambda)x_{i1} + \lambda x_{i2},\ 0 \leq \lambda \leq 1,\ i = 1, 2, 3\}$, and let us call S the set of consequent points of G_1. The mapping $\pi: S \to Z'_\delta \cap S$ defined by $\pi(P) = P_1$ if P verifies (8.2) and $\pi(P) = P_2$ if P verifies (8.3) is a retraction. On the other hand, $\{P_1\} \cup \{P_2\}$ is not a retract of Z'_δ. All the hypotheses of Theorem 4.7 are satisfied, and therefore G_1 is a sector.

The only thing which remains to be proved is that G_1 is an absolute expeller. But this follows from 5.5 considered along with the function $V = x_1^2$. The proof is therefore complete. Q.E.D.

9. Exercises

9.1. Consider the non autonomous linear system:

$$\begin{aligned}\dot{x}_1 &= (-b + a\cos^2 bt)x_1 + (b - a\sin bt\cos bt)x_2,\\ \dot{x}_2 &= (-b - a\sin bt\cos bt)x_1 \\ &\qquad\qquad + (-b + a\sin^2 bt)x_2,\end{aligned} \tag{9.1}$$

where a and b are two real constants with $b < a < 2b$. Show that the matrix of the system (9.1) has two eigenvalues with the same time-independent strictly negative real part, and, however, the origin is unstable.

Hint: One computes easily that the common real part of the eigenvalues is: $\frac{a - 2b}{2}$.

The set $G = \{(t,x,y): t \in \mathscr{R},\ x_2 \sin bt - x_1 \cos bt > 0\} \cap C_\varepsilon$ is an absolute sector and an absolute expeller. To prove it, use the function $V = x_2 \sin bt - x_1 \cos bt$; one has $\dot{V} = (a - b)V$. Then, apply Lemma 6.2 and Theorems 4.2 and 5.2.

A particular case of (9.1) was dealt with by L. Markus and H. Yamabe [1960] who proved the instability by exhibiting a class of unbounded solutions. The proof suggested above is due to M. Laloy $[1974]_2$.

9.2. For the system:

$$\begin{aligned}\dot{x}_1 &= 2x_1x_2 - x_2^2,\\ \dot{x}_2 &= 4x_1x_2 - x_2^2 - 2x_1^2,\end{aligned}$$

the origin is unstable (K. P. Persidski [1947]).

Hint: Consider the set $G = \mathscr{R} \times \psi$, where $\psi = \{(x_1,x_2): 0 < x_2 < 3x_1\}$. Using Lemma 8.2 and the function $W_1 = x_2$ and $W_2 = 3x_1 - x_2$, one can see that all the hypotheses of Theorem 4.6 are satisfied. Then, by means of the function $V = x_1^2 + (x_1 - x_2)^2$, and applying Theorem 5.2, one proves that G is an absolute expeller. In fact, on G, $\dot{V} \geq \sqrt{\frac{4}{5} V^3}$.

9.3. Consider the system:

$$\begin{aligned} \dot{x}_1 &= a_1x_1^2 + b_1x_1x_2 + c_1x_2^2, \\ \dot{x}_2 &= a_2x_1^2 + b_2x_1x_2 + c_2x_2^2; \end{aligned} \tag{9.2}$$

suppose the trinomials $g_i(\lambda) = a_i + b_i\lambda + c_i\lambda^2$ $(i = 1, 2)$ have real roots $\lambda_{i1}, \lambda_{i2}$. Suppose moreover that $a_2 < 0$ and $g_1(\lambda)$ has a root $\lambda_{11} > 0$ such that $g_2(\lambda_{11}) > 0$ (then, as $g_2(0) < 0$, $g_2(\lambda)$ has a root $\lambda_{21} > 0$). Then, each of the following conditions is sufficient for the instability of the origin

(i) $c_1 > 0$ and $g_1(\lambda)$ has a second root $\lambda_{12} \geq \lambda_{11}$;

(ii) $c_1 > 0$ and $\lambda_{21} < \lambda_{12} < \lambda_{11}$;

(iii) $c_1 < 0$ and $\lambda_{12} \leq 0$.

Hint: Consider the set $G = \mathscr{R} \times \psi$, where $\psi = \{(x_1,x_2) \in \mathscr{R}^2: 0 < x_2 < \lambda_{11}x_1\}$. By means of the functions $W_1 = \lambda_{11}x_1 - x_2$ and $W_2 = x_2$, and using Lemma 8.2 and Theorem 4.6, prove that G is a sector. In fact, for $x_2 = \lambda_{11}x_1$, one gets, for $x_1 \neq 0$:

$$\dot{W}_1 = -x_1g_2(\lambda_{11}) < 0, \tag{9.3}$$

and, for $x_2 = 0$,

$$\dot{W}_2 = a_2x_1^2.$$

Moreover, G is an absolute expeller. Indeed,

a) in the cases (i) and (iii), use the function $V = x_1$, with $\dot{V} = a_1x_1^2 + b_1x_1x_2 + c_1x_2^2$. For $x \in \bar{\Psi}$, one has $\dot{V}_1 \geq 0$, and $\dot{V}_1 = 0$ if and only if $x_2 = \lambda_{11}x_1$. Moreover, in view of (9.3), the set defined by $x_2 = \lambda_{11}x_1$, $x_1 \neq 0$, does not contain any invariant set. So, all the hypotheses of Theorem 5.6 are verified.

b) in case (ii), use Corollary 5.5 with the function $V(x)$ defined as follows: $V(x) = x_2$ if $x_2 > \lambda_{21}x_1$, and, if $x_2 \leq \lambda_{21}x_1$, $V(x) = \lambda_{21}x_1$.

More complete stability results for Equation (9.2) (and, more generally, for equations with homogeneous right member) can be found, e.g., in W. Hahn [1967]).

9.4. Consider the third order scalar equation

$$\frac{d^3y}{dt^3} = -a\frac{d^2y}{dt^2} - b\frac{dy}{dt} - k(y), \tag{9.4}$$

where a and b are real constants and $k(y)$ is defined on an open interval containing 0. We suppose $k(0) = 0$ and k continuous and regular enough to ensure uniqueness of solutions. We write $h(\lambda)$ for $\lambda(\lambda^2 + a\lambda + b)$. Equation (9.4) is equivalent to the system:

$$\begin{aligned}\dot{x}_1 &= x_2,\\ \dot{x}_2 &= x_3,\\ \dot{x}_3 &= -ax_3 - bx_2 - k(x_1).\end{aligned}$$

Suppose $a < 0$, $a^2 \geq 4b$ and, for $y < 0$:

$$0 < k(y) < -yh(\tfrac{1}{3}\ |a| - \sqrt{a^2 - 3b}). \tag{9.5}$$

Then, the origin is unstable (M. Laloy [1974]$_2$; this completes some results of A. Huaux [1964]; see also R. Reissig, G. Sansone and R. Conti [1969]).

Hint: Notice that the function $h(\lambda)$ has a relative maximum for $\lambda_1 = \frac{1}{3}(|a| - \sqrt{a^2 - 3b})$ and has a root $\lambda_2 > \lambda_1$. Then, taking $\alpha = \lambda_1$ and $\beta = \lambda_2$, the hypotheses relative to system (8.1) are verified if (9.5) is satisfied.

9.5. Prove the following generalization of Theorem III.3.2. (cf. M. Laloy [1975]): if there exists an $\varepsilon > 0$ (with $\overline{B}_\varepsilon \subset \Omega$) such that:

(i) $\Theta = \{q \in B_\varepsilon, \Pi(q) < 0\} \neq \phi$;

(ii) $0 \in \partial\Theta$;

(iii) $(\frac{\partial \Pi}{\partial q} \mid q) + \Pi(q) \leq 0, \ \forall q \in \Theta$,

then the origin $q = p = 0$ is unstable.

Hint: Use the absolute sector $G = \mathscr{R} \times \psi$, with $\psi = \{(q,p): 0 < ||q|| + ||p|| < \varepsilon, H(q,p) = 0\}$, and the auxiliary function $V = (q \mid p)$; a trivial modification of Corollary 5.5 implies that G is an expeller.

10. Bibliographical Note

As already noticed in Chapter I, the classical instability theorem of N. G. Chetaev [1934] generalized two previous results of A. M. Liapunov [1892]. It already used, at least implicitly, the concept of a sector. But the method of sectors as such was developed first by K. P. Persidski [1946], [1947] and later by S. K. Persidski [1961], [1968],

[1970]. The main contribution was due to the former of these authors, when he introduced a sector for which all points of the side-boundary are egress points. Some difficulties arose concerning the definition of a sector and the proof of sufficient conditions for a set to be a sector. They have been discussed by A. D. Myshkis [1947] using arguments of topological algebra, as well as by J. L. Massera [1956] and M. Laloy $[1973]_1$. The definitions of a sector and an absolute sector as given in the present text differ slightly from the original ones and are those of M. Laloy $[1973]_1$, $[1973]_2$.

As compared to those of N. G. Chetaev, K. P. and S. K. Persidski, the sectors considered here are more general in two respects: first they are not necessarily open sets (see J. A. Yorke [1968] and M. Laloy $[1973]_1$), and further they admit in their side-boundary, the coexistence of ingress and egress points (M. Laloy $[1973]_1$), a result which has been obtained by using the so-called topological principle of T. Wazewski [1947]. Notice also that the sectors considered here are subsets not of $\mathscr{R}^n$, but of $\mathscr{R} \times \mathscr{R}^n$. It would have been possible to think of a sector as a set $G(t)$ of $\mathscr{R}^n$, variable with t. This point of view leads to a specific difficulty: the necessity to define some kind of continuity for $G(t)$. On this point, cf. Dang Chau Phien and N. Rouche [1970].

On the other hand, the conditions for a set to be an expeller, as they appear in the original theorems, have been improved in various ways. In particular, the requirements put on the auxiliary function (or functions) have been weakened, for example by Kh. I. Ibrachev [1947], N. N.

Krasovski [1959], V. M. Matrosov $[1962]_1$, N. Rouche [1968], L. Salvadori [1971], J. A. Yorke [1968]. It is shown in N. Rouche [1969] that the upper bound on the V functions can be chosen as a function of t.

Case a) of the problem of an equation of nth order dealt with in Section 6 was already treated by D. W. Muller [1965] with more stringent regularity hypotheses. The same author considers also cases b_1) and b_2). These results have been generalized, in M. Laloy $[1973]_3$, to the case of a non autonomous bounded second member. In this paper, the case $n = 2m$ for m odd is also considered. The necessity of a bounded second member was dispensed with in B. D'Onofrio, R. Sarno and M. Laloy [1974]. At last in M. Laloy $[1974]_2$, the hypothesis of an isolated root for f is weakened and one considers second members depending on t and on all coordinates.

The instability of the betatron is established in M. Laloy $[1974]_2$. The example of Section 8 has been inspired by J. D. Schuur [1967]. Many more references, including the contributions of the Alma-Ata school, appear in V. M. Matrosov [1965].

CHAPTER VI

A SURVEY OF QUALITATIVE CONCEPTS

1. Introduction

1.1. Up to this point, we have studied a small number of concepts such as stability, attractivity, asymptotic stability, etc. They all pertain to the origin of $\mathscr{R}^n$ for a differential equation

$$\dot{x} = f(t,x) \tag{1.1}$$

with a continuous second member defined on $I \times \Omega$ with values in $\mathscr{R}^n$ (the symbols I and Ω have the same meaning here as in Section IV.2). An exception was the orbital stability, as described in Section I.1.4: this is properly the stability of a set. It appears that the attractivity or asymptotic stability of a set are natural concepts fitting many practical applications: several examples will be seen in Chapter VII. But as soon as a set, and especially an unbounded set, is substituted to a point, namely the origin, the number of potentially useful concepts increases rapidly:

in fact several types of "uniformities" with respect to spatial coordinates have to be considered in addition to the familiar uniformity with respect to t_0 (the one which makes the distinction between stability and uniform stability). The number of concepts increases to such an extent that one may well wonder which ones should be studied, and which ones shouldn't. The situation appears even worse when observing that some other qualitative concepts such as boundedness, dissipativity, ... possess also many variants corresponding to different types of uniformities. The purpose of this chapter is to throw some light, and, if possible, to produce some order in this forest of concepts. The interest of such an undertaking is partly theoretical and, as will be seen, it can be carried through only at the expense of introducing some new extremely concise symbolism. This is why, from Section 4 below to the end of the chapter, the familiar aspect of our statements and formulas suddenly changes. But except for Section IX.6, the rest of the book can be read without knowledge of this new language. If he is interested mainly in the applications, the reader can skip Sections 3 to 6 and only cast a glance at Section 7.

The expression qualitative concept designates hereafter any concept such as stability, attractivity, boundedness, etc. "Qualitative" is meant as an acceptable substitute for "stabilitylike": both adjectives have a rather loose signification, but as they are, they will be found useful.

The qualitative concepts studied below pertain to bounded or unbounded closed sets. An important feature is that these sets are considered as subsets not of Ω, but of $\overline{\Omega}$: in other words, they can overlap the frontier of Ω. Several examples in Chapter VII will show why and how this is relevant to many a practical application. For the sake of brevity, we do not consider time-varying sets: on this point see T. Yoshizawa [1966] and P. Habets and K. Peiffer [1971]. For the same reason, we decided to leave outside the scope of this chapter any qualitative property of a set "with respect to another set". Thus partial stability, and similar "partial" properties, will be absent because, as is known, they are irreducible to any kind of stability of a set (see K. Peiffer and N. Rouche [1969]).

The terminology adopted is, as far as possible, in agreement with that of H. A. Antosiewicz [1958], J. L. Massera [1956-1958], T. Yoshizawa [1966] and N. P. Bhatia and G. P. Szegö [1967].

1.2. Our general hypotheses in this chapter are those of Section IV.2, with the exceptions that uniqueness is assumed throughout and that the origin is no more necessarily a point of Ω, nor a critical point for Equation (1.1). As usual, the expression $x(t;t_0,x_0)$ will often be abbreviated as $x(t)$.

The following notations will remain in force throughout the chapter: M is a closed non-empty subset of $\overline{\Omega}$. For any $\delta > 0$,

$$B(M,\delta) = \{x \in \mathscr{R}^n\colon\ d(x,M) < \delta\},$$

$$M_\delta = \{x \in \Omega\colon\ d(x,M) < \delta\},$$

$$\mathscr{A} = M_\delta \setminus M.$$

Remember that $d(x,M)$ is the distance from x to M.

2. A View of Stability and Attractivity Concepts

2.1. Table 6.1 exhibits the two stability definitions we are interested in.

	Definition	M is
S_1	$(\forall t_0\in I)(\forall\varepsilon>0)(\exists\delta>0)\quad(\forall x_0\in\mathscr{A})(\forall t\in J^+)\ x(t)\in M_\varepsilon$	S
S_2	$(\forall\varepsilon>0)(\exists\delta>0)(\forall t_0\in I)(\forall x_0\in\mathscr{A})(\forall t\in J^+)\ x(t)\in M_\varepsilon$	U S

Table 6.1. Stability concepts.

In this table, S means <u>stable</u>, US <u>uniformly stable</u>. The adverb <u>uniformly</u> obviously means that δ can be chosen independent of t_0. Two remarks are appropriate here:

a) Definitions S_1 and S_2 are respectively equivalent to those obtained by writing M_δ instead of $\mathscr{A}$.

b) If M is uniformly stable, M is stable. Conversely, if Equation (1.1) is autonomous and M is stable, then M is uniformly stable. This statement will be extended later, mutatis mutandis, (cf. Section 4.10) to periodic second members.

2.2. Six types of attractivities will be considered, as listed in Table 6.2. Some common denominations are:

A: M is an <u>attractor</u>,

t_0UA: M is a t_0-<u>uniform attractor</u>,

EA: M is an <u>equi-attractor</u>.

UA: M is a <u>uniform attractor</u>.

Thus, t_0-uniformity corresponds to δ <u>and</u> σ independent of t_0, the prefix <u>equi</u> to uniformity with respect to x_0, and

	Definition	M is
A_1	$(\forall t_0 \in I)(\exists \delta > 0)(\forall \varepsilon > 0)(\forall x_0 \in \mathscr{A})(\exists \sigma > 0)$ $(t_0 + \sigma \in J)$ and $(\forall t \geq t_0 + \sigma, t \in J)$ $x(t) \in M_\varepsilon$	A
A_2	$(\exists \delta > 0)(\forall t_0 \in I)(\forall \varepsilon > 0)(\forall x_0 \in \mathscr{A})(\exists \sigma > 0)$ " " "	
A_3	$(\exists \delta > 0)(\forall \varepsilon > 0)(\forall x_0 \in \mathscr{A})(\exists \sigma > 0)(\forall t_0 \in I)$ " " "	t_0 UA
A_4	$(\forall t_0 \in I)(\exists \delta > 0)(\forall \varepsilon > 0)(\exists \sigma > 0)(\forall x_0 \in \mathscr{A})$ " " "	EA
A_5	$(\exists \delta > 0)(\forall t_0 \in I)(\forall \varepsilon > 0)(\exists \sigma > 0)(\forall x_0 \in \mathscr{A})$ " " "	
A_6	$(\exists \delta > 0)(\forall \varepsilon > 0)(\exists \sigma > 0)(\forall x_0 \in \mathscr{A})(\forall t_0 \in I)$ " " "	UA

Table 6.2. Attractivity concepts.

uniform without any further qualification to the strongest concept: δ and σ both independent of t_0 and x_0. We notice the following:

c) If $M \cap \Omega$ is positively invariant, definitions A_1 to A_6 are equivalent to those obtained by substituting M_δ to $\mathscr{A}$.

d) The diagram of implications presented in Fig. 6.1 is obvious.

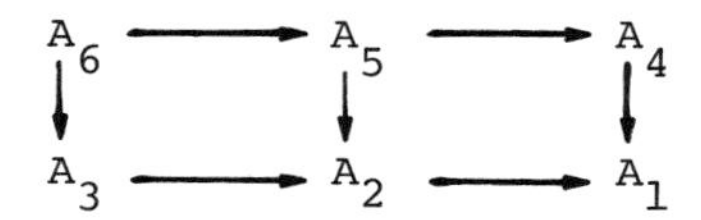

Fig. 6.1. Implications between attractivity concepts.

If Equation (1.1) is autonomous, every horizontal arrow may be reversed and the number of distinct concepts reduces to 2. The extension of this property to periodic equations is considered in Sections 4.11 to 4.14: it is no trivial matter.

e) To each attractivity concept A_i, there corresponds a global attractivity concept (GA_i), obtained by substituting $(\forall\delta > 0)$ to $(\exists\delta > 0)$.

2.3. The following observations pertain to the concepts defined in both Tables 6.1 and 6.2.

f) If $M \subset \Omega$ is compact and ε is chosen such that $\overline{B(M,\varepsilon)} \subset \Omega$, every solution mentioned in definitions S_1, S_2 and A_1 to A_6 exist for all $t \geq t_0$.

g) If $M \subset \Omega$ is a closed set with a compact boundary (for instance the complement of an open ball in $\Omega = \mathscr{R}^n$) and is negatively invariant, and if $\overline{B(M,\varepsilon)} \subset \Omega$, then every solution mentioned in definitions S_1, S_2 and A_1 to A_6 exists for all $t \geq t_0$.

Along with I.2.12 and I.2.13, the following exercises show that A_1 to A_6 are meaningful and distinct concepts, even if M is compact.

2.4. <u>Exercise</u>. For the equation $\dot{x} = -x/(1 + t)$, with $x \in \mathscr{R}$ and $t > 0$, the origin verifies A_5, but not A_3.

2.5. <u>Exercise</u>. For the equation $\dot{x} = (1 + x)^2(1 - x)$ with $x \in \mathscr{R}$ and for $M = \{-1\} \cup \{+1\}$, A_3 is verified but A_4 is not. Show in addition that the same conclusion holds for Example I.2.7 if M is the origin.

2.6. <u>Exercise</u>. For the equation $\dot{x} = (1 + x)^2(1 - x)/(1 + t)$ with $x \in \mathscr{R}$ and $t > 0$, the set $M = \{-1\} \cup \{+1\}$ verifies A_2, but not A_3 nor A_4.

2.7. <u>Exercise</u>. Let $\dot{x} = f(t,x)$ be a scalar differential equation where, for $t > 0$

(1) f is defined as in I.2.13 for $x < 0$;

(2) $f(t,x) = (\text{th}^{-1}x)(1 - x^2)/t$ for $0 \leq x < 1$;

(3) $f(t,x) = 1 - x$ for $1 \leq x$.

Then $M = \{0\} \cup \{1\}$ verifies A_1 but not A_2 nor A_4.

3. <u>Qualitative Concepts in General</u>

3.1. Tables 6.1 and 6.2 exhibit some kind of logical structure common to all concepts of stability and attractivity. This structure might be roughly depicted as follows: a given concept is applicable to a set M if, given a set $\mathscr{A} \subset \Omega$ described by a sequence of quantified variables (existential or universal) and a set $\mathscr{B} = M_\varepsilon$, the solutions starting from $\mathscr{A}$ remain in $\mathscr{B}$ in the future, or enter $\mathscr{B}$ after a given lapse of time and remain in it thereafter.

Now we want to play the following game: after having specified the sets $\mathscr{A}$ and $\mathscr{B}$ in some reasonable way, to generate a maximum number of concepts by interchanging the quantified variables in every possible manner. The result will be a huge number of concepts, and it will make sense to try some classification, to recognize the most familiar amongst them and to wonder why some aroused interest and many others did not.

The game will be considerably eased by reducing our symbolism to as few letters and signs as possible: we follow D. Bushaw [1969]. By doing this, we deviate from the usual mathematical symbolism, but it will be realized that the conciseness thus obtained more than compensates for the disadvantage.

3.2. Definition of a general qualitative concept. A set M will be said to verify a qualitative concept with respect to Equation (1.1) if the following proposition

$$W,\ \gamma \ast(\lambda,\beta) \tag{3.1}$$

is true. The rest of this section is mainly devoted to defining the symbols in (3.1).

(1) W is an abbreviation for a word whose letters are borrowed from the following alphabet $\{d,e,s,r,\tau_0,\xi_0,D,E,S,R,T_0,\Xi_0\}$. Each letter represents a quantified variable, as shown in Table 6.3; lower-case letters correspond to the

existential quantifiers, and upper-case letters to the universal ones. As compared with Tables 6.1 and 6.2, Table 6.3 contains a new variable, namely ρ, the <u>upper bound of sample interval</u>. It will be needed below to cover the cases of weak attractivity and boundedness.

Name of the variable	Variable	Associated letters	Meaning of these letters
Estimate of the initial perturbation	δ	d D	$\exists\delta > 0$ $\forall\delta > 0$
Estimate of the perturbation at time t	ε	e E	$\exists\varepsilon > 0$ $\forall\varepsilon > 0$
Delay of sample interval	σ	s S	$\exists\sigma \geq 0$ $\forall\sigma \geq 0$
Upper bound of sample interval	ρ	r R	$\exists\rho \geq \sigma$ $\forall\rho \geq \sigma$
Initial time	t_0	τ_0 T_0	$\exists t_0 \in I$ $\forall t_0 \in I$
Initial position	x_0	ξ_0 Ξ_0	$\exists x_0 \in \mathscr{A}$ $\forall x_0 \in \mathscr{A}$

Table 6.3. Variables for a qualitative concept. $I =]\tau,\infty[$, $\mathscr{A} = M_\delta \setminus M$.

The word W is subject to the following obvious constraints:

(i) no letter is repeated in W; any letter may

appear as lower- or upper-case, not both;

(ii) a variable used in the definition of the domain of another variable must appear, in the word, before this one: for instance, s or S has to appear before r or R.

A word verifying these two conditions will be called <u>well-formed</u>. An example of a well formed word is $(\forall t_0 \in I)$ $(\forall\varepsilon > 0)(\exists\delta > 0)(\forall x_0 \in M_\delta \setminus M)(\forall\sigma \geq 0)$. It will be written $T_0 E d \Xi_0 S$.

Let us now proceed with the explanation of the symbols appearing in (3.1). As will be seen, the meaning of γ, * and λ will be different according as σ and ρ are present or not in W, and according as σ appears under s (existential) or S (universal). Further, β will admit two values also, but this time independently of what precedes in the proposition. In detail, we assume the following

(2) γ stands for "$t_0 + \sigma \in J$" if σ appears in W, and for "$t_0 \in J$" if it does not.

(3) The asterisk * stands for the implication "$\Longrightarrow$" if W contains the upper-case letter S, and for the logical conjunction "&" in every other case.

(4) The letter λ can be freely chosen as representing τ or T, where τ stands for $(\exists t \in [t_0 + \sigma, t_0 + \rho] \cap J)$ and of course T for $(\forall t \in [t_0 + \sigma, t_0 + \rho] \cap J)$. Notice however that if σ does not appear in W, σ in the parentheses above should be replaced by 0. Similarly, if ρ is absent from W, "$t_0 + \rho]$" should be replaced above by "$\infty[$".

(5) The letter β stands for β^+ or β^-, where β^+ means "$x(t) \in \mathscr{B} = M_\varepsilon$" and β^- means "$x(t) \in \mathscr{B} = \Omega \setminus M_\varepsilon$". The symbol β^+ corresponds to stability and β^- to instability.

The meanings of the various symbols have to be kept in mind to understand what follows. They are summarized in Tables 6.4 and 6.5.

3.3. Remark. To every concept C constructed as above, there corresponds another concept C' derived from C by writing $\mathscr{A}'$ instead of $\mathscr{A}$, where $\mathscr{A}' = (\mathscr{N}(M) \setminus M) \cap \Omega$, with $\mathscr{N}(M)$ an open neighborhood of M. Such concepts are studied in detail by N. P. Bhatia and G. P. Szegö [1967]. They are associated with the prefix "semi". Clearly, if δ is existential and C is verified, so is C'. It can be shown that, if furthermore M is closed with a compact boundary and if x_0 is quantified universally, C and C' are equivalent. For lack of space, C' concepts will be considered outside the scope of our study.

3.4. Remark. Consider the following sequence of parentheses:

$$(\forall t_0 \in I)(\exists\delta > 0)(\forall x_0 \in \mathscr{A})(\exists\varepsilon > 0) \ldots$$

Normally, δ varies with t_0 and ε with t_0 and x_0. It is conceivable however to imagine a concept where ε would depend on x_0, but not on t_0. Such a feature is also beyond the scope of this study. It has been considered by C. Avramescu [1973].

3.5. Some helpful familiarity with the symbolism introduced above may be gained by transcribing a few well-known

Symbols	Particular symbols	Conditions	Meaning
γ			$t_0 \in J$
*			&
λ	τ	if r or R appears in W	$\exists t \in [t_0, t_0+\rho] \cap J$
		otherwise	$\exists t \in [t_0, \infty[\cap J$
	T	if r or R appears in W	$\forall t \in [t_0, t_0+\rho] \cap J$
		otherwise	$\forall t \in [t_0, \infty[\cap J$
β	β^+		$x(t) \in \mathcal{B} = M_\varepsilon$
	β^-		$x(t) \in \mathcal{B} = \Omega \setminus M_\varepsilon$

Table 6.4. Meaning of $\gamma * (\lambda, \beta)$ if neither s nor S appears in W.

Symbols	Particular symbols	Conditions	Meaning
γ			$t_0 + \sigma \in J$
*		if s appears in W	&
		if S appears in W	$\Longrightarrow$
λ	τ	if r or R appears in W	$\exists t \in [t_0+\sigma, t_0+\rho] \cap J$
		otherwise	$\exists t \in [t_0+\sigma, \infty[\cap J$
	T	if r or R appears in W	$\forall t \in [t_0+\sigma, t_0+\rho] \cap J$
		otherwise	$\forall t \in [t_0+\sigma, \infty[\cap J$
β	β^+		$x(t) \in \mathcal{B} = M_\varepsilon$
	β^-		$x(t) \in \mathcal{B} = \Omega \setminus M_\varepsilon$

Table 6.5. Meaning of $\gamma * (\lambda, \beta)$ if s or S appears in W.

definitions into the new system. This is done in Table 6.6. In order to make the comparison easier, the formulas in the second column have been given in a "complete" form, i.e. using every variable. The simplified formulas of the third column are obtained by using one or more of the equivalence theorems of the next section.

3.6. Exercise. If a concept C_1 derives from a concept C_2 by moving a parenthesis with a universal quantifier to the right, or with an existential one to the left, or by substituting a universal quantifier to an existential one, then $C_1 \Longrightarrow C_2$.

Terminology	Formula	Simplified formula
1. M is uniformly stable	$EdSRT_0\Xi_0,\gamma^*(T,\beta^+)$	$EdT_0\Xi_0T\beta^+$
2. Solutions uniformly bounded with respect to M	$DeSRT_0\Xi_0,\gamma^*(T,\beta^+)$	$DeT_0\Xi_0T\beta^+$
3. Solutions uniformly locally bounded with respect to M	$edSRT_0\Xi_0,\gamma^*(T,\beta^+)$	$edT_0\Xi_0T\beta^+$
4. M is a uniform attractor	$dEsRT_0\Xi_0,\gamma^*(T,\beta^+)$	$dEsT_0\Xi_0,\gamma^*(T,\beta^+)$
5. M is a uniform global attractor	$EDsRT_0\Xi_0,\gamma^*(T,\beta^+)$	$EDsT_0\Xi_0,\gamma^*(T,\beta^+)$
6. Solutions uniformly ultimately bounded with respect to M	$eDsRT_0\Xi_0,\gamma^*(T,\beta^+)$	$eDsT_0\Xi_0,\gamma^*(T,\beta^+)$
7. M is a uniform weak attractor	$dEsrT_0\Xi_0,\gamma^*(\tau,\beta^+)$	$dErT_0\Xi_0\tau\beta^+$
8. M is a uniform global weak attractor	$EDsrT_0\Xi_0,\gamma^*(\tau,\beta^+)$	$EDrT_0\Xi_0\tau\beta^+$
9. M is unstable	$\tau_0eD\xi_0sr,\gamma^*(\tau,\beta^-)$	$\tau_0eD\xi_0\tau\beta^-$

Table 6.6. Some examples of concepts.

4. Equivalence Theorems for Qualitative Concepts

4.1. Various examples of two different, but essentially equivalent formulations of a concept are given in Table 6.6. Such equivalences, proved by purely logical considerations, will be established in the present section. But it happens also that two concepts are equivalent if some further assumption is made concerning the differential equation (for example that it is autonomous, or periodic, or that its solutions are continuable up to $+\infty$) or the set M (for example that it is compact, or has a compact boundary). The present section deals also with such conditional equivalences.

The only two propositions of the first type which we shall give are the following ones, and they are obvious.

4.2. Theorem. If W_1 and W_2 are two well-formed words obtained from each other by permutation of two adjacent lower-case (upper-case) letters, the corresponding concepts are equivalent, i.e.

$$[W_1, \gamma * (\lambda,\beta)] \iff [W_2, \gamma * (\lambda,\beta)].$$

4.3. Theorem. If W_1 s W_2 is a well-formed word, then

(a) $W_1 \text{ s } W_2, \gamma * (\tau,\beta) \iff W_1W_2\tau\beta.$

If W_1 SR W_2 is a well-formed word and if W_2 contains only universal quantifiers or is empty (contains no letter), then

(b) $W_1 \text{ SR } W_2, \gamma * (T,\beta) \iff W_1W_2T\beta;$

(c) $W_1 \text{ Sr } W_2, \gamma * (T,\beta) \iff W_1W_2T\beta;$

(d) $W_1 \text{ SR } W_2, \gamma * (\tau,\beta) \iff W_1W_2T\beta;$

(e) $W_1 \text{ sR } W_2, \gamma * (T,\beta) \iff W_1 \text{ s } W_2, \gamma * (T,\beta).$

4.4. <u>Exercise</u>. Prove a dual proposition to 4.3 for the case where W_2 contains only existential quantifiers or is empty.

Let us now proceed to a few theorems of the second type, considering successively solutions which can be continued up to $+\infty$, autonomous equations, periodic equations and, finally periodic equations associated with a set M having a compact boundary. Most proofs are simple and will be either omitted or merely outlined. However, in order to illustrate clearly the type of proofs which are relevant in this context, Theorems 4.10 and 4.11 are demonstrated in detail.

4.5. <u>Theorem</u>. If every solution of the differential equation can be continued to $+\infty$ and if W is a well-formed word, then

$$[W,\gamma * (\lambda,\beta)] \iff [W,\lambda,\beta].$$

4.6. <u>Theorem</u>. If $I \supset [0,\infty[$, M is a compact set, every solution of the differential equation can be continued to $+\infty$ and $W_1 T_0 D W_2 \Xi_0 W_3$ is a well-formed word, then

$$(\text{for } t_0 = 0) \quad W_1 D W_2 (\forall x_0 \in M_\delta) W_3, \gamma * (\lambda,\beta^+) \tag{4.1}$$

if and only if

$$W_1 (\forall t_0 \geq 0) D W_2 (\forall x_0 \in M_\delta) W_3, \gamma * (\lambda,\beta^+). \tag{4.2}$$

<u>Proof</u>. It is sufficient to prove the "only if". From the continuity of the solutions, the continuability hypothesis and the fact that M is compact, one deduces that

$$(\forall t_0 \geq 0) D (\exists \eta \geq 0) \Xi_0,\ x'_0 = x(0;t_0,x_0) \in M_\eta. \tag{4.3}$$

(4.1) can be rewritten as

$$W_1'(\forall\eta > 0)W_2'(\forall x_0' \in M_\eta)W_3,\gamma * (\lambda, x(t;0,x_0') \in M_\varepsilon). \quad (4.4)$$

Combining (4.3) and (4.4), one gets

$$W_1(\forall t_0 \geq 0)DW_2\exists_0 W_3,\gamma * (\lambda, x(t;t_0,x_0) \in M_\varepsilon). \quad (4.5)$$

Indeed, let us in W_1 choose the universal variables arbitrarily and the existential ones as for (4.4). Choose then t_0 and δ arbitrarily. Then we adopt for η in (4.4) a value coming from (4.3). Further, we choose in W_2 the universal variables arbitrarily and the existential ones as for (4.4). Then, for every $x_0 \in M_\delta$, (4.3) yields $x_0' = x(0;t_0,x_0) \in M_\eta$. But then by (4.4): $W_3,\gamma * (\lambda, x(t;0,x_0') \in M_\varepsilon)$, which completes the proof of (4.5), since $x(t;0,x_0') = x(t;t_0,x_0)$. Q.E.D.

Remark. The same theorem holds for T_0 located anywhere in the word in (4.2).

4.7. Theorem. If the differential equation is autonomous and if the word $W_1T_0W_2$ is well-formed, then

$$\tau_0 W_1 W_2,\gamma * (\lambda,\beta) \iff W_1 T_0 W_2,\gamma * (\lambda,\beta)$$

where W_1 or W_2 can be empty.

The proof, which is left as an exercise, is based on the fact that, if (1.1) is autonomous, one has for every $t_0 \in I$ and $t_0' \in I$ that

(1) $t' \in J(t_0',x_0)$ is equivalent to: $t = t' + t_0 - t_0' \in J(t_0,x_0)$;

(2) $x(t';t_0',x_0) = x(t' + t_0 - t_0';t_0,x_0) = x(t;t_0,x_0)$.

4.8. Corollary. If the differential equation is autonomous and if $T_0W_1W_2$ is well-formed, then

$$T_0W_1W_2,\gamma * (\lambda,\beta) \iff W_1T_0W_2,\gamma * (\lambda,\beta) \iff W_1W_2T_0,\gamma * (\lambda,\beta).$$

If (1.1) is ω-periodic for some $\omega \neq 0$, obtaining results similar to 4.7 and 4.8 requires some further hypotheses on M, W_1 or W_2. The following lemma is essential in deriving such results. By way of exception, it mentions another domain than I for the variable t_0.

4.9. Lemma. If the differential equation is ω-periodic with $\omega > 0$, if $W_1\tau_0W_2$ is a well-formed word and if there exists a $\tilde{t}_0$ such that the concept $W_1(\forall t_0 \in [\tilde{t}_0,\tilde{t}_0+\omega])W_2$, $\gamma * (\lambda,\beta)$ is verified, so is $W_1T_0W_2,\gamma * (\lambda,\beta)$.

This can be proved by observing that, for all $t_0' \in I$ and any integer m, if $t_0 = t_0' + m\omega \in I$, then:

(i) $t' \in J(t_0',x_0)$ is equivalent to $t = t' + t_0 - t_0' \in J(t_0,x_0)$;

(2) $x(t';t_0',x_0) = x(t;t_0,x_0)$.

The following results concerning periodic differential equations pertain to the concepts S_1, S_2 and A_1 to A_6 of Section 2, i.e. to the various kinds of stability and attractivity.

4.10. Theorem. If $M \subset \Omega$ is closed with a compact boundary and is positively invariant, if the differential equation is ω-periodic and if each solution $x(t;t_0,x_0)$ with $x_0 \in \partial M$ is defined at least on $[t_0,t_0 + \omega]$, then stability (S_1) is equivalent to uniform stability (S_2).

Proof. It is sufficient to show that S_1 implies S_2. Choose any $\tilde{t}_0 \in I$ and an arbitrary $\varepsilon > 0$. Let then δ_0 be a value of δ suiting S_1 where t_0 is replaced by $\tilde{t}_0 + \omega$. The general solution $x(t;t_0,x_0)$ being continuous with respect to the initial conditions, there exists a $\delta_1 > 0$, $\overline{B(M,\delta_1)} \subset \Omega$ such that, for every $t_0 \in [\tilde{t}_0,\tilde{t}_0 + \omega]$, every $x_0 \in B(M,\delta_1) \setminus M$ and every $t \in [t_0,\tilde{t}_0 + \omega]$: $x(t;t_0,x_0) \in B(M,\delta_0)$. But then, because S_1 is verified, $x(t;t_0,x_0) \in M_\varepsilon$ for every $t \geq \tilde{t}_0 + \omega$. Lemma 4.9 then shows that S_2 is also verified. Q.E.D.

4.11. Theorem. If the hypotheses of Theorem 4.10 are satisfied, one has $A_1 \Longleftrightarrow A_2$ and $A_4 \Longleftrightarrow A_5 \Longleftrightarrow A_6$.

Proof. Consider an arbitrary $\tilde{t}_0 \in I$ and let δ_0 be a value of δ suiting A_4 for $t_0 = \tilde{t}_0 + \omega$. One may find, as in the proof of 4.10, a δ_1 with $\overline{B(M,\delta_1)} \subset \Omega$ such that

$$(\forall x_0 \in B(M,\delta_1) \setminus M)(\forall t_0 \in [\tilde{t}_0,\tilde{t}_0 + \omega]) \quad x(\tilde{t}_0 + \omega;t_0,x_0) \in B(M,\delta_0).$$

We then choose an arbitrary $\varepsilon > 0$ and determine a value of σ suiting A_4 (in A_6, we shall adopt the value $\sigma + \omega$). To the possible values (4.1) of t_0 and x_0, there corresponds initial values $x(\tilde{t}_0 + \omega;t_0,x_0)$ suiting A_4. Hence A_6 is verified with $(\forall t_0 \in I)$ replaced by $(\forall t_0 \in [\tilde{t}_0,\tilde{t}_0 + \omega])$ and the thesis follows from 4.9. Q.E.D.

4.12. Exercise. Prove the other implications appearing in the statement of Theorem 4.11.

4.13. The following example shows that, even in the hypotheses of Theorem 4.11, A_2 does not imply A_3. Consider an autonomous system in $\mathscr{R}^2$, the orbits of which are represented in Fig. 6.2.

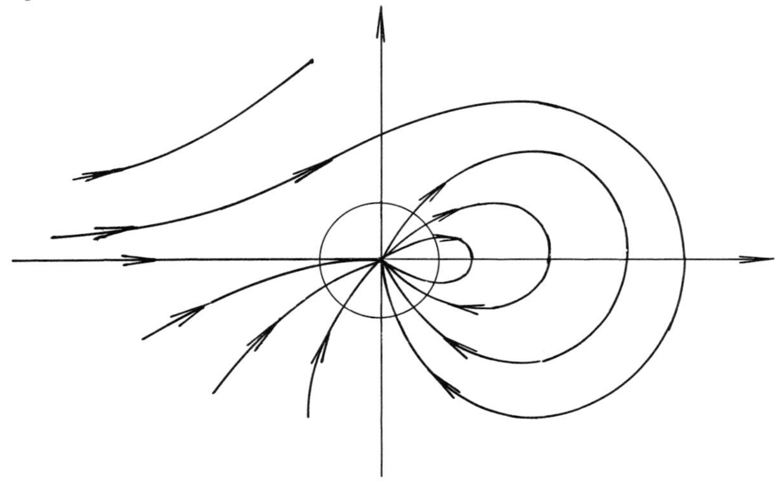

Fig. 6.2. Phase plane for Equations (4.6).

We are interested in the equations of this system written in cartesian coordinates. Let us however consider these equations written as follows in polar coordinates:

$$\begin{aligned} \dot{r} &= R(t,\theta) \\ \dot{\theta} &= \psi(r,\theta) \end{aligned} \tag{4.6}$$

with R and ψ 2π-periodic in θ. It will not be necessary to specify these two functions any further. The origin 0 verifies A_1. The transformation $\bar{r} = r$, $\bar{\theta} = \theta + t$, changes (4.6) into a 2π-periodic differential system. It corresponds to a rotation of the plane around 0 with angular velocity equal to 1. For the new equations, A_1 and A_2 are clearly satisfied whereas A_3 is not. Indeed, given any $\varepsilon > 0$ and $\delta > 0$, one has for each point $(\bar{r}_0, \bar{\theta}_0) \in B_\delta$ with $\bar{r}_0 \neq 0$ that $\sup \{\sigma(t_0): t_0 \in [0,2\pi]\} = +\infty$.

4.14. Exercise. Suppose we try to use a reasoning similar to that of 4.11 to prove that, under the hypotheses of 4.10, $A_2 \Longrightarrow A_3$. Where shall we fail?

4.15. In Theorem 4.5 and in several other theorems to be proved in the rest of this chapter, there appears the hypothesis that every solution of (1.1) may be continued up to $+\infty$. Sufficient conditions to get this are given by the following lemma.

4.16. Lemma. Consider a concept such that

(i) ρ does not appear in W;

(ii) t is a universal variable, or $W = W_1 S$;

(iii) $\beta = \beta^+$;

suppose further that M, either is compact, or is closed with compact boundary and negatively invariant; if moreover ε is chosen such that $\overline{B(M,\varepsilon)} \subset \Omega$, then every solution mentioned in this concept can be continued up to $+\infty$.

Proof. Hypotheses (i) to (iii) imply that $\gamma * (\lambda,\beta)$ has to be written in one of the three following explicit forms:

(1) $(t_0 + \sigma \in J)$ & $[(\forall t \in [t_0 + \sigma, \infty[\cap J)\ x(t) \in M_\varepsilon]$

where σ can vanish, or

(2) $(t_0 + \sigma \in J) \Longrightarrow [(\forall t \in [t_0 + \sigma, \infty[\cap J)\ x(t) \in M_\varepsilon]$

or

(3) $(t_0 + \sigma \in J) \Longrightarrow [(\exists t \in [t_0 + \sigma, \infty[\cap J)\ x(t) \in M_\varepsilon]$.

In the first two cases, $x(t)$ remains, for $t \geq t_0 + \sigma$, $t \in J$, in $B(M,\varepsilon)$ if M is compact or in $B(M,\varepsilon) \setminus M$ otherwise. Both sets are bounded and do not touch $\partial\Omega$. Hence the thesis. In the third case, suppose a solution $x(t)$ is defined on the right maximal interval $[t_0,t^*[$ for some $t^* \in \mathscr{R}$. This implies that $x(t)$ ultimately leaves $B(M,\varepsilon)$ (or $B(M,\varepsilon) \setminus M$) and never comes back. So there exists a $\sigma \geq 0$ such that $t_0 + \sigma \in J$ and for every $t \geq t_0 + \sigma$, $t \in J$: $x(t) \notin M_\varepsilon$. This contradicts (3). Q.E.D.

5. A Tentative Classification of Concepts

Let us attempt to classify the concepts whose word W is made up of letters borrowed from Table 6.3. While retaining the definitions of the sets $\mathscr{A}$ and $\mathscr{B}$, we generate, by mere substitutions and permutations, 46.080 formally different concepts, all well-formed! Hence the necessity to rule out all uninteresting items. We propose the following reasonable restrictions:

(1) t_0 and x_0 will always be preceded by the same quantifier: one is generally interested in the behavior either of all solutions starting near M, or of only one;

(2) ε and δ will appear before σ: the reason is that we normally want to fix an estimate of the initial perturbation and then choose a sample delay to satisfy the final perturbation.

Respecting both conditions reduces to 5.376 the number of formally different concepts. Indeed, for δ, ε, x_0, σ and ρ, the only possibilities are those listed in Table 6.7.

I	II	III	IV	V
δ	x_0	ε	σ	ρ
δ	ε	x_0	σ	ρ
δ	ε	σ	x_0	ρ
δ	ε	σ	ρ	x_0
ε	δ	x_0	σ	ρ
ε	δ	σ	x_0	ρ
ε	δ	σ	ρ	x_0

Table 6.7. Admissible orders for the variables δ, ε, x_0,σ,ρ.

As t_0 may be inserted at any place, one gets $7 \times 6 = 42$ possibilities. With all possible choices of quantifiers for (t_0,x_0), δ, ε, σ, ρ and t, and the binary choice between β^+ and β^-, one gets the announced 5.376 well-formed concepts.

They can first be grouped as suggested in Table 6.8 in stability-like (β^+) or instability-like (β^-) properties concerning all (T_0,Ξ_0) or one (τ_0,ξ_0) solutions near M. Four families are obtained, each of 1.344 concepts.

1.	Stability-like concepts	$T_0 \Xi_0 \beta^+$
2.	Instability-like concepts	$\tau_0 \xi_0 \beta^-$
3.	Complete instability-like concepts	$T_0 \Xi_0 \beta^-$
4.	Incomplete stability-like concepts	$\tau_0 \xi_0 \beta^+$

Table 6.8. Fundamental families of concepts.

Each family can be divided into natural equivalence classes, two concepts being considered "equivalent" if they differ only by the positions of t_0 and x_0 in W: in this case, they do not differ by more than some spatial or time uniformity conditions. This divides the 1.344 concepts into 32 classes of 24 concepts, plus 32 classes of 18 concepts (Exercise: why 24 and why 18?). At last, in any listing of concepts, advantage can be taken of the partial order induced between them by the implication relation: a concept will be said to be stronger than another one if the former implies the latter.

Let us now examine the important family of stability-like concepts. In Table 6.9, each class of the family is represented by its maximal element in the sense of partial order: T_0 and Ξ_0 occupy the utmost right position. The table begins by the strongest concepts, so strong indeed as to be meaningless, and it ends with the weakest ones, so weak as to be trivial. It should contain 64 rows, but it has only 42 because the following simplifications have been introduced:

(1) owing to Theorem 4.3, SRT, SrT and SRτ are equivalent;

	Formula	Terminology
1.	E D S R $T_0\ \Xi_0,\ \gamma * (T,\beta^+)$	meaningless
2.	e D " " "	"
3.	d E " " "	"
4.	D e " " "	solutions around M are bounded
5.	E d " " "	M is stable
6.	d e " " "	(locally bounded)
7.	e d " " "	(locally bounded)
8.	E D s R " T	M is a global attractor
9.	e D " " "	(ultimately bounded)
10.	d E " " "	M is an attractor
11.	D e " " "	
12.	E d " " "	
13.	d e " " "	
14.	e d " " "	
15.	E D s r " T	
16.	e D " " "	
17.	d E " " "	
18.	D e " " "	
19.	E d " " "	
20.	d e " " "	
21.	e d " " "	
22.	E D s R $T_0\ \Xi_0,\ \gamma * (\tau,\beta^+)$	
23.	e D " " "	
24.	d E " " "	
25.	D e " " "	
26.	E d " " "	
27.	d e " " "	
28.	e d " " "	
29.	E D S r " τ	
30.	e D " " "	
31.	d E " " "	
32.	D e " " "	
33.	E d " " "	
34.	d e " " "	
35.	e d " " "	
36.	E D s r " τ	M is a global weak attractor
37.	e D " " "	(weakly ultimately bounded)
38.	d E " " "	M is a weak attractor
39.	D e " " "	trivial
40.	E d " " "	"
41.	d e " " "	"
42.	e d " " "	"

Table 6.9. Family of stability-like concepts.

(2) E and D can be interchanged. This is not true for e and d because T_0, or Ξ_0, or both, may appear between e and d.

Table 6.9 also gives the common denominations of some concepts. A point of divergence with previously known definitions is weak ultimate boundedness which, for N. Pavel [1972], corresponds to number 10 and for us to 37. For lack of space, the adverb "uniformly" has been omitted throughout in Table 6.9.

Similar tables can be set up for each of the four families of Table 6.8. For example, Table 6.10 presents, for instability-like concepts, the ten definitions corresponding to the first ten of Table 6.9. Here, each class is represented by its minimal element in sense of partial order, so that each concept of Table 6.10 is the negation of the corresponding one of Table 6.9.

	Formula				Terminology
1.	e d	s r	τ_0 ξ_0, γ	* (τ,β^-)	trivial
2.	E d	"	"	"	"
3.	D e	"	"	"	"
4.	d E	"	"	"	(unbounded)
5.	e D	"	"	"	M is unstable
6.	D E	"	"	"	
7.	E D	"	"	"	
8.	e d	S r	"	"	
9.	E d	"	"	"	
10.	D e	"	"	"	

Table 6.10. Some classes of instability-like concepts.

6. Weak Attractivity, Boundedness, Ultimate Boundedness

6.1. The purpose of this section is to examine in some detail those classes of Table 6.9 which aroused most interest amongst scientists. To avoid useless repetitions, we use of course the equivalence theorems of Section 4. Stability and attractivity has been studied already and were presented in Tables 6.1 and 6.2. The variable ρ is absent there in virtue of Theorem 4.3 (b) and (e) respectively. Weak attractivity is presented in Table 6.11, where σ does not appear owing to Theorem 4.3 (a).

	Formula									M is
WA_1	T_0	d	E		Ξ_0	r			τ,β^+	a weak attractor
WA_2		d	E	T_0	Ξ_0	r			τ,β^+	
WA_3		d	E		Ξ_0	r		T_0	τ,β^+	
WA_4	T_0	d	E			r	Ξ_0		τ,β^+	an equi-weak attractor
WA_5		d	E	T_0		r	Ξ_0		τ,β^+	
WA_6		d	E			r	Ξ_0	T_0	τ,β^+	a uniform weak attractor

Table 6.11. Weak attractivity.

The implications between these concepts are those of Fig. 6.3, where the horizontal arrows can be reversed if Equation (1.1) is autonomous.

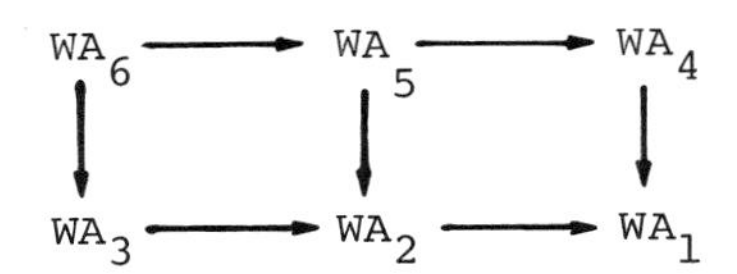

Fig. 6.3. Implications between weak attractivity concepts.

6.2. <u>Theorem</u>. If $M \cup \Omega$ is a neighborhood of M and if M is closed with a compact boundary, then

(a) weak attractivity $(WA_1) \Longleftrightarrow$ equi-weak attractivity (WA_4):

(b) $WA_2 \Longleftrightarrow WA_5$.

<u>Proof</u>. We prove the first equivalence only: the reasoning would be substantially the same for the second one. Only the direct implication has to be established, namely

$$WA_1 \equiv T_0 dE\Xi_0 \tau, \beta^+ \Longrightarrow WA_4 \equiv T_0 dEr\Xi_0 \tau, \beta^+.$$

Suppose WA_1 is satisfied and WA_4 is not, i.e.

$$(\exists t_0 \in I)(\forall \delta > 0(\exists \varepsilon > 0)(\forall \rho \geq 0)(\exists x_0 \in \mathscr{A})$$

$$(\forall t \in [t_0, t_0+\rho] \cap J) \quad x(t) \notin M_\varepsilon. \tag{6.1}$$

Choose a t_0 suiting (6.1) and then a δ as for WA_1, such moreover that $\overline{M}_\delta \subset \Omega \cup M$. Choose some $\delta^* < \delta$ to be used in (6.1) and an ε^* suiting (6.1) in these conditions. Then by (6.1), there exists a sequence $\rho_i \to \infty$ and a sequence of points $x_{0i} \in M_{\delta*} \setminus M_{\varepsilon*}$. Since $M_{\delta*} \setminus M_{\varepsilon*}$ is bounded, the sequence $\{x_{0i}\}$ has a cluster point $x_0 \in \overline{M_{\delta*} \setminus M_{\varepsilon*}} \subset M_\delta \setminus M$. Let us use this x_0 in WA_1. Then there exists a $t \in J^+(t_0,x_0)$ such that $x(t) \in M_{\varepsilon*}$. By continuity of the solutions with respect to the initial conditions, one gets, for i large enough, that

$$t \in J^+(t_0,x_{0i}) \cap [t_0,t_0 + \rho_i],$$

and since, for every i, $x(t;t_0,x_{0i}) \notin M_{\varepsilon*}$, then also $x(t;t_0,x_0) \notin M_{\varepsilon*}$, which is a contradiction. Q.E.D.

6.3. Exercise. If f is ω-periodic, $WA_2 \iff WA_3$. If $M \cup \Omega$ is a neighborhood of M and M has a compact boundary, and if f is ω-periodic, $WA_2 \iff WA_6$. If $M \subset \Omega$ has a compact boundary and is positively invariant and if, for $x_0 \in \partial M$, $x(t;t_0,x_0)$ is defined at least on $[t_0,t_0+\omega]$, then $WA_1 \iff WA_2$. If f does not depend on t and if M satisfies the hypotheses of Theorem 6.2, all concepts of weak attractivity are equivalent.

6.4. Remark. An alternative way to introduce a class of concepts also likely to be grouped under the heading of "weak attractivity" is to consider the variations of

$$dEST_0\Xi_0,\gamma * (\tau,\beta^+).$$

The following relation between both kinds of concepts is proved in P. Habets and K. Peiffer [1973]: if M is negatively invariant and if WEr and WES are well-formed, then

$$WEr,\gamma * (\tau,\beta^+) \iff WES,\gamma * (\tau,\beta^+).$$

The following theorem links attractivity and weak attractivity.

6.5. Theorem. $A_i \implies WA_i$, $1 \leq i \leq 6$.

Proof. Each A_i has the form

$$W_1sW_2,[t_0 + \sigma \in J] \text{ and } [\forall t \in [t_0 + \sigma,\infty[\cap J,\beta^+],$$

where W_1 and W_2 are well-formed and do not contain ρ. But this implies that $W_1sW_2,(\exists t = t_0 + \sigma \in J)\beta^+$ and, taking $\rho = \sigma$, that $W_1rW_2,(\exists t \in [t_0,t_0 + \rho] \cap J)\beta^+$, which is the corresponding weak attractivity concept. Q.E.D.

6.6. Boundedness, ultimate boundedness and weak ultimate boundedness concepts are presented in Tables 6.12, 6.13 and 6.14. They are studied here for their own interest and further because they lead naturally to the important notion of a <u>dissipative system</u> (cf. 6.18 and Chapter VIII).

	Formula							The solutions are ... with respect to M
B_1	T_0	D	Ξ_0	e			T,β^+	bounded
B_2		D	Ξ_0	e	T_0		T,β^+	bounded uniformly in t_0
B_3	T_0	D		e		Ξ_0	T,β^+	equi-bounded
B_4		D		e	T_0	Ξ_0	T,β^+	uniformly bounded

Table 6.12. Boundedness.

	Formula										The solutions are ... with respect to M
UB_1	T_0	e	D		Ξ_0	s		$,\gamma$	$*$	(T,β^+)	
UB_2		e	D	T_0	Ξ_0	s		$,\gamma$	$*$	(T,β^+)	ultimately bounded
UB_3		e	D		Ξ_0	s		T_0,γ	$*$	(T,β^+)	
UB_4	T_0	e	D			s	Ξ_0	$,\gamma$	$*$	(T,β^+)	
UB_5		e	D	T_0		s	Ξ_0	$,\gamma$	$*$	(T,β^+)	equi-ultimately bounded
UB_6		e	D			s	Ξ_0	T_0,γ	$*$	(T,β^+)	uniformly ultimately bounded

Table 6.13. Ultimate boundedness.

	Formula								
WUB_1	T_0	e	D		Ξ_0	r			τ,β^+
WUB_2		e	D	T_0	Ξ_0	r			τ,β^+
WUB_3		e	D		Ξ_0	r		T_0	τ,β^+
WUB_4	T_0	e	D			r	Ξ_0		τ,β^+
WUB_5		e	D	T_0		r	Ξ_0		τ,β^+
WUB_6		e	D			r	Ξ_0	T_0	τ,β^+

Table 6.14. Weak ultimate boundedness.

Owing to Theorem 4.3 (b), Table 6.12 contains neither σ nor ρ. In the same way, and due to Theorem 4.3 (e) and (a) respectively, Table 6.13 contains no r and Table 6.14 no σ. Some immediate implications are given in Fig. 6.4. If (1.1) is autonomous, the horizontal arrows can be reversed.

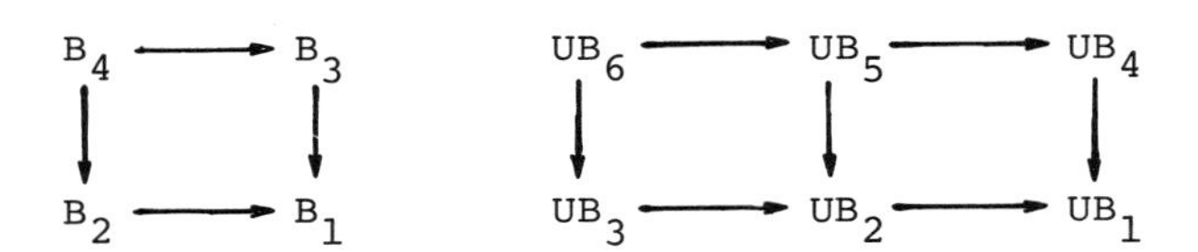

Fig. 6.4. Implications between boundedness and ultimate boundedness concepts.

6.7. Exercise. For $\Omega = \mathscr{R}^2$ and $M = \{0\}$, show, by using the following system of equations:

$$\dot{x} = y^2(1 - xy),$$
$$\dot{y} = 0,$$

that even in the autonomous case, B_2 does not imply B_3.

6.8. <u>Exercise</u>. For $\Omega = \mathscr{R}^2$, $M = \{0\}$ and f ω-periodic, one does not necessarily have that $B_1 \Longrightarrow B_2$.

Hint: Use the following change of variables in the differential system of Exercise 6.7: $X = x$, $Y = y + \cos t$.

6.9. <u>Exercise</u>. Show that for an unbounded set M with compact boundary, all concepts B_i, $i = 1, 2, 3, 4$, are trivially verified.

6.10. <u>Exercise</u> (T. Yoshizawa [1959]). If f is ω-periodic and $M = \{0\} \subset \Omega = \mathscr{R}^n$, equi-boundedness ($B_3$) is equivalent to uniform boundedness (B_4).

The following theorem exhibits some relations between boundedness and ultimate boundedness concepts. Proposition 6.11 (b) is due to T. Yoshizawa [1959].

6.11. <u>Theorem</u>. If $M = \{0\} \subset \Omega = \mathscr{R}^n$:

(a) $UB_4 \Longrightarrow$ equi-boundedness (B_3);

(b) ultimate boundedness (UB_2) plus uniform boundedness (B_4) imply equi-ultimate boundedness (UB_5).

<u>Proof</u>. (a) Let t_0 be chosen in I. If UB_4 is verified for $\varepsilon = \varepsilon_1$, then for each $\delta > 0$, B_3 is satisfied with $\varepsilon = \max(\varepsilon_1, \varepsilon_2)$, where $\varepsilon_2 = \sup\{||x(t;t_0,x_0)||: x_0 \in \mathscr{A}, t \in [t_0, t_0 + s_1]\}$, s_1 being the value of σ corresponding to δ in UB_4. This proves (a).

(b) Let $\varepsilon = \varepsilon_1$ be chosen as for UB_2 and let $\varepsilon = \varepsilon_2$ correspond to $\delta = \varepsilon_1$ in B_4. Then, in virtue of UB_2, for every $(t_0,x_0) \in \mathscr{R}^n$, there exists an $s_1 > 0$ such that $x(t_0 + s_1;t_0,x_0) \in M_{\varepsilon_1}$. Let N_1 be a neighborhood of $x(t_0 + s_1;t_0,x_0)$ such that $N_1 \subset M_{\varepsilon_1}$. Then, for some neighborhood N^* of x_0, $x(t_0 + s_1;t_0,x_0^*) \in N_1$ for every $x_0^* \in N^*$, and, by B_4, one gets that

$$(\forall x_0^* \in N^*)(\forall t \geq t_0 + s_1) \quad x(t;t_0,x_0^*) \in M_{\varepsilon_2}. \tag{6.2}$$

Let us write $s_1 = s_1(N^*)$ and choose δ arbitrary. Then $B(M,\delta)$ is covered by a finite number of such neighborhoods, say N_i^*, $(1 \leq i \leq m)$. If $s = \max s(N_i^*)$, one gets from (6.2) that

$$(\exists \varepsilon = \varepsilon_2 > 0)(\forall \delta > 0)(\forall t_0 \in I)(\exists \sigma = s \geq 0)(\forall x_0 \in B(M,\delta))$$

$$(\forall t \geq t_0 + s) \; t_0 + s \in J \text{ and } x(t;t_0,x_0) \in M_{\varepsilon_2},$$

which is nothing but UB_5. Q.E.D.

6.12. Exercise (N. Pavel $[1972]_2$). If f is ω-periodic, $M = \{0\} \subset \Omega = \mathscr{R}^n$, then

(a) ultimate boundedness (UB_2) implies uniform boundedness (B_4);

(b) ultimate boundedness (UB_2) implies uniform ultimate boundedness (UB_6).

6.13. Exercise. Use the equation $\dot{x} = \sin x$ with $\Omega = \mathscr{R}$ and $M = \{0\}$ to prove that B_4 does not imply UB_1.

6.14. Theorem. WUB_i + uniform boundedness (B_4) $\Longrightarrow$ UB_i, $1 \leq i \leq 6$.

Proof. Suppose WUB_i is satisfied for $\varepsilon = \varepsilon_1$ and $\rho = \rho_1$. Then, due to B_4, UB_i will be satisfied for $\sigma = \rho_1$ and $\varepsilon = \varepsilon_2$, where ε_2 is, in B_4, the value of ε corresponding to $\delta = \varepsilon_1$. Q.E.D.

6.15. Theorem (V. A. Pliss [1964]). If f is ω-periodic and $M = \{0\} \subset \Omega = \mathscr{R}^n$, then weak ultimate boundedness (WUB_2) implies ultimate boundedness (UB_2).

6.16. Theorem (N. Pavel [1972]). If f is ω-periodic, $M = \{0\} \subset \Omega = \mathscr{R}^n$ and every solution can be continued to $+\infty$, all concepts of weak ultimate boundedness and ultimate boundedness are equivalent.

Proof. It will be enough to prove that WUB_1 implies UB_6. But this implication will be established if we show that WUB_1 implies WUB_2, for then, using 6.15 and 6.12 (b), one gets the chain $WUB_1 \Longrightarrow WUB_2 \Longrightarrow UB_2 \Longrightarrow UB_6$. Let then $\tilde{t}_0$ be fixed in I and let $\varepsilon = \varepsilon_1$ correspond to $t_0 = \tilde{t}_0 + \omega$ in WUB_1. By continuity, for every $\delta > 0$, there is a $\delta_1 > 0$ such that

$$(\forall t_0 \in [\tilde{t}_0, \tilde{t}_0 + \omega])(\forall x_0 \in M_\delta) \quad x(\tilde{t}_0 + \omega; t_0, x_0) \in M_{\delta_1}.$$

Then there exists a $t \in [t_0, \infty[$ with $x(t) \in M_\varepsilon$ and WUB_2 is verified with I replaced by $[\tilde{t}_0, \tilde{t}_0 + \omega]$. Therefore, WUB_2 is verified, owing to Lemma 4.9. Q.E.D.

6.17. Exercise. If f is ω-periodic, $M = \{0\} \subset \Omega = \mathscr{R}^n$ and every solution can be continued to $+\infty$, then all concepts of class 30 in Table 6.9 are equivalent to all concepts of ultimate boundedness and weak ultimate boundedness.

6.18. An equation like (1.1) will be called dissipative if, for $M = \{0\}$, all concepts of weak ultimate boundedness, of ultimate boundedness and of class 30 (in Table 6.9) are verified for the equation. In N. Levinson [1944], (1.1) was called dissipative if it verified UB_2. Theorem 6.16 and Exercise 6.17 yield sufficient conditions for an equation to be dissipative. The following exercise gives particularly simple conditions to get the same result.

6.19. Exercise. If f is ω-periodic, $M = \{0\} \subset \Omega = \mathscr{R}^n$, $I \supset [0,\infty[$ and every solution can be continued to $+\infty$, the differential equation is dissipative if and only if

$$(\exists\varepsilon > 0)(\forall x_0 \in \mathscr{R}^n)(\exists t > 0) \quad x(t;0,x_0) \in B_\varepsilon.$$

Hint: Using Theorem 4.6, one can replace in all concepts of weak ultimate boundedness, of ultimate boundedness and of class 30, $(\forall t_0 \in \mathscr{R})$ by (for $t_0 = 0$).

7. Asymptotic Stability

7.1. A set M is said to verify a concept of asymptotic stability if it possesses some kind of attractivity along with some kind of stability. It results from the number of concepts listed in Tables 6.1 and 6.2 that there are, formally, 12 types of asymptotic stability. Some of them are equivalent however.

A set M is said to be

(i) asymptotically stable (AS) if it is stable (S_1) and attractive (A_1);

(ii) equi-asymptotically stable (EAS) if it is stable (S_1) and equi-attractive (A_4);

(iii) uniformly asymptotically stable (UAS) if it is uniformly stable (S_2) and uniformly attractive (A_6).

Global asymptotic stability corresponds to asymptotic stability like global attractivity to attractivity. For instance, M is said to be

(iv) globally asymptotically stable (GAS) if it is stable (S_1) and globally attractive (GA_1);

(v) uniformly globally asymptotically stable if it is uniformly stable (S_2) and uniformly globally attractive (GA_6).

Let us now prove a frequently used partial converse of Theorem 6.5.

7.2. Theorem. If

$$(\exists\delta_1 > 0)(\forall x_0 \in M_{\delta_1} \setminus M)(\forall t_0 \in I) \quad J^+(t_0,x_0) \supset [t_0,+\infty[,$$

then $S_2 + WA_i \Longrightarrow A_i$, $(1 \leq i \leq 6)$.

Proof. It is possible, with some care, to prove all six implications at a time. Let us, for simplicity, limit ourselves to the sixth one. More precisely, we shall show that if S_2 and WA_6 are satisfied, then A_6 is satisfied with the same choice $\delta = \delta_1$ in A_6 as in WA_6. We first choose a $\delta = \delta_1$ suiting WA_6 and then select, to prove A_6, an arbitrary $\varepsilon = \varepsilon'$. Let $\delta_2 < \delta_1$ correspond to ε' in S_2. Selecting $\varepsilon = \delta_2$ in WA_6, we get that for $\delta = \delta_1$, $\varepsilon = \delta_2$ and some $\rho = \rho'$:

$$\begin{gathered}(\forall x_0 \in M_{\delta_1} \setminus M)(\forall t_0 \in I)(\exists t_a \in [t_0,t_0+\rho']) \\ x(t_a) \in M_{\delta_2}.\end{gathered} \tag{7.1}$$

Clearly t_a can be chosen such that $x(t_a) \notin M$ and so

$x(t_a) \in M_{\delta_2} \setminus M$. But then combining (7.1) and S_2 (with $\varepsilon = \varepsilon'$, $\delta = \delta_2$, $t_0 = t_a$, $x_0 = x(t_a)$), one gets that for $\delta = \delta_1$ and $\varepsilon = \varepsilon'$

$$(\forall t_0 \in I)(\forall x_0 \in M_{\delta_1} \setminus M)(\forall t \in [t_a,\infty[) \quad x(t) \in M_\varepsilon.$$

Finally, since $t_a \leq t_0 + \rho$, one obtains for $\delta = \delta_1$, $\varepsilon = \varepsilon'$ and $\sigma = \rho$ that

$$(\forall t_0 \in I)(\forall x_0 \in M_{\delta_1} \setminus M)(\forall t \in [t_0 + \sigma,\infty[) \quad x(t) \in M_\varepsilon.$$

Q.E.D.

7.3. Exercise. Prove that, for $i = 1$ or 2, the unique (continuability) hypothesis of Theorem 7.2 can be dispensed with.

Owing to Lemma 4.16, if $M \subset \Omega$ either is compact, or is closed with compact boundary and negatively invariant, then the (continuability) hypothesis of Theorem 7.2 can be dispensed with.

We have already observed that attractivity does not imply stability (cf. I.2.7). However, we can prove the following.

7.4. Theorem. If $M \subset \Omega$ is positively invariant, equi-attractivity (A_4) implies stability (S_1).

The proof is left as an exercise (see Exercise I.2.9).

7.5. Exercise. Suppose f is ω-periodic and $M \subset \Omega$ either is compact and positively invariant, or is closed with a compact boundary and invariant. Suppose further that each solution $x(t;t_0,x_0)$ with $x_0 \in \partial M$ is defined at least on

$[t_0, t_0 + \omega]$. Then asymptotic stability is equivalent to uniform asymptotic stability.

Hint: Applying Theorem 4.10, one has $S_1 \Longleftrightarrow S_2$. Then show that, even if f is not periodic, S_2 along with A_1 imply A_4. Using Theorem 4.11, one gets that $A_4 \Longrightarrow A_6$. (The case $M = \{0\}$ has been treated in Theorem I.2.14.)

7.6. Exercise. If the hypotheses of Exercise 7.5 are satisfied, global asymptotic stability is equivalent to global uniform asymptotic stability.

Hint: One may reason as in 7.5, while changing the implication "S_2 and A_1 imply A_4" into "S_2 and GA_1 imply GA_4". The case $M = \{0\}$ appears in T. Yoshizawa [1966].

8. Bibliographical Note

In his paper of 1892, A. M. Liapunov proposed a fairly general definition for the stability of a solution of a differential equation. Roughly speaking, he said that a solution $x^*(t)$ is stable with respect to some function $h(x)$, if the norm $||h(x(t)) - h(x^*(t))||$ remains small whenever $||x(t_0) - x^*(t_0)||$ is chosen small enough. But he studied the simplest case only: $h(x) \equiv x$ and $x^*(t) \equiv 0$.

Later, K. P. Persidski [1933] introduced the uniformity in t_0, and I. G. Malkin [1954], for asymptotic stability, the uniformity in x_0. The example due to R. E. Vinograd [1957] (cf. I.2.7) of an unstable attractor led naturally to splitting asymptotic stability into two simpler concepts, namely stability and attractivity. Later, the notion of weak attractivity appeared as a natural one, when one is trying to prove some kind of asymptotic stability by using a theorem like 7.2 (see also I.6.8). The definition of weak attractivity as proposed here is slightly different from the one of N. P. Bhatia [1966]. Boundedness was studied in detail by T. Yoshizawa [1966]. On weak ultimate boundedness, see P. Habets and K. Peiffer [1973].

The extension of these definitions to sets is due to several authors: V. I. Zubov [1957], T. Yoshizawa [1966], N. P. Bhatia and G. P. Szegö [1967]. Time-varying sets were considered by T. Yoshizawa, and later, in an effort to encompass partial stability, P. Habets and K. Peiffer [1973] mentioned the stability of a set with respect to another set. An early work in the same direction is A. N. Michel [1969]. A number of other concepts have been studied, as for example, exponential stability by many authors, stability in tube-like domains, stability of conditionnally invariant sets, by V. Lakshmikantham and S. Leela, etc.

Systematic presentations of the most classical concepts with an effort to unify the terminology appear in J. L. Massera [1956] and H. A. Antosiewicz [1958]. Rather extensive sets of definitions can be found in the books of V. I. Zubov [1957], L. Cesari [1959], W. Hahn [1959], [1967], T. Yoshizawa [1966], N. P. Bhatia and G. P. Szegö [1967], V. Lakshmikantham and S. Leela [1969] and A. Halanay [1966].

A classification of concepts by means of various types of bounds on the solutions has been presented by W. Hahn [1967]. A different approach was adopted by D. Bushaw [1969] who defines a large number of concepts by combining in various ways a sequence of quantified variables. The same type of symbolism was used later by P. Habets and K. Peiffer [1976], whose general presentation has been adopted here. Other contributions of the same type are those of M. Dana [1972], V. M. Matrosov [1973] and C. Avramescu [1973].

CHAPTER VII

ATTRACTIVITY FOR AUTONOMOUS EQUATIONS

1. Introduction

Attractivity properties are studied in this chapter and the next one. As contrasted with Chapters I to V, the origin will play no particular role here: it will not be assumed that it is a critical point of the differential equation on hand, nor even that it belongs to Ω. The reason will become clear as soon as we tackle some applications: the attractivity of a set is frequently encountered as a natural question.

The autonomous case is treated separately in this chapter for several reasons. First, strong results can be obtained rather simply in this setting. Further, an important use can be made of the invariance property of limit sets: in the non autonomous case, this property can be only partially preserved, at the expense of some supplementary hypothesis. Finally, there exist very good and important examples to illustrate the autonomous case, whereas this is less true for the non autonomous one. Many significant theoretical results concerning the latter are more recent indeed.

2. General Hypotheses

Let Ω be some open set of $\mathscr{R}^n$, $f\colon \Omega \to \mathscr{R}^n$, $x \to f(x)$ a continuous function and x_0 a point of Ω. The initial time will always be chosen equal to 0. Thus, our general Cauchy problem will be:

$$\dot{x} = f(x) \tag{2.1}$$

$$x(0) = x_0 \tag{2.2}$$

By the way, the uniqueness of solutions is not assumed. A non-continuable solution will be written $x\colon]\alpha,\omega[\to \mathscr{R}^n$, and we shall often write $J =]\alpha,\omega[$ and $J^+ = [0,\omega[$. The positive limit set of a solution x will be designated by $\Lambda^+(x)$. The properties of limit sets are studied in Appendix III.

3. The Invariance Principle

3.1. The following theorem is important from a theoretical point of view: it shows in fact which conclusions remain, in Liapunov's direct method, when a single function is considered along with hypotheses as weak as possible, not much more indeed than a negative derivative. Every inessential consideration being discarded in this way, there remains something like the essence of the method. Further, this theorem, which is known under the name of <u>invariance principle</u>, is amongst the most useful ones in the applications.

3.2. Theorem (J. P. LaSalle [1968]). Let x be a solution of (2.1), (2.2) and $V: \Omega \to \mathscr{R}$ a locally lipschitzian function such that $D^+V(x) \leq 0$ on $x(J^+)$. Then $\Lambda^+ \cap \Omega \subset M$ where M is the union of all non-continuable orbits, each of which is a subset of $E = \{x \in \Omega: D^+V(x) = 0\}$.

Proof. If $||x(t)|| \to \infty$ as $t \to \infty$, then $\Lambda^+ = \phi$ and the thesis is trivially satisfied. Otherwise, one has either $\Lambda^+ \subset \partial\Omega$ and the thesis is again trivially satisfied, or $\Lambda^+ \not\subset \partial\Omega$. In the latter case, consider some point $x^* \in \Lambda^+ \cap \Omega$. There exists an increasing sequence $\{t_i\} \subset J^+$ such that $t_i \to \omega$ and $x(t_i) \to x^*$ as $i \to \infty$. But $\{V(t_i)\}$ is bounded from below and decreasing. Therefore, for some $c \in \mathscr{R}$, $V(t_i) \to c$ as $i \to \infty$, and, as V is continuous, $V(x^*) = c$. But V is decreasing, and therefore $V(t) \to c$ as $t \to \omega$. Therefore the limiting value of V is the same for every point $x^* \in \Lambda^+ \cap \Omega$. In other words: $V(x) = c$ on $\Lambda^+ \cap \Omega$. But as $\Lambda^+ \cap \Omega$ is semi-invariant, $D^+V(x) = 0$ on $\Lambda^+ \cap \Omega$ (on the concept of semi-invariance, cf. Appendix III). Therefore $\Lambda^+ \cap \Omega \subset E$ and the thesis is obtained by invoking again the semi-invariance of $\Lambda^+ \cap \Omega$. Q.E.D.

More precise and practically important conclusions concerning the asymptotic behavior of $x(t)$ are given in the following corollary.

3.3. Corollary. If $x(t)$ does not approach infinity when $t \to \omega$, then $x(t) \to M \cup \partial\Omega$ on every compact set when $t \to \omega$. If $x(J^+)$ is bounded, $x(t) \to M \cup \partial\Omega$ when $t \to \infty$.

In this statement, the expression "$x(t)$ tends to

$M \cup \partial\Omega$ on every compact set" means that for every compact set K, if J_K is defined as $\{t \in J: x(t) \in K\}$, then $x(t) \to M \cup \partial\Omega$ for values of t contained in J_K.

3.4. In connection with Theorem 3.2, it is important to be able to identify the set M. It is often easier to recognize $E \setminus M$, i.e. the largest subset of E no point of which belongs to a non-continuable orbit entirely contained in E. This is usually possible by considering the vector field $x \to f(x)$. But auxiliary functions can be used also as is shown by the following two propositions. The first one, which is rather crude, is obvious; the second one is more refined.

3.5. <u>Proposition</u>. Let $P \subset \Omega$ be compact, $P' \subset \Omega$ an open neighborhood of P and $W: P' \to \mathscr{R}$ a $\mathscr{C}^1$ function such that $\dot{W}(x) \neq 0$ on P. Then, if $x(t)$ is a solution such that $x(0) \in P$: $x(J^+) \not\subset P$.

Unfortunately, $E \setminus M$ is not often, if ever, compact. This is because, in many applications, M is compact: think of the simplest case, where M is a critical point. Of course, Proposition 3.5 is no more true if the hypothesis of compacity of P is removed. In a way, it's a pity, because one would have liked to draw some useful conclusions from the very simple assumption that $\dot{W}(x) \neq 0$. (See however Section 4.4.)

3.6. <u>Proposition</u>. For some sets P and P^* with $P \subset P^* \subset \Omega$, let $\{P_i\} \subset \Omega$ and $\{Q_j\} \subset \Omega$ be two sequences of open sets such that

$$(\bigcup_i P_i) \cup (\bigcup_j Q_j) \supset P.$$

Let us write $P'_j = P_j \cap P$ and $Q'_j = Q_j \cap P$. Assume that the closure of Q'_j (in $\mathscr{R}^n$) is disjoint from $\partial\Omega$. Assume that, in P^* considered as a metric space, the Q'_j are pairwise disjoint and such that $\overline{Q}'_j \cap \partial P = \phi$. If there exist two sequences of $\mathscr{C}^1$ functions $W_i: P_i \to \mathscr{R}$ and $W^*_j: Q_j \to \mathscr{R}$ such that

(i) on P'_i: $W_i(x) = 0$, $\dot{W}_i(x) \neq 0$;

(ii) on Q'_j: $W^*_j(x)$ is bounded, $|\dot{W}^*_j(x)| > \alpha_j$ for some $\alpha_j > 0$;

then, for every solution $x(t)$ such that $x(0) \in P$: $x(J^+) \not\subset P$.

Proof. (a) If $x(0) \in P'_i$ for some i, there is no $\tau > 0$ such that $x(t) \in P'_i$ for every $t \in [0,\tau[$. Otherwise, one would get, for all these t, that $W_i(x(t)) = 0$, and therefore $\dot{W}_i(x(0)) = 0$, which is excluded by (i). But P'_i is open with respect to P: as a consequence, $x(t)$ leaves P immediately.

(b) If $x(0) \in Q'_j$ for some j and $x(t)$ remains in P for every $t \in [0,\omega[$, then either $x(t)$ remains in Q'_j for every $t \in [0,\infty[$ and it follows that $\omega = \infty$, or $x(t)$ comes out of Q'_j. In the former case, $x(t)$ remains in some connected component of Q'_j upon which $|\dot{W}^*_j(x)| > \alpha_j$, whereas W^*_j is bounded: this is a contradiction. In the latter case, $x(t)$ comes out of Q'_j while remaining in P. Therefore it touches one of the P'_i and, as proved above, comes out of P, which is another contradiction. Q.E.D.

The importance of P^* will be better understood if one knows that in practical situations, the interior of P^*

will often be empty. Notice also, that if one assumes further that W_j^* is defined and continuous on some compact set containing Q_j, then (ii) can be replaced by: $\dot{W}_j^*(x) \neq 0$ on this set.

4. An Attractivity and a Weak Attractivity Theorem

4.1. Corollary 3.3 implies that if $x(J^+)$ is bounded, and bounded away from $\partial\Omega$, then $x(t) \to M$ as $t \to \infty$. On the other hand, if $x(J^+)$ is unbounded, there is no indication in this corollary that $x(t)$ might approach some set. The theorem to follow yields sufficient conditions for a set, possibly an unbounded set, to be attractive. By the way, the function $\phi(x)$ was introduced in this theorem to avoid imposing a bound on $f(x)$.

4.2. Theorem. Let $S \subset \Omega$ be closed with respect to Ω and put $M = \partial S \cap \partial\Omega$. Suppose there exists a locally Lipschitzian function $V: \Omega \to \mathscr{R}$ such that $D^+V(x) \leq 0$ on Ω. Suppose that $E = \{x \in \Omega: D^+V(x) = 0\}$ contains no non-continuable orbit. Suppose at last that there exist a continuous, strictly positive function $\phi: \Omega \to \mathscr{R}$ and, for every $\rho > 0$, four numbers $A, B, C, D > 0$ such that, for every $x \in S \setminus B(M,\rho)$:

(i) $\phi(x)||f(x)|| < A$;

(ii) $(||x|| > B) \Longrightarrow [\phi(x)D^+V(x) \leq -C]$;

(iii) $V(x) \geq -D$;

then, every solution $x(t)$ such that $x(J^+) \subset S$ tends to M as $t \to \omega$.

Proof. If the thesis is wrong, there exists a sequence

$\{t_i\} \subset J^+$ and a $\rho > 0$ such that $t_i \to \omega$ as $i \to \infty$ and $d(x(t_i),M) \geq 2\rho$ for every i. But $x(t_i) \to \infty$ for otherwise, there would exist a point of Λ^+ outside M, which is excluded by Theorem 3.2. For an appropriate subsequence, again noted $\{t_i\}$, one has $||x(t_{i+1}) - x(t_i)|| > \rho$. Further, for every i there is a $\tau_i \in]t_i,t_{i+1}[$ such that $d(x(t),M) \geq \rho$ for every $t \in [t_i,\tau_i[$ and $||x(\tau_i) - x(t_i)|| = \rho$. Therefore, for i large enough, one has

$$V(x(\tau_i)) - V(x(t_i)) \leq \int_{t_i}^{\tau_i} D^+V(x(s))ds \leq -C\int_{t_i}^{\tau_i} \frac{ds}{\phi(x(s))}$$

$$\leq - \frac{C}{A} \int_{t_i}^{\tau_i} ||f(x(s))||ds \leq - \frac{C}{A} \left|\left|\int_{t_i}^{\tau_i} \dot{x}(s)ds\right|\right|$$

$$= - \frac{C}{A}||x(\tau_i) - x(t_i)||.$$

At last, since $V(x(t))$ is decreasing, $V(x(t)) \to -\infty$ as $t \to \infty$, which is excluded by (iii). Q.E.D.

4.3. Remarks. (a) In the applications, any suitable criterion can be used to recognize that the set E contains no non-continuable orbit. In particular Proposition 3.6 can prove helpful.

(b) One might well wonder why the attractive set of Theorem 4.2 has been chosen as a subset not of $\overline{\Omega}$ but of $\partial\Omega$. The reason was one of convenience, and it will be no limitation in any case, for it is always possible to decide that $f(x)$ is not defined on the attractive set.

(c) It has been proved that $x(t) \to M$ as $t \to \omega$, and ω may be finite. The reader who is familiar with Chapters I and II knows that in a proof ab absurdo like this one, one always shows that a function $V(x(t)) \to -\infty$, whereas on some other ground, it is bounded from below. But usually, to prove that $V(x(t)) \to -\infty$, one uses the fact that $\omega = \infty$ and that $V(x(t))$ is bounded from above by a strictly negative quantity multiplied by t. This reasoning was impossible here. This is why one has used, to "push" $V(x(t))$ towards $-\infty$, the fact that $x(t) \to \infty$.

4.4. A theorem like VI.7.2 has shown that weak attractivity is a useful concept. The following theorem leads obviously to some sufficient conditions for a compact set to be weakly attractive. It gives us an opportunity to draw some interesting conclusion from the main hypothesis of Proposition 3.5 (namely that $\dot{W}(x) \neq 0$), which, as will be remembered, led to a rather disappointing result.

4.5. We shall say that a solution $x\colon]\alpha,\omega[\to \mathscr{R}^n$ <u>tends weakly</u> to a set M as $t \to \omega$, if there exists a sequence $\{t_i\} \subset]\alpha,\omega[$ such that $t_i \to \omega$ and $x(t_i) \to M$ as $i \to \infty$.

4.6. <u>Theorem</u>. Let $S \subset \Omega$ be closed with respect to Ω and bounded. Let us put $M = \partial S \cap \partial\Omega$. Suppose there exist a locally lipschitzian function $V\colon \Omega \to \mathscr{R}$ and a $\mathscr{C}^1$ function $W\colon \Omega \to \mathscr{R}$ such that:

(i) $(\forall x \in \Omega)\ D^+V(x) \leq 0$;

(ii) $(\forall x \in E)\ \dot{W}(x) \neq 0$ where $E = \{x \in S\colon D^+V(x) = 0\}$;

then, for every solution $x(t)$ such that $x(J^+) \subset S$, $x(t)$ tends weakly towards M as $t \to \infty$.

<u>Proof</u>. If the thesis is wrong, there exists a solution $x(t)$ such that $x(J^+) \subset S$ and which does not tend weakly to M. Then there exists an $\varepsilon > 0$ and a $\tau \in J^+$ such that

$x(t) \in S \setminus B(M,\varepsilon)$ for every $t \in [\tau,\omega[$, and therefore $\omega = \infty$. But one knows, by Theorem 3.2, that $\Lambda^+ \subset E$ and further, since Λ^+ is compact, that $x(t) \to \Lambda^+$ as $t \to \infty$. On the other hand, there exists a sequence $\{t_i\}$ such that $t_i \to \infty$ and $\dot{W}(x(t_i)) \to 0$ as $i \to \infty$: indeed, $W(x)$ being bounded on the compact set $S \setminus B(M,\varepsilon)$, $|\dot{W}(x)|$ cannot be bounded away from 0 when $t \to \infty$. But this is absurd, for Λ^+ and $\{x \in S \setminus B(M,\varepsilon): \dot{W}(x) = 0\}$ are two disjoint compact sets.

Q.E.D.

5. Attraction of a Particle by a Fixed Center

5.1. Let 0 be the origin of some inertial frame of reference and assume a particle of mass equal to 1 is subject to two forces: an attraction towards 0 and some kind of friction. Let $\underline{r}$ and $\underline{v}$ be respectively the position vector and the velocity of the particle. For simplicity, we shall write r and v for the euclidean norms of the two vectors. Further, for the vector $(\underline{r},\underline{v})$ of phase space, we shall use the norm $(r^2 + v^2)^{1/2}$.

The attraction towards 0 will derive from a $\mathscr{C}^1$ potential function $P(r)$, defined for $r > 0$ and such that $\frac{dP}{dr} > 0$ for every $r > 0$. The friction force will be directed along $\underline{v}$ with opposite direction and an amplitude given by a function $k(v)$ defined for $v \geq 0$, with $k(0) = 0$, $k(v) > 0$ and a convergent integral

$$\int_0^V \frac{v}{k(v)}\, dv.$$

The last hypothesis is verified for instance if $k(v)$ is proportional to v^α for some $\alpha \in [0,2[$, which might be acceptable for a physicist in many circumstances.

5.2. The equations of motion read

$$\begin{aligned} \dot{\underline{r}} &= \underline{v}, \\ \dot{\underline{v}} &= -\frac{dP}{dr}\frac{\underline{r}}{r} - k(v)\frac{\underline{v}}{v}. \end{aligned} \tag{5.1}$$

We choose as Liapunov function the total energy

$$V(r,v) = T(v) + P(r),$$

where $T(v) = v^2/2$ is the kinetic energy. One computes readily

$$\dot{V} = -k(v)v \leq 0$$

where the equality obtains if and only if $v = 0$.

5.3. Let us now prove that <u>for every solution of</u> (5.1), $\underline{r}(t)$ <u>is bounded in the future</u>. More precisely, if a motion $(\underline{r}(t),\underline{v}(t))$ is such that, for some t_0, $(\underline{r}(t_0),\underline{v}(t_0)) = (\underline{r}_0,\underline{v}_0)$, then, for every $t \in [t_0,\omega[$

$$r(t) \leq r_0 + \int_0^{v_0} \frac{v}{k(v)}\, dv,$$

the second member being finite by hypothesis. Let us write

$$\phi(r,v) = r + \int_0^{v} \frac{v}{k(v)}\, dv,$$

and assume there exists a $t_1 > t_0$ such that $r(t_1) = r_1 > \phi(r_0,v_0)$. Then, one would get

$$\phi(r_1,v_1) \geq r_1 > \phi(r_0,v_0) \tag{5.2}$$

which is impossible, as we shall prove now.

Observe that if $\dot{r} = \dfrac{(\underline{r}|\underline{v})}{r}$ is positive,

$$\dot{\phi}(r(t),v(t)) = -v + \frac{(\underline{r}|\underline{v})}{r} - \frac{1}{k(v)} \frac{dP}{dr} \frac{(\underline{r}|\underline{v})}{r}$$

is negative, because the last term is negative and $\dot{r} \leq v$. Further, if there exist two values t' and t'' such that $t_0 \leq t' \leq t'' < t_1$ with $r(t') = r(t'')$, one gets successively $V(r(t'),v(t')) \geq V(r(t''),v(t''))$, whence $v(t') \geq v(t'')$ and at last $\phi(r(t'),v(t')) \geq \phi(r(t''),v(t''))$. Every hypothesis of Lemma 5.6 below is verified if we identify r to y, $-\phi(r(t),v(t))$ to $z(t)$, t_0 to a and t_1 to b. Therefore we get that $\phi(r_1,v_1) \leq \phi(r_0,v_0)$, which contradicts (5.2). We have thus proved that $r(t)$ is bounded in the future. Next we consider successively two types of potential functions.

5.4. Assume first that $P(r)$ does not tend to $-\infty$ as $r \to 0$. Since $P(r)$ is increasing, it tends to a limit when $r \to 0$, and we may suppose, without loss of generality, that $P(0) = 0$. With a view to apply Theorem 4.2, let us identify Ω with the phase space $\mathscr{R}^6 \setminus \{0\}$ of the non-zero vectors $(\underline{r},\underline{v})$ and let M be precisely the origin. Every solution of (5.1) is bounded in the future. This has just been proved for $r(t)$. As for $v(t)$, since V is decreasing and P is bounded from below, $T(v)$ and therefore v are bounded in the future. The set E is constituted of those points of the phase space where $v = 0$ and $r \neq 0$. At every point of E, one component at least of $\underline{r}$, say r_j (for $j = 1$, 2 or 3) is non vanishing. Let us choose, for every such point $(\overline{r},0)$, an open neighborhood $N(\overline{r},0)$ where $r_j \neq 0$, and let us define the auxiliary function $W_{(\overline{r},0)}(\underline{r},\underline{v}) = v_j$. But, owing to (5.1),

$$\dot{W}_{(\bar{r},0)}(\underline{r},\underline{v}) = -\frac{dP}{dr}\frac{r_j}{r}$$

a quantity which doesn't vanish on $N(\bar{r},0) \cap E$ for $N(\bar{r},0)$ small enough. It follows then from Proposition 3.6 that E contains no non-continuable orbit. Therefore, all the hypotheses of Theorem 4.2 are satisfied for any motion, if one chooses $\phi(x)$ identically equal to 1 and a set S large enough to encompass the orbit of the motion: the origin of the phase space is a global attractor.

5.5. Removing now the restriction that $P(r)$ does not tend to $-\infty$ when $r \to 0$, we identify the set M of Theorem 4.2 with the hyperplane $\underline{r} = 0$ and Ω with $\mathscr{R}^6 \setminus M$. Let V and W be as above. At last, let

$$\phi(\underline{r},\underline{v}) = \frac{1}{(1 + v)(1 + k(v))} .$$

One computes easily that

$$||\phi(\underline{r},\underline{v})f(\underline{r},\underline{v})|| \leq [1 + (1 + \frac{dP}{dr})^2]^{1/2}.$$

We know that for every solution $(\underline{r}(t),\underline{v}(t))$, there exists an $r_1 > 0$ such that $(\underline{r}(t),\underline{v}(t)) \in B[M,r_1]$. The set $B[M,r_1] \cap \Omega$ will play the role of S in 4.2. Then all the hypotheses of 4.2 are satisfied obviously, except perhaps (ii) which we verify now. Choosing $B > r_1$, we see that $||(\underline{r},\underline{v})|| > B$ implies that $v > (B^2 - r_1^2)^{1/2} > 0$. If we limit our study to those functions $k(v)$ which are, for v large, greater than some strictly positive constant, we can prove that

$$\phi(\underline{r},\underline{v})\dot{V}(\underline{r},\underline{v}) = \frac{-k(v)v}{(1 + k(v))(1 + v)}$$

is smaller than some strictly negative constant in the appropriate region of phase space. Theorem 4.2 applies, and as it applies to every solution, M is a global attractor.

5.6. <u>Lemma</u>. Let $y(t)$ and $z(t)$ be two real $\mathscr{C}^1$ functions defined on some non-empty compact interval $[a,b] \subset \mathscr{R}_0$. Assume that

(i) $y(a) < y(b)$;

(ii) for every $t \in [a,b]$ such that $y'(t) \geq 0$: $z'(t) \geq 0$;

(iii) for every pair $t', t'' \in [a,b]$ such that $t' \leq t''$ and $y(t') = y(t'')$: $z(t') \leq z(t'')$;

then $z(a) \leq z(b)$.

<u>Proof</u>. Let $I^* = \{t^* \in [a,b]: (\forall t \in [a,t^*])\ y(t) \leq y(t^*)\}$. This set is closed. Indeed, let $\{t_i^*\} \subset I^*$ be a sequence approaching some t^* and let t be fixed in $[a,t^*[$. For every i large enough, $y(t) \leq y(t_i^*)$ and thus $y(t) \leq y(t^*)$. Therefore $t^* \in I^*$. Further, for any $t^* \in I^*$: $y'(t^*) \geq 0$ and, owing to (ii),

$$\int_{I^*} z'(\tau)d\tau \geq 0. \tag{5.3}$$

The set $I^{**} = [a,b] \setminus I^*$ is open and is even a denumerable union of open pairwise disjoint intervals:

$$I^{**} = \bigcup_{1 \leq i < p}]a_i,b_i[$$

where p may equal ∞. Next $y(a_i) = y(b_i)$ for every i. To prove this, suppose first that $y(a_i) < y(b_i)$ and let

$$a_i' = \inf \{t \in [a_i,b_i]: y(t) = \frac{y(a_i) + y(b_i)}{2}\}.$$

One shows easily that $a_i' \in]a_i,b_i[$ and also that $a_i' \in I^*$, which is absurd. Neither can one have $y(a_i) > y(b_i)$, for then b_i wouldn't belong to I^*, and this is absurd again. We conclude that for every i

$$\int_{]a_i,b_i[} z'(\tau)d\tau = z(b_i) - z(a_i) \geq 0,$$

and

$$\int_{I^{**}} z'(\tau)d\tau \geq 0. \tag{5.4}$$

It results from (5.3) and (5.4) that

$$\int_{[a,b]} z'(\tau)d\tau = z(b) - z(a) \geq 0.$$

Q.E.D.

6. A Class of Nonlinear Electrical Networks

6.1. In electrical engineering, there are many examples of networks whose normal operation consists in some kind of switching from one equilibrium to another, and which should be prevented from oscillating in any manner: think for instance of the bistable devices used in computer technology. In this section, we establish sufficient conditions for a certain class of networks to be such that, starting from any initial conditions, they approach an equilibrium as $t \to \infty$.

6.2. The networks are those whose equations, following R. K. Brayton and J. Moser [1964] can be derived from the knowledge of a single state function called the <u>mixed potential</u> (or by some others the "hybrid dissipation function"). More precisely, the variables describing the state of the network will be supposed to be, for some positive integers r and s, a current vector $i \in \mathscr{R}^r$ and a voltage vector $v \in \mathscr{R}^s$.

The components of v are voltages across capacitors, and the components of i are currents through inductors. The mixed potential is a real function P of the form

$$P(v,i) = -v^T\chi i - G(v) + F(i),$$

where χ is a real $s \times r$ matrix (whose elements, by the way, are equal either to 1 or to -1, but this fact will be immaterial for what follows) and where G and F are two real $\mathscr{C}^1$ functions, defined on $\mathscr{R}^s$ and $\mathscr{R}^r$ respectively. The equations of the network read

$$C\frac{dv}{dt} = \frac{\partial P}{\partial v}, \quad L\frac{di}{dt} = -\frac{\partial P}{\partial i}, \tag{6.1}$$

where C and L are two square matrices, respectively of order s and r, usually functions of v and i. A description of the type of networks admitting equations of the form (6.1) and the actual derivation of these equations appear in the above mentioned paper by K. K. Brayton and J. Moser. The following theorem is an interesting example of construction of a Liapunov function.

6.3. <u>Theorem</u>. Assume that

(i) C is a function of v, symmetric and positive definite for $v \in \mathscr{R}^s$;

(ii) L is a function of i, symmetric and positive definite for $i \in \mathscr{R}^r$;

(iii) for some $a \in \mathscr{R}^r$ and some constant, symmetric and positive definite matrix R of order r: $F(i) = \frac{1}{2}\, i^T R i + a^T i$;

(iv) $G(v) + ||\chi^T v|| \to \infty$ as $v \to \infty$;

(v) for any $x \in \mathscr{R}^s$, any $(i,v) \in \mathscr{R}^r \times \mathscr{R}^s$ and some $\theta \in]0,1[$:

$$(Kx)^T Kx \leq x^T x(1 - \theta)$$

where

$$K = L^{1/2} R^{-1} \chi^T C^{-1/2};$$

(vi) the system possesses a finite number of equilibriums;

then every solution of (6.1) tends to an equilibrium as $t \to \infty$.

Proof. Since

$$P(v,i) = -v^T \chi i - G(v) + \frac{1}{2} i^T R i + a^T i, \qquad (6.2)$$

the equations of the network read

$$C \frac{dv}{dt} = -\chi i - \frac{\partial G}{\partial v},$$

$$L \frac{di}{dt} = \chi^T v - Ri - a.$$

To get simpler notations, let us define the functions f and g by the following equations:

$$f = \frac{\partial P}{\partial v} = -\chi i - \frac{\partial G}{\partial v},$$

$$g = -\frac{\partial P}{\partial i} = \chi^T v - Ri - a. \qquad (6.3)$$

Our auxiliary function will be

$$V = g^T R^{-1} g - P, \qquad (6.4)$$

with a derivative, computed along the solutions of (6.1), given by

$$\frac{dV}{dt} = 2g^T R^{-1} \frac{dg}{dt} - f^T C^{-1} f + g^T L^{-1} g.$$

But

$$\frac{dg}{dt} = \chi^T C^{-1} f - RL^{-1} g,$$

and therefore

$$\frac{dV}{dt} = -f^T C^{-1} f + 2g^T R^{-1} \chi^T C^{-1} f - g^T L^{-1} g.$$

This derivative vanishes at every equilibrium point, i.e. at every point where $f = g = 0$. Let us show next that it is strictly negative everywhere else. Since C and L are symmetric and positive definite, the matrices $C^{1/2}$ and $L^{1/2}$ exist, are real, regular and symmetric. Therefore we may write

$$-\frac{dV}{dt} = (C^{-1/2} f)^T (C^{-1/2} f) - 2(L^{-1/2} g)^T L^{1/2} R^{-1} \chi^T C^{-1/2} (C^{-1/2} f)$$
$$+ (L^{-1/2} g)^T (L^{-1/2} g),$$

or

$$-\frac{dV}{dt} = x^T x - 2y^T Kx + y^T y,$$

with

$$x = C^{-1/2} f, \quad y = L^{-1/2} g \quad \text{and} \quad K = L^{1/2} R^{-1} \chi^T C^{-1/2}.$$

From this result and Hypothesis (v), one deduces that

$$-\frac{dV}{dt} = (y - Kx)^T (y - Kx) + x^T x - (Kx)^T Kx \geq ||y - Kx||^2$$
$$+ \theta ||x||^2.$$

But Hypothesis (i) and (ii) imply that x vanishes when and only when f vanishes, and y when and only when g vanishes. Therefore $\frac{dV}{dt} < 0$ everywhere, except at the equilibrium points.

Let us show at last that $V \to \infty$ when $||i|| + ||v|| \to \infty$. In order to achieve this, we first write P as a function of v and g. We substitute to i in (6.2) its expression

$$i = R^{-1}(-g + \chi^T v - a).$$

One obtains after some easy calculation, that

$$P = \frac{1}{2} g^T R^{-1} g - U(v)$$

where $U(v) = \frac{1}{2} [v^T\chi - a^T]R^{-1}[\chi^T v - a] + G(v)$. Therefore, by (6.4),

$$V = \frac{1}{2} g^T R^{-1} g + U(v).$$

But R being positive definite, the same is true of R^{-1}. Then Hypothesis (iv) implies that $V \to \infty$ when $||g|| + ||v|| \to \infty$. But $||g|| \to \infty$ if $||i|| \to \infty$ and $v \not\to \infty$, as is evident from (6.3) and the fact that R is regular.

Getting back now to Theorem 3.2 and Corollary 3.3, we observe first that $E = M =$ the set of equilibrium points. Further, the fact that V is decreasing and tends to infinity with $||i|| + ||v||$ implies that every solution is bounded. Therefore every solution tends to the set of equilibrium points. But since there is a finite number of such points, every solution tends to some equilibrium point as $t \to \infty$.

Q.E.D.

6.4. As an elementary illustration, consider the circuit of Fig. 7.1, where the rectangle represents a nonlinear resistor, and where the capacitance C, the inductance L and the resistance R are constant. It appears from the theory of the mixed potential that here χ is of order 1 and is equal to 1. One obtains

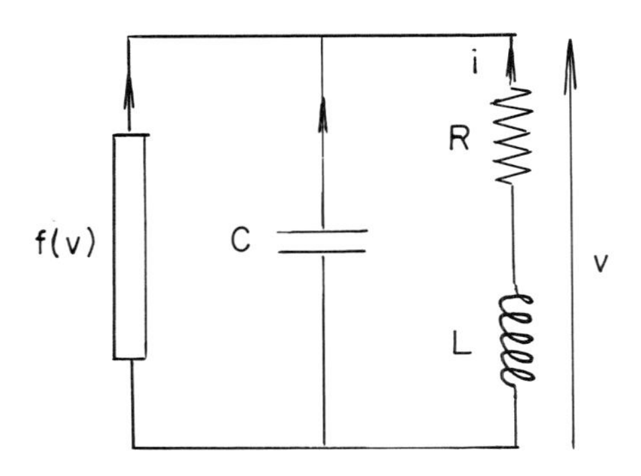

Fig. 7.1. RLC circuit.

$$P = - vi - \int_0^v f(v)dv + \frac{1}{2} Ri^2,$$

whence the equations of the circuit,

$$C \frac{dv}{dt} = - i - f(v),$$

$$L \frac{di}{dt} = v - Ri.$$

We assume that the characteristic function f is such that there is a finite number of equilibriums and that $\int_0^v f(v)dv + |v|$ tends to ∞ with $|v|$. Then the circuit will always approach an equilibrium if

$$L^{1/2}R^{-1}C^{-1/2} < 1,$$

or in other words, if $L < R^2C$.

More involved illustrations of Theorem 6.3 will be found for instance in R. K. Brayton and J. Moser [1964] or in T. E. Stern [1965].

7. The Ecological Problem of Interacting Populations

7.1. The problem dealt with in this section yields a nice illustration of Theorem 4.2 but it touches also several other questions of stability theory, for example partial asymptotic stability and the use of first integrals. It was first considered by A. J. Lotka [1920] (cited by N. S. Goel et altr. [1971]) as a problem of chemical reactions or competing species and then by V. Volterra [1931] when he tried to explain some cyclic variations of fish catches in the Adriatic. Let us quote N. S. Goel et altr.: "It was apparently observed that the populations of two species of fish commonly found in these catches varied with the same period, but somewhat out of phase. One of these was a species of small fish ... and the other was a species of a larger fish ... It seemed as though the large fish ate the small ones, grew, and multiplied until the population of small ones diminished to such a level that there were insufficient numbers for the survival of the large ones. As the population of the large one declined, that of the small species prospered to the degree that a larger number of large fish could be supported, etc.".

7.2. Let us call N_1 the number of fish preyed upon and N_2 the number of predators. It will appear natural to assume that in the absence of predators, the birth rate $(dN_1/dt)/N_1$ amongst the first population is a constant, which we write α_1. Assume similarly that in the absence of the first population, the death rate amongst predators is a negative constant, which we write $-\alpha_2$. In the presence of predators, the birth rate of fish preyed upon has to be corrected by a negative number, which we suppose to be proportional to the number of predators. It may be considered that the probability of encounter between fish of the first population and predators varies like the product $N_1 N_2$. Introducing a symmetric correction for the death rate of predators, one gets the following pair of equations:

$$\begin{aligned} \dot{N}_1 &= N_1(\alpha_1 - \lambda_1 N_2), \\ \dot{N}_2 &= N_2(-\alpha_2 + \lambda_2 N_1) \end{aligned} \tag{7.1}$$

where α_1, α_2, λ_1 and λ_2 are > 0. They make sense only for $N_1 \geq 0$ or $N_2 \geq 0$ since they are concentrations of animals or chemical products. From a mathematical point of view, they can of course be studied in the entire (N_1,N_2)-plane. One verifies that both N_1 and N_2 axes are invariant. There are two critical points, namely the origin $(N_1,N_2) = (0,0)$, and the point $(N_1,N_2) = \left(\frac{\alpha_2}{\lambda_2}, \frac{\alpha_1}{\lambda_1}\right) = (n_1,n_2)$.

The former is unstable, as is shown by the existence of an exponentially increasing N_1, for N_2 identically equal to zero. The latter is stable, because one verifies readily that the function

$$V = N_1 - \frac{\alpha_2}{\lambda_2} - \frac{\alpha_2}{\lambda_2} \ln \frac{\lambda_2}{\alpha_2} N_1 + \frac{\lambda_1}{\lambda_2} [N_2 - \frac{\alpha_1}{\lambda_1} - \frac{\alpha_1}{\lambda_1} \ln \frac{\lambda_1}{\alpha_1} N_2]$$

is a first integral for the Equations (7.1) and that is is positive definite around (n_1,n_2). The last property is proved by noticing that the Taylor series of V around (n_1,n_2) begins by the terms

$$\frac{1}{2} \frac{\lambda_2}{\alpha_2} (N_1 - \frac{\alpha_2}{\lambda_2})^2 + \frac{1}{2} \frac{\lambda_1}{\lambda_2} \frac{\lambda_1}{\alpha_1} (N_2 - \frac{\alpha_1}{\lambda_1})^2 .$$

7.3. Following Volterra, we now generalize the problem to the case of n species, for which the equations read

$$\frac{dN_i}{dt} = N_i(k_i + \beta_i^{-1} \sum_{1 \le j \le n} a_{ij}N_j), \quad 1 \le i \le n. \tag{7.2}$$

Here k_i is the difference between birth and death rate of the corresponding species, when it is supposed to be left to itself. When a parameter a_{ij} is > 0, it means that the ith species increases at the expense of the jth one, whereas when it is < 0, the ith one diminishes on behalf of the jth one. The a_{ij} form an antisymmetric matrix. The parameters β_i are > 0 and take into account the fact that, for instance, to generate one predator, usually more than one prey has to disappear. Practically, Equations (7.2) make sense for $N_i \ge 0$, $1 \le i \le n$. Mathematically, they can be studied in the whole of $\mathscr{R}^n$. Moreover, one observes that any of the subsets of $\mathscr{R}^n$ characterized by an arbitrary number of the N_i equal to zero, is invariant.

7.4. The equilibriums are solutions of the equations

$$N_i(\beta_i k_i + \sum_{1 \le j \le n} a_{ij}N_j) = 0.$$

The origin of $\mathscr{R}^n$ is one of them, and it is unstable as soon as one of the k_i is > 0. Let $N_i = n_i$ be any equilibrium, and suppose the subscripts have been arranged in such a way that $n_i > 0$ for $i = 1,\ldots,k$ and $n_i = 0$ for $i = k + 1,\ldots,n$, where k is some number equal to $1, 2, \ldots,$ or n. Then of course

$$\beta_i k_i + \sum_{1 \le j \le k} a_{ij}n_j = 0 \qquad 1 \le i \le k,$$

$$n_i = 0 \qquad k + 1 \le i \le n.$$

7.5. Consider now the auxiliary function

$$V = \sum_{1 \le i \le k} \beta_i n_i \left[\frac{N_i}{n_i} - \ln \frac{N_i}{n_i}\right] + \sum_{k+1 \le i \le n} \beta_i N_i,$$

which should be considered as defined on the set

$$\Psi = \{(N_1,\ldots,N_n) : N_i > 0,\ 1 \le i \le k,\ N_i \ge 0,\ k+1 \le i \le n\}.$$

Its time derivative along the solutions of (7.2) is computed as follows:

$$\begin{aligned}
\dot{V} &= \sum_{\substack{1 \le i \le k \\ 1 \le j \le n}} a_{ij}(N_i - n_i)(N_j - n_j) \\
&\qquad + \sum_{k+1 \le i \le n} \Big(k_i \beta_i N_i + \sum_{1 \le j \le n} a_{ij} N_i N_j\Big) \\
&= \sum_{1 \le i,j \le n} a_{ij}(N_i - n_i)(N_j - n_j) \\
&\qquad + \sum_{k+1 \le i \le n} \Big(k_i \beta_i + \sum_{1 \le j \le n} a_{ij} n_j\Big) N_i \\
&= \sum_{k+1 \le i \le n} \Big(k_i \beta_i + \sum_{1 \le j \le n} a_{ij} n_j\Big) N_i .
\end{aligned}$$

The antisymmetric character of the a_{ij} has been used to derive the final expression of the derivative. There is a function V for every equilibrium $(n_1,\ldots,n_n)$. Now if $n_i > 0$ for every i, one gets $\dot{V} = 0$, and V is a first integral. Further, let us subtract from V a constant equal to its value at $(n_1,\ldots,n_n)$. The function thus obtained is positive definite. Therefore, the corresponding equilibrium is stable.

Consider next an equilibrium for which $k < n$, i.e. belonging to the frontier of Ψ. One verifies that, when corrected by a constant term chosen as above, the function V is again positive definite around the equilibrium. Of course, the positive-definiteness referred to here is relative to a neighborhood (of the equilibrium) in Ψ, not in $\mathscr{R}^n$. On the other hand, $\dot{V} \leq 0$ if the following conditions are satisfied:

$$k_i\beta_i + \sum_{1 \leq j \leq k} a_{ij}n_j \leq 0, \quad k+1 \leq i \leq n. \tag{7.3}$$

The function V is no more a first integral in this case. A simple generalization of Theorems I.4.2 and I.6.33 to the case of a critical point located, as above, in $\partial\Psi$, yields parts (b) and (c) of the following proposition, which sums up the stability informations derivable from the properties of the function V.

7.6. Proposition. (a) Any equilibrium $(n_1,\ldots,n_n)$ of (7.2) such that $n_i > 0$, $1 \leq i \leq n$, is stable;

(b) for any equilibrium $(n_1,\ldots,n_n)$ of (7.2) for which $n_i > 0$, $1 \leq i \leq k$, $n_i = 0$, $k+1 \leq i \leq n$, $k < n$, (7.3) is a sufficient stability condition;

(c) in the same conditions, if the inequalities (7.3) are strict for some values of i, the stability is asymptotic with respect to the variables having these i's as subscripts.

7.7. Exercise. State and prove the extensions of Theorems I.4.2 and I.6.33 referred to in Section 7.5.

7.8. As an interesting particular case, let us consider a ternary system where two species live on a third one. The equations read

$$\dot{N}_1 = (k_1 + \beta_1^{-1}a_{13}N_3)N_1,$$

$$\dot{N}_2 = (k_2 + \beta_2^{-1}a_{23}N_3)N_2,$$

$$\dot{N}_3 = (k_3 - \beta_3^{-1}(a_{i3}N_1 + a_{23}N_2))N_3,$$

where $k_1 < 0$, $k_2 < 0$, $k_3 > 0$, $a_{13} > 0$, $a_{23} > 0$ and $\beta_i > 0$, $i = 1, 2, 3$. In general, one has $k_1\beta_1/a_{13} \neq k_2\beta_2/a_{23}$. If we disregard this case, we find three and only three critical points, namely

$$(P_1) \quad N_1 = N_2 = N_3 = 0;$$

$$(P_2) \quad N_1 = 0,\ N_2 = \frac{k_3\beta_3}{a_{23}} > 0,\ N_3 = -\frac{k_2\beta_2}{a_{23}} > 0;$$

$$(P_3) \quad N_1 = \frac{k_3\beta_3}{a_{13}} > 0,\ N_2 = 0,\ N_3 = -\frac{k_1\beta_1}{a_{13}} > 0.$$

We know that (P_1) is unstable. As for (P_2), the criterion (7.3) yields

$$k_1\beta_1 - a_{13}\frac{k_2\beta_2}{a_{23}} < 0, \tag{7.4}$$

with a $<$ sign because the equality has been discarded by hypothesis. The stability is asymptotic with respect to N_1. On the other hand, if we have, instead of (7.4), the inequality

$$k_1\beta_1 - a_{13}\frac{k_2\beta_2}{a_{23}} > 0,$$

(P_2) is unstable. Indeed, since the hyperplane $N_1 = 0$ is invariant, the set $\{(N_1,N_2,N_3)\colon N_1 > 0\} \cap B(P_2,\varepsilon)$ is a sector for any $\varepsilon > 0$. But on such a set, and if ε is chosen small enough, the auxiliary function $V(N_1,N_2,N_3) = N_1$ is strictly positive and $\dot{N}_1 = (k_1 + \beta_1^{-1}a_{13}N_3)\ N_1 > 0$. Similarly, the stability condition for (P_3) reads

$$k_2\beta_2 - a_{23}\frac{k_1\beta_1}{a_{13}} < 0. \tag{7.5}$$

The inequalities (7.4) and (7.5) are mutually exclusive. If we continue to disregard the exceptional case where $k_1\beta_1/a_{13} = k_2\beta_2/a_{23}$, we therefore obtain two unstable and one stable equilibriums. Further, if the system starts in some neighborhood of the stable one, one of the species disappears.

7.9. Exercise. Prove that this conclusion can be extended to any initial point. More explicitly, prove that if the system starts from any point in the set $\{(N_1,N_2,N_3): N_i > 0, 1 \le i \le 3\}$, then one of the species disappears asymptotically. This is a precise statement of what is known as "the ecological principle of exclusion of Volterra-Lotka".

7.10. One of the most conspicuous drawbacks of Volterra's model is that the birth rate of the species preyed upon left alone is a strictly positive constant, in such a way that the corresponding population increases beyond any bound. Of course, in any real situation, there will occur a saturation effect due to the limited resources of the ecological environment. This observation was made by P. Verhulst [1845] when discussing the Malthus theory of exponential population growth. This effect is taken into account in the following expression for the birth rate: $\frac{dN}{dt}\frac{1}{N} = k(\theta - N)/\theta$, where θ is some positive real quantity. The Equations (7.2) are now replaced by the following ones:

$$\frac{dN_i}{dt} = N_i\left[\frac{k_i}{\theta_i}\left(\theta_i - \frac{1}{2}(1 + \operatorname{sign} k_i)N_i\right) + \beta_i^{-1}\sum_{1 \le j \le n} a_{ij}N_j\right].$$

The equilibriums are obtained by equating the second members to zero. Let $N_i = n_i$, $1 \le i \le n$, be such an equilibrium, and assume as above that $n_i > 0$ for $i = 1,\ldots,k$, and $n_i = 0$ for $i = k+1,\ldots,n$. Then of course

$$\beta_i k_i \frac{\theta_i - \frac{1}{2}(1 + \operatorname{sign} k_i)n_i}{\theta_i} + \sum_{1 \leq j \leq k} a_{ij} n_j = 0, \quad i = 1,\dots,k$$

$$n_i = 0, \qquad\qquad i = k+1,\dots,n.$$

The derivative of the function V defined as above reads

$$\dot{V} = -\sum_{1 \leq i \leq n} \beta_i k_i \frac{(1 + \operatorname{sign} k_i)}{2\theta_i} (N_i - n_i)^2 + \sum_{k+1 \leq i \leq n} (\beta_i k_i + \sum_{1 \leq j \leq n} a_{ij} n_j) N_i .$$

A first observation is that if the equilibrium is such that $n_i > 0$ for $i = 1,\dots,n$, the associated function V is no more a first integral of the differential equations. Further, the same type of stability conclusions hold as for the simplest model, the one without saturation effect, and the stability conditions remain unchanged. One observes however that in the present case, the stability is necessarily asymptotic with respect to those variables for which $i \leq k$ and $k_i > 0$.

7.11. Let us now try to get sufficient conditions of asymptotic stability with respect to all variables, for the case where $n_i > 0$ for all i (i.e. $k = n$). Applying Theorem II.1.3, the only thing to prove is that the set $E \setminus \{(n_1,\dots,n_n)\}$ contains no non-continuable orbit, where

$$E = \{(N_1,\dots,N_n): \quad (\forall i: k_i > 0) N_i = n_i\}.$$

Assume that we renumber the N_i in such a way that

$$k_i > 0 \quad \text{if} \quad i = 1,\dots,l,$$
$$k_i \leq 0 \quad \text{if} \quad i = l+1,\dots,n.$$

Then of course every point of E is of the form $(n_1,\ldots,n_l,$ $N_{l+1},\ldots,N_n)$ and, owing to Proposition 3.6, it will be enough to prove that the vector function $W = (\beta_1 N_1,\ldots,\beta_l N_l)$ has a non vanishing derivative on $E \setminus \{(n_1,\ldots,n_n)\}$. One computes that, on E,

$$\beta_i \frac{dN_i}{dt} = k_i \beta_i n_i \frac{\theta_i - n_i}{\theta_i} + \sum_{1 \le j \le l} a_{ij} n_i n_j + \sum_{l+1 \le j \le n} a_{ij} n_i N_j$$

$$= \sum_{l+1 \le j \le n} a_{ij} n_i (N_j - n_j). \qquad (i = 1,\ldots,l)$$

Therefore, the stability of <u>the equilibrium is asymptotic whenever the rank of the matrix</u>

$$\begin{pmatrix} n_1 & & \\ & \ddots & \\ & & n_l \end{pmatrix} \begin{pmatrix} a_{1,l+1} & \cdots & a_{1,n} \\ \cdots & \cdots & \cdots \\ a_{l,l+1} & \cdots & a_{l,n} \end{pmatrix}$$

<u>equals</u> $n - l$. Of course, this condition is never satisfied if $l < n/2$.

7.12. As an example, consider the quaternary system where the fourth species subsists on the third and the second, while the third subsists on the first. The corresponding differential equation are

$$\dot N_1 = k_1 N_1 \frac{\theta_1 - N_1}{\theta_1} - \beta_1^{-1} a_{31} N_3 N_1,$$

$$\dot N_2 = k_2 N_2 \frac{\theta_2 - N_2}{\theta_2} - \beta_2^{-1} a_{42} N_4 N_2,$$

$$\dot N_3 = k_3 N_3 + \beta_3^{-1} (a_{31} N_1 - a_{43} N_4) N_3,$$

$$\dot N_4 = k_4 N_4 + \beta_4^{-1} (a_{42} N_2 + a_{43} N_3) N_4,$$

where $k_1 > 0$, $k_2 > 0$, $k_3 < 0$, $k_4 < 0$, $\theta_i > 0$, $\beta_i > 0$ and

$a_{ij} > 0$. Under suitable conditions, there exists an equilibrium $N_i = n_i > 0$ for $i = 1,\ldots,4$. On the set

$$E = \{(n_1,n_2,N_3,N_4): \quad (N_3,N_4) \neq (n_3,n_4)\},$$

the derivative of the function

$$W = (\beta_1 N_1, \beta_2 N_2)$$

doesn't vanish. Indeed

$$\dot{W} = -(a_{31}n_1(N_3 - n_3),\ a_{42}n_2(N_4 - n_4)) \neq 0.$$

Hence the equilibrium mentioned above is asymptotically stable.

8. Bibliographical Note

The invariance principle for bounded solutions of autonomous equations is due to J. P. LaSalle [1960]. Antecedent to the theory are the theorems of E. A. Barbashin and N. N. Krasovski appearing in Section II.1 of this book and which dealt already with periodic equations. The invariance principle for unbounded solutions comes from J. P. LaSalle [1968]. A particular case of Proposition 3.6 on non invariant sets was proved in N. Rouche [1968] and subsequently generalized by M. Laloy [1974] and N. Rouche [1974]. The attractivity Theorem 4.2 comes from J. L. Corne and N. Rouche [1973], where it is applied to the fixed center problem treated in Section 5. The problem of electrical network of Section 6 will be found, as already mentioned, in R. K. Brayton and J. K. Moser [1964], whereas a good review of the ecological problems of interacting populations appears in N. S. Goel et altr. [1971].

CHAPTER VIII

ATTRACTIVITY FOR NON AUTONOMOUS EQUATIONS

1. Introduction, General Hypotheses

Proving attractivity or asymptotic stability is more difficult in the non autonomous case than in the autonomous one, because in the former, one cannot rely in general on any invariance property of the limit sets. The situation is more complex and the asymptotic properties which can be proved with substantially equivalent hypotheses are weaker. In Section 2, we introduce the one-parameter families of Liapunov functions of L. Salvadori: they do not appear in the statements of the theorems, but are powerful tools of demonstration. We use them to prove a significant extension of Matrosov's theorem II.2.5, yielding a new and interesting characterization of the uniform asymptotic stability of the origin. By the way, in order to grade the difficulties, the origin, instead of a set, is studied in this Section 2. Section 3 gives another useful extension of Matrosov's theorem.

With Section 4, we come back to the attractivity of sets and examine what kind of generalizations of LaSalle's Theorem VII.3.2 can be proved when starting from the most natural and simple hypotheses in Liapunov's direct method: some kind of lower bound on an auxiliary function $V(t,x)$, and some kind of upper bound on its derivative $\dot{V}(t,x)$. Next we extend the attractivity Theorem VII.4.2 to the non autonomous case, and it proves helpful, in this setting, to generalize also Proposition VII.3.6 on the expulsion of the solutions from a given set. Section 5 is devoted to those particular types of non autonomous equations for which the limit sets admit some kind of invariance property: for such equations, which are called asymptotically autonomous, asymptotically almost periodic, etc., there exist interesting particular extensions of LaSalle's theorem. Finally, Section 6 is an introduction to dissipative systems.

Our general hypotheses are again here those of Chapter IV. For reference purposes, let us recall the Cauchy problem on hand:

$$\dot{x} = f(t,x), \tag{1.1}$$

$$x(t_0) = x_0. \tag{1.2}$$

2. The Families of Auxiliary Functions

2.1. Let us show how a family of auxiliary functions in the sense of L. Salvadori [1969] [1971] can be used to prove a theorem of weak attractivity. These families are non denumerable: there corresponds a function to each possible choice of the quantity ε in the definition of stability, or of a similar quantity in another definition. A lemma and two definitions are needed before stating the theorem.

2.2. Lemma (L. Salvadori [1969]). Let S be a set and f, g two functions on S into $\mathscr{R}$. Assume there exist three constants $\beta_1, \beta_2, \beta_3 > 0$ such that, for every $z \in S$

$$f(z) \leq 0, \quad g(z) \leq \beta_1,$$
$$[f(z) > -\beta_2] \Longrightarrow [g(z) < -\beta_3];$$

then, there exist two constants $\mu, l > 0$ such that, for every $z \in S$:

$$f(z) + \mu g(z) < -l.$$

Proof. One only has to choose μ such that $0 < \mu < \beta_2/\beta_1$ and $l < \min(\beta_2 - \beta_1\mu, \mu\beta_3)$. Q.E.D.

2.3. If N is a compact neighborhood of the origin of $\mathscr{R}^n$ and if $E \subset N \subset \Omega$, we say with V. M. Matrosov $[1962]_1$ that a function $V\colon I \times N \to \mathscr{R}$ is non-vanishing definite with respect to E on N if

$$(\forall\varepsilon > 0)(\exists\eta > 0)(\exists\xi > 0)(\forall t \in I)(\forall x \in N \setminus B_\varepsilon)$$
$$[d(x,E) < \eta] \Longrightarrow [|V(t,x)| > \xi]$$

2.4. Exercise. Show that $V(t,x)$ is non-vanishing definite with respect to E on N if and only if there exists a neighborhood $\mathscr{N}$ of $E\setminus\{0\}$ and a function $a \in \mathscr{K}$ such that

$$(\forall t \in I)(\forall x \in \mathscr{N} \cap N) \quad |V(t,x)| \geq a(||x||);$$

also if and only if there exists a function $a \in \mathscr{K}$ such that

$$(\forall t \in I)(\forall x \in N) \quad \max(d(x,E), |V(t,x)|) \geq a(||x||).$$

2.5. The following definition is akin to the one of weak attractivity (cf. VI.6). A solution $x:]\alpha,\omega[\to \mathscr{R}^n$ of (1.1) is said to <u>tend weakly</u> towards the origin as $t \to \omega$, if there exists a sequence $\{t_i\}$ such that $t_i \to \omega$ and $x(t_i) \to 0$ as $i \to \infty$. Our first theorem refers to this concept and can be compared to Theorem II.2.5. As it is rather difficult, we shall avoid some inessential difficulties by assuming that both auxiliary functions V and W are $\mathscr{C}^1$.

2.6. <u>Theorem</u>. Assume there exist three constants M, a, $b > 0$, two $\mathscr{C}^1$ functions $V(t,x)$ and $W(t,x)$ on $I \times \Omega$ into $\mathscr{R}$, a continuous function $V^*(x)$ on Ω into $\mathscr{R}$ and a compact neighborhood N of the origin, $N \subset \Omega$, such that, for every $(t,x) \in I \times N$:

(i) $||f(t,x)|| \leq M$;

(ii) $-a \leq V(t,x)$;

(iii) $\dot{V}(t,x) \leq V^*(x) \leq 0$; we put $E = \{x \in N: V^*(x) = 0\}$;

(iv) $|W(t,x)| < b$;

(v) if moreover $\dot{W}$ is non-vanishing definite with respect to E on N,

then every solution x such that $x(J^+) \subset N$ tends weakly to the origin when $t \to \infty$.

<u>Proof</u>. The solution x on hand cannot approach the frontier of Ω, and therefore for $x: J^+ = [t_0,\infty[$. To prove the theorem, one only has to show that there exists, for every $\varepsilon > 0$, a $T > 0$ such that $||x(t)||$ cannot remain $\geq \varepsilon$ on a time interval of duration T. Owing to (v), there exist η and ξ such that for every $t \in I$ and every

$$x \in G(\varepsilon) = B(E,\eta) \cap (N \setminus B_\varepsilon),$$

one gets $|\dot{W}(t,x)| > \xi$. Defining the compact set $H(\varepsilon)$ by

$$H(\varepsilon) = (N \setminus B_\varepsilon) \setminus B(E,\eta/2),$$

we observe that, by (iii),

$$(\exists \beta > 0)(\forall (t,x) \in I \times H(\varepsilon)) \quad \dot{V}(t,x) < -\beta. \qquad (2.1)$$

Let us now use V and W to construct a function $v_\varepsilon: J \times (N \setminus B_\varepsilon) \to \mathscr{R}$ with a strictly negative derivative. Notice first that $G(\varepsilon)$ can be split into two disjoint parts, namely

$$G^{(1)}(\varepsilon) = \{x \in G(\varepsilon): \quad (\forall t \in I) \quad \dot{W}(t,x) < -\xi\},$$

$$G^{(2)}(\varepsilon) = \{x \in G(\varepsilon): \quad (\forall t \in I) \quad \dot{W}(t,x) > \xi\}.$$

Let $\nu_\varepsilon: \mathscr{R}^+ \to \mathscr{R}^+$ be a $\mathscr{C}^1$ function such that

$$\nu_\varepsilon(\tau) = 1 \quad \text{for} \quad \tau \in [0,\eta/2],$$
$$\nu_\varepsilon(\tau) = 0 \quad \text{for} \quad \tau \geq \eta.$$

Its derivative is obviously bounded. For $i = 1, 2$, let us define

$$\alpha_\varepsilon^{(i)}(x) = \nu_\varepsilon(d(x,E)) \qquad \text{for} \quad x \in G^{(i)}(\varepsilon),$$

$$\alpha_\varepsilon^{(i)}(x) = 0 \qquad \text{for} \quad x \notin G^{(i)}(\varepsilon).$$

These functions will not usually be $\mathscr{C}^1$, because $d(x,E)$ is not $\mathscr{C}^1$. However, they have bounded Dini derivatives, for the derivative of ν_ε is bounded, and one gets further that

$$|D^+ d(x(t),E)| \leq ||\frac{dx}{dt}(t)|| = ||f(t,x(t))|| < M,$$

for $(t,x) \in I \times (N \setminus B_\varepsilon)$, as well as similar inequalities for the other Dini derivatives. Define next h_ε:

$I \times (N \setminus B_\varepsilon) \to \mathscr{R}$ by the equation

$$h_\varepsilon(t,x) = (\alpha_\varepsilon^{(1)}(x) - \alpha_\varepsilon^{(2)}(x))\, W(t,x).$$

And now let us prove that if one chooses $S = I \times (N \setminus B_\varepsilon)$, the functions $\dot{V}(t,x)$ and $D^+h_\varepsilon(t,x)$ can be identified respectively with the functions f and g of Lemma 2.2. Indeed (see Appendix I.1.2 e):

(1) $\dot{V}(t,x) \leq 0$;

(2) $D^+h_\varepsilon(t,x) = (\alpha_\varepsilon^{(1)}(x) - \alpha_\varepsilon^{(2)}(x))\, \dot{W}(t,x) + (D\alpha_\varepsilon^{(1)}(x) - D\alpha_\varepsilon^{(2)}(x))\, W(t,x)$, where D means D^+ or D_+ according to the sign of W. Owing to the construction mode of the $\alpha_\varepsilon^{(i)}$, the first term is ≤ 0. The second one is bounded, for it is constituted of three bounded functions;

(3) At last, $D^+h_\varepsilon(t,x) < -\xi$ for every $t \in I$ and every $x \notin H(\varepsilon)$, for $D\alpha_\varepsilon^{(1)}$ and $D\alpha_\varepsilon^{(2)}$ vanish when $d(x,E) \leq \eta/2$, and, on the other hand, $\dot{V}(t,x) \leq -\beta$ for $(t,x) \in I \times H(\varepsilon)$.

Since the hypotheses of Lemma 2.2 are verified, there exist two quantities $\mu > 0$ and $l > 0$ such that, if $v_\varepsilon(t,x) = V(t,x) + \mu h_\varepsilon(t,x)$, one gets

$$D^+v_\varepsilon(t,x) = \dot{V}(t,x) + \mu D^+h_\varepsilon(t,x) < -l$$

for $(t,x) \in I \times (N \setminus B_\varepsilon)$. But on this set, v_ε is bounded from below by $-a - \mu b$. Let us put

$$T(t_0) = \sup \left\{\frac{v_\varepsilon(t_0,x_0) + a + \mu b}{l} : x_0 \in N \setminus B_\varepsilon\right\}.$$

No solution can remain in $N \setminus B_\varepsilon$ during an interval of time of length T, for if this were the case, one would get for this solution that

$$-(v_\varepsilon(t_0,x_0) + a + \mu b) < v_\varepsilon(t_0 + T) - v_\varepsilon(t_0) \leq \int_{t_0}^{t_0+T} D^+v_\varepsilon(\tau)d\tau$$

$$< - lT \leq - (v_\varepsilon(t_0,x_0) + a + \mu b),$$

which is absurd. Q.E.D.

2.7. Another proof of the same theorem. With a view to compare the one-parameter families of auxiliary functions to the more classical types of proof, let us outline now another demonstration of the same theorem, this time inspired by II. 2.5. We retain the definitions of E, $G(\varepsilon)$ and $H(\varepsilon)$ and start anew from (2.1). We first show that no solution can remain in $G(\varepsilon)$ for a period equal to or longer than $\tau = 2b/\xi$. Indeed, $|\dot{W}(t,x)|$ being $\geq \xi$ on $G(\varepsilon)$, and since $\dot{W}$ is continuous, one would get in the opposite case

$$2b > |W(t + \tau) - W(t)| = \int_t^{t+\tau} |\dot{W}(\sigma)|d\sigma \geq \xi\tau = 2b$$

which is absurd.

Let us put $\gamma = \min \{\beta\tau,\beta\eta/2M\} > 0$, $k = \min \{i \in N: \gamma i \geq a + \sup [V(t_0,x_0): x_0 \in N \setminus B_\varepsilon]\}$ and $T(t_0) = k\tau$. Assume, ab absurdo, that the solution may remain in $N \setminus B_\varepsilon$ during the time interval $[t_0,t_0+T]$. Let us cut this interval in k sub-intervals of length τ:

$$I_i = [t_0 + (i - 1)\tau, t_0 + i\tau] \quad i = 1,\dots,k.$$

Then, for each i:

(1) either $x(t) \in H(\varepsilon)$ for every $t \in I$, and owing to (2.1),

$$V(t_0 + i\tau) - V(t_0 + (i - 1)\tau) < -\beta\tau < -\gamma$$

(2) <u>or</u> $x(t) \in G(\varepsilon)$ for some $t' \in I_i$. In this case, there exists a value $t \in I_i$ for which $x(t) \notin G(\varepsilon)$, and therefore two values t_i and $t_i^* \in I_i$ (and we assume, without loss of generality, that $t_i < t_i^*$), such that

$$d(x(t_i),E) = \eta/2, \quad d(x(t_i^*),E) = \eta,$$

and $\eta/2 < d(x(t),E) < \eta$ for every $t \in]t_i,t_i^*[$. If this is the case, $|t_i^* - t_i| > \eta/2M$. Indeed,

$$\frac{\eta}{2} \leq ||x(t_i^*) - x(t_i)|| \leq ||\int_{t_i}^{t_i^*} \dot{x}(\sigma)d\sigma|| \leq \int_{t_i}^{t_i^*}||f(\sigma,x(\sigma))||d\sigma$$

$$\leq M|t_i^* - t_i|.$$

Since $\dot{V}$ is negative everywhere and smaller than $-\beta$ on $H(\varepsilon)$,

$$V(t_0 + i\tau) - V(t_0 + (i - 1)\tau) < V(t_i^*) - V(t_i) \leq \frac{\beta\eta}{2M} \leq -\gamma,$$

and therefore $V(t_0 + T) - V(t_0) < -k\gamma \leq -V(t_0) - a$, whence at last $V(t_0 + T) < -a$, which is absurd. Q.E.D.

2.8. The differences between these two proofs are interesting to notice. In the former, W is changed into h_ε to get a strictly negative derivative, then combined with v to yield v_ε, and using v_ε, the solutions are shown to be expelled from $N \setminus B_\varepsilon$. In the latter, W is used to prove that the solution is expelled from an appropriate neighborhood of $E(\varepsilon)$; after which V is shown to decrease at a sufficient pace for the solutions to be expelled from $N \setminus B_\varepsilon$. Observe the use of Hypothesis (i), i.e. the bound on $f(t,x)$, in both demonstrations. In the former, it yields an upper bound on the derivative of h_ε. It is proved below that a bound on

Wf would be enough. But this new hypothesis would not suffice in the latter proof, where the bound on f gives a sufficient transit time of the solution through $H(\varepsilon) \cap G(\varepsilon)$, and therefore a sufficient decrease of V. At last, the hypotheses on W might be weakened for the second proof, where they are used to show that the solutions are expelled from $G(\varepsilon)$, a result which can be obtained in various ways.

2.9. Corollary. The thesis of Theorem 2.6 remains true if Hypothesis (i) is replaced by

$$\text{(i')} \quad (\forall (t,x) \in I \times N) \quad ||W(t,x)f(t,x)|| \leq M.$$

Proof. One only has to observe that, for $i = 1, 2$,

$$|D\alpha_\varepsilon^{(i)}(x)| = \left| \left.\frac{d\nu_\varepsilon}{d\tau}\right|_{d(x(t),E)} Dd(x(t),E) \right| \leq c||\dot{x}(t)|| = c||f(t,x(t))||$$

where c is the bound on the derivative of ν_ε. In these conditions

$$D^+h_\varepsilon(t,x) \leq 2c||f(t,x)W(t,x)||.$$

Q.E.D.

2.10. Corollary. In Theorem 2.6 as well as in its version modified by Corollary 2.9, the quantity T may be determined independent of t_0 provided that V is bounded (and no more bounded from below only).

The thesis of the theorem will then be expressed saying that the solutions starting from some point in N tend weakly to the origin, uniformly with respect to t_0, when $t \to \infty$. Notice that T was, from the beginning, independent of x_0.

We know that uniform stability plus weak attractivity entails attractivity. Therefore, if we add to the hypotheses of Theorem 2.6 any conditions of uniform stability, we get a theorem of equi-asymptotic stability. Further, if Corollary 2.10 applies, one obtains uniform asymptotic stability. This leads to the following theorem, generalizing Theorem II.2.5.

2.11. Theorem. Assume there exist two constants M and $c > 0$, two $\mathscr{C}^1$ functions $V(t,x)$ and $W(t,x)$ on $I \times \Omega$ into $\mathscr{R}$, a continuous function $V^*(x)$ on Ω into $\mathscr{R}$, two functions a and $b \in \mathscr{K}$ and at last a compact neighborhood N of the origin, $N \subset \Omega$, such that, for every $(t,x) \in I \times \Omega$:

(i) $||f(t,x)W(t,x)|| \leq M$;

(ii) $a(||x||) \leq V(t,x) \leq b(||x||)$;

(iii) $\dot{V}(t,x) \leq V^*(x) \leq 0$; we put $E = \{x \in \Omega: V^*(x) = 0\}$;

(iv) $|W(t,x)| \leq c$;

(v) if moreover $\dot{W}$ is non-vanishing definite with respect to E on N,

then the origin is uniformly asymptotically stable.

Assume for a moment that we include in our general hypotheses a local Lipschitz condition for f on $I \times \Omega$, entailing uniqueness of the solutions. One knows then (cf. I.7.4) that the classical Liapunov conditions for uniform asymptotic stability (i.e. (ii) hereabove along with a negative definite $\dot{V}$) are also necessary. Clearly, any other sufficient condition ensuring this property can be but stronger than this one, or at least equivalent.

In the case of Theorem 2.11, we really have an equivalence. Indeed, the Liapunov hypotheses imply those of 2.11

with E reduced to the origin and an identically vanishing function W. Notice that if we had kept, instead of (i) the hypothesis (i') $||f(t,x)|| \leq M$, the Liapunov hypotheses would not have implied those of 2.11! The method of one-parameter families of auxiliary functions led us therefore to an original characterization of uniform asymptotic stability of the origin.

3. Another Asymptotic Stability Theorem

3.1. Theorem 2.11 in the last section was an extension of Matrosov's Theorem II.2.5, to the case of possibly unbounded second members $f(t,x)$. Another extension of the same theorem is considered here, which consists in replacing the second auxiliary function, the one written W, by some more elaborate means of proving that the solution is repelled by a given compact set. This is the object of the following lemma, to be compared with Proposition VII.3.6. In addition to its use in the asymptotic stability theorem in question, this lemma can serve other purposes, for instance to prove that a set is an absolute expeller in the sense of Chapter V (see M. Laloy $[1974]_2$).

3.2. Lemma. Let K be a compact subset of Ω and let $\{M_i: 1 \leq i \leq n_1\}$, $\{N_j: 1 \leq j \leq n_2\}$ be two sequences of open subsets of Ω such that $M = (\bigcup_i M_i) \cup (\bigcup_j N_j) \supset K$. Assume the N_j are pairwise disjoint and there exist two sequences of $\mathscr{C}^1$ functions

$$U_i: I \times M_i \to \mathscr{R}, \quad W_j: I \times N_j \to \mathscr{R}$$

such that

(i) $(\forall i: 1 \leq i \leq n_1)\ U_i(t,x) \to 0$ uniformly in t as $x \to K,\ x \in M_i$;

(ii) $(\exists \beta_1 > 0)(\forall j: 1 \leq j \leq n_2)(\forall (t,x) \in I \times N_j)\ |W_j(t,x)| < \beta_1$;

(iii) $(\exists \xi > 0) \begin{cases} (\forall i: 1 \leq i \leq n_1)(\forall (t,x) \in I \times M_i)\ |\dot{U}_i(t,x)| > \xi; \\ (\forall j: 1 \leq j \leq n_2)(\forall (t,x) \in I \times N_j)\ |\dot{W}_j(t,x)| > \xi; \end{cases}$

assume finally that

(iv) $(\exists A > 0)(\forall (t,x) \in I \times \Omega)\quad \|f(t,x)\| \leq A$;

then there exist two numbers T and σ such that a solution $x(t)$ of (1.1) cannot remain in $D = \{x: d(x,K) \leq \sigma\}$ on a time-interval of duration T.

Proof. One shows easily (compare with VII.3.6) the existence of an $r > 0$ such that, if we put

$$M_i^* = M_i \setminus \{x: d(x,\partial M_i) \leq r\},$$
$$N_j^* = N_j \setminus \{x: d(x,\partial N_j) \leq r\},$$

the whole family of the M_i^* and N_j^* still constitutes an open covering of K. Let us write

$$M^* = (\bigcup_i M_i^*) \cup (\bigcup_j N_j^*).$$

Choose now β_2 such that

$$0 < \beta_2 < \frac{\xi r}{2A} \tag{3.1}$$

and $\sigma > 0$ such that $\sigma < d(K,\partial M^*)$ and further that, owing to (i),

$$(\forall (t,x) \in I \times (M_i \cap D))\quad |U_i(t,x)| \leq \beta_2,\quad 1 \leq i \leq n_1, \tag{3.2}$$

where $D = \{x: d(x,K) \leq \sigma\}$. Let us put further

$$P_i = M_i \cap D, \quad P_i^* = M_i^* \cap D, \quad 1 \le i \le n_1,$$
$$Q_j = N_j \cap D, \quad Q_j^* = N_j^* \cap D, \quad 1 \le j \le n_2.$$

Let us now show that a solution $x(t)$ of (1.1) cannot remain in any of the P_i or Q_j for a period of duration greater than $T = 2\beta/\xi$, where $\beta = \max\{\beta_1,\beta_2\}$. Let then x be a solution such that $x(t_0) \in P_i$ (one would reason alike for one of the Q_j's). If $x(t)$ remains in P_i from t_0 up to $t_1 = t_0 + \frac{2\beta}{\xi}$, one gets, owing to (iii) and (3.2) that $2\beta \ge 2\beta_2 \ge |U_i(t_1,x(t_1)) - U_i(t_0,x(t_0))| > \xi(t_1 - t_0) = 2\beta$, which is absurd.

Assume now that $x(t_0)$ be in P_i^* for some i. If $x(t)$ leaves P_i without leaving D, it has to cross the "barrier" $P_i \setminus P_i^*$ and therefore to travel a distance at least equal to r. If t_1 and t_2 are respectively the instants where $x(t)$ enters and leaves the barrier, one gets that

$$2\beta_2 \ge |U_i(t_2,x(t_2)) - U_i(t_1,x(t_1))| > \xi(t_2-t_1) \ge \frac{\xi r}{A} \quad (3.3)$$

which contradicts (3.1). Assume on the contrary that $x(t_0)$ be in Q_j^* for some j. If the solution leaves Q_i without leaving D, it has to cross the barrier $Q_i \setminus Q_i^*$ and therefore again to travel a distance at least equal to r. If t_1 and t_2 are here also the instants where $x(t)$ enters and leaves the barrier, one knows that $x(t_1)$ belongs to some P_i^*. But then <u>either</u> $x(t)$ remains in P_i between t_1 and t_2 and one gets the inequalities (3.3), <u>or</u> $x(t)$ comes out of P_i, but then it crosses $P_i \setminus P_i^*$, and one obtains (3.3) again.

Q.E.D.

3.3. <u>Exercise</u>. Let K be a compact subset of Ω and let, for some integer $k > 0$ and a neighborhood N of K, $N \subset \Omega$, a $\mathscr{C}^1$ function $W: N \to \mathscr{R}^k$ be such that its derivative along the solutions of (1.1), written $\dot{W}(x)$, doesn't depend on t and that

(i) $(\forall x \in K)\ \dot{W}(x) \neq 0$;

(ii) at least $k - 1$ components W_i of W vanish identically on K;

then, except for (iv), all the hypotheses of Lemma 3.2 are verified.

<u>Hint</u>: use Hypothesis (i) and the compactness of K to construct two sequences $\{M_i\}$ and $\{N_j\}$ containing respectively $k - m$ and m elements, where $m = 0$ if all components of W vanish on K and $m = 1$ if one component of W doesn't vanish identically on K.

Lemma 3.2 enables one to render Theorem II.2.5 on asymptotic stability more versatile by merely changing the hypotheses concerning W. The generalized theorem reads explicitly as follows.

3.4. <u>Theorem</u>. Let there exist a $\mathscr{C}^1$ function $V: I \times \Omega \to \mathscr{R}$, a $\mathscr{C}^0$ function $V^*: \Omega \to \mathscr{R}$, two functions a and $b \in \mathscr{K}$ and a constant $A > 0$ such that, for every $(t,x) \in I \times \Omega$:

(i) $a(||x||) \le V(t,x) \le b(||x||)$;

(ii) $\dot{V}(t,x) \le V^*(x) \le 0$; put $E = \{x \in \Omega: V^*(x) = 0\}$;

(iii) $||f(t,x)|| \le A$;

suppose further that for every pair of numbers ν and ε, $0 < \nu < \varepsilon$ with $B_\varepsilon \subset \Omega$, the set $K = \{x \in E: \nu \le ||x|| \le \varepsilon\}$ verifies the hypotheses of Lemma 3.2 (or of course of Exercise

3.3); choosing $\alpha > 0$ such that $\bar{B}_\alpha \subset \Omega$, let us put for every $t \in I$,

$$V_{t,\alpha}^{-1} = \{x \in \Omega\colon V(t,x) \leq a(\alpha)\}.$$

Then

(a) for any $t_0 \in I$ and any $x_0 \in V_{t_0,\alpha}^{-1}$: $x(t;t_0,x_0) \to 0$ uniformly in (t_0,x_0), when $t \to \infty$;

(b) the origin is uniformly asymptotically stable.

The proof is left as an exercise.

3.5. Example of a general Lagrangian system. Consider a mechanical system such as described at Section 13 of Appendix II, i.e. with a kinetic energy $T(q,\dot{q})$ which is $\mathscr{C}^1$ on some domain $\mathscr{R}^n \times \Omega$, a potential function $\Pi(q)$ also $\mathscr{C}^1$ on Ω, stationary at the origin, with $\Pi(0) = 0$. Assume further that there exist Lagrangian forces $\tilde{Q}(t,q,\dot{q})$, also expressible in terms of t, p, q as $Q(t,p,q)$. The equations of motion for such a system are Hamiltonian equations modified by the adjunction of generalized forces. They read

$$\dot{p} = -\frac{\partial H}{\partial q}(p,q) + Q(t,p,q), \quad \dot{q} = \frac{\partial H}{\partial p}(p,q).$$

Assume that $\Pi(q)$ admits an isolated minimum at $q = 0$ and that

$$\dot{H}(t,p,q) = Q^T(t,p,q)\frac{\partial H}{\partial p}(p,q) \leq -a(||p||)$$

for some function $a \in \mathscr{K}$. Assume further that $Q(t,p,q)$ approaches 0 uniformly with respect to t when $p \to 0$. The set E of Theorem II.2.5 as modified in Theorem 3.4 above is $\{(p,q)\colon p = 0\}$. The p_i will play the role of the U_i (with appropriate sets M_i) in Lemma 3.2. Of course every component of p vanishes on E. As for $\dot{p}$, observe first that, owing to Proposition 7 of Appendix II and the fact that p and $\dot{q}$ vanish simultaneously, $Q(t,0,q) = 0$ for every q and t. Therefore, on E,

$$\dot{p} = -\frac{\partial H}{\partial q}(0,q) = -\frac{\partial \Pi}{\partial q},$$

a vector which, by hypothesis, vanishes at $q = 0$ only. Then the p_i possess all the properties required from the U_i in 3.4 and 3.2. The full power of Lemma 3.2 will not be used here, since there is in our example, no function like the W_j. We conclude that the origin $p = q = 0$ of the phase space is uniformly asymptotically stable. This extends to some types of non autonomous mechanical systems the result embodied in Theorem III.5.2. It is pleasant to observe that the auxiliary functions used in connection with this mechanical problem are the total energy and the vector of conjugate momenta, two functions which are obviously meaningful to the physicist.

4. Extensions of the Invariance Principle and Related Questions

4.1. We now come back to the attractivity of sets and remove from our general hypotheses the requirement for the origin of $\mathscr{R}^n$ to belong to Ω and to be a critical point of (1.1).

4.2. In the autonomous case, LaSalle's theorem proved that $\Lambda^+ \cap \Omega$ was a subset of M, where M is the largest semi-invariant set of $E = \{x \in \Omega: D^+V(x) = 0\}$. In the absence of a semi-invariance property, all that can be proved in the non autonomous case is that $\Lambda^+ \cap \Omega$ is a subset of E. But how shall we define E in the present case where D^+V is a function of x but also of t? In his paper of 1968, LaSalle supposed the existence of a function $V^*(x)$ such that $D^+V(t,x) \leq V^*(x) \leq 0$ and defined E as $\{x: V^*(x) = 0\}$. Here, we shall find interesting to assume that $D^+V(t,x)$ is bounded above by a function of t and x, which by the way will be identified sometimes with $D^+V(t,x)$ itself. Further, we shall not immediately show that $x(t)$ approaches some set, but rather that some function of t and $x(t)$ tends to zero. Later, this function will be identified with the distance from $x(t)$ to some set. Most results in this section are not very far from being variants of one another: they form altogether a set of tools enabling one to cope with rather varied situations. A final remark: a majority of the theorems below are consequences of a unique fundamental lemma, which by the way formalizes a type of reasoning already encountered in this book (see Theorem I.6.23).

4.3. <u>Fundamental Lemma</u>. For some $\alpha \in \mathscr{R}$, let $\phi:]\alpha,\infty[\to \mathscr{R}$, $\psi:]\alpha,\infty[\to \mathscr{R}$ be two continuous functions such that

(i) ϕ is bounded from below;

(ii) $(\forall t \in]\alpha,\infty[)\ \psi(t) \geq 0$;

assume further that $(\forall \varepsilon > 0)(\exists a \in \mathscr{K})(\exists A > 0)(\forall t: \psi(t) \geq \varepsilon)$

(iii) $D^+\phi(t) \leq -a(\psi(t))$;

(iv) $D^+\psi(t) \leq A$ (or $D^+\psi(t) \geq -A$);

then $\psi(t) \to 0$ as $t \to \infty$.

<u>Proof</u>. If $\psi(t)$ does not approach 0, there exist an $\varepsilon > 0$ and an increasing sequence $\{t_n\}$ of time-values such that $t_n \to \infty$ and for every n: $t_{n+1} - t_n \geq \varepsilon/A$ and $\psi(t_n) \geq 2\varepsilon$.

But then by (iv), $\psi(t) \geq \varepsilon$ on $[t_n, t_n + \frac{\varepsilon}{A}]$ or on $[t_n - \frac{\varepsilon}{A}, t_n]$. Then, as the case may be, $\phi(t_n + \frac{\varepsilon}{A}) - \phi(t_n) \leq -\frac{\varepsilon}{A} a(\varepsilon)$ or $\phi(t_n) - \phi(t_n - \frac{\varepsilon}{A}) \leq -\frac{\varepsilon}{A} a(\varepsilon)$. In both cases, $\phi(t) \to -\infty$ as $t \to \infty$, which contradicts (i). Q.E.D.

4.4. The properties which can be proved are of a different character according as, for the solution on hand, $\omega = \infty$, in which case there is no need for the second member to be bounded, or ω is arbitrary, and then some kind of boundedness has to be assumed for $f(t,x)$. Let us first study _the case where_ $\omega = \infty$ _and no bound is assumed for the second member_.

4.5. _Theorem_ (_arbitrary solutions_). Let S be a subset of Ω and let V and ψ be two functions on $I \times \Omega$ into $\mathscr{R}$, locally lipschitzian in x and continuous. If there exist two numbers $A, B > 0$ and a function $a \in \mathscr{K}$ such that, for every $(t,x) \in I \times S$:

(i) $V(t,x) \geq -A$;

(ii) $\psi(t,x) \geq 0$;

(iii) $D^+V(t,x) \leq -a(\psi(t,x))$;

(iv) $D^+\psi(t,x) \geq -B$ (or $D^+\psi(t,x) \leq B$);

then $\psi(t,x(t)) \to 0$ as $t \to \infty$ for every solution $x(t)$ for which $\omega = \infty$ and $x(J^+) \subset S$.

This is an immediate consequence of Lemma 4.3. The requirement of a lower bound on V may be relaxed, but only for those solutions which don't tend to ∞ as $t \to \infty$. This is shown in the next theorem.

4.6. _Theorem_ (_solutions which don't tend to_ ∞). Let S be a subset of Ω and let V and ψ be two functions on $I \times \Omega$ into $\mathcal{R}$, locally lipschitzian in x and continuous. If there exists a number $B > 0$ and a function $a \in \mathcal{K}$ such that, for every $(t,x) \in I \times S$:

(ii) $\psi(t,x) \geq 0$;

(iii) $D^+V(t,x) \leq -a(\psi(t,x))$;

(iv) $D^+\psi(t,x) \geq -B$ (or $D^+\psi(t,x) \leq B$);

if moreover

(i') for every compact set C, there exists an $A > 0$ such that $V(t,x) \geq -A$ on $I \times (C \cap \Omega)$;

then $\psi(t,x(t)) \to 0$ as $t \to \infty$ for every solution $x(t)$ which does not tend to ∞ as t tends to ∞ and for which $\omega = \infty$ and $x(J^+) \subset S$.

<u>Proof</u>. For every solution of the mentioned type, there exists a compact set C and a sequence $\{t_n\}$ such that $t_n \to \infty$ as $n \to \infty$, and for every n: $x(t_n) \in C$. Then, owing to (i'), $V(t_n)$ doesn't tend to $-\infty$ as $t_n \to \infty$. But since $V(t)$ is decreasing, it tends to a limit as $t \to \infty$, and in particular, it admits a lower bound. Therefore Lemma 4.3 again entails the announced result. Q.E.D.

4.7. <u>Exercise</u>. The following statement is interesting in that, like Theorem VII.3.2, it draws some conclusion of asymptotic behavior from the consideration of a single auxiliary function.

Let S be a subset of Ω and V a function on $I \times \Omega$ into $\mathcal{R}$ such that $\dot{V}(t,x)$ exists, be locally lipschitzian in x and continuous; if there exist two numbers A and B such that, for every $(t,x) \in I \times S$:

(i) $\dot{V}(t,x) \le 0$;

(ii) $V(t,x) \ge -A$;

(iii) $D^+\dot{V}(t,x) \ge -B$ (or $D^+\dot{V}(t,x) \le B$);

then $\dot{V}(t,x(t)) \to 0$ as $t \to \infty$ for every solution $x(t)$ for which $\omega = \infty$ and $x(J^+) \subset S$.

4.8. For ψ independent of t, Theorem 4.6 admits the following corollary, which is akin to Theorem 1 (b) of J. P. LaSalle [1968].

4.9. Corollary. Let S be a subset of Ω, closed with respect to Ω; let $V: I \times \Omega \to \mathscr{R}$ be a function which is locally lipschitzian in x and continuous, and let $\psi: \Omega \to \mathscr{R}$ be a locally lipschitzian function; if there exists a number $B > 0$ such that, for every $(t,x) \in I \times S$:

(ii) $D^+V(t,x) \le -\psi(x)$:

(iii) $\psi(x) \ge 0$;

(iv) $D^+\psi(t,x) \ge -B$ (or $D^+\psi(t,x) \le B$);

if moreover

(i') for every compact set C, there exists an $A > 0$ such that $V(t,x) \ge -A$ on $I \times (C \cap \Omega)$:

then $\Lambda^+ \cap \Omega \subset E = \{x \in S: \psi(x) = 0\}$, for every solution $x(t)$ such that $\omega = \infty$ and $x(J^+) \subset S$.

Proof. If the solution in question tends to ∞ as $t \to \infty$, its limit set is empty and the thesis is trivially verified. Otherwise, the solution is such, owing to Theorem 4.6, that $\psi(x(t)) \to 0$ as $t \to \infty$. Since S is closed in Ω, E is closed in Ω. If there existed a point x^* of Λ^+ in $\Omega \setminus E$, its distance to E would be strictly positive. Therefore $\psi(x(t))$ couldn't approach 0 as $t \to \infty$. Q.E.D.

4.10. One deduces from this corollary that $\Lambda^+ \subset E \cup (\partial S \cap \partial\Omega)$. In particular, if $\overline{S} \subset \Omega$, then $\Lambda^+ \subset E$, which is the situation described in Theorem 1 (b) of LaSalle [1968]. Further, in this paper, the regularity conditions imposed on ψ are somewhat different from those adopted here. The thesis of Corollary 4.9 will appear insufficiently precise in many practical circumstances. The following corollary gives conclusions which are stronger in some sense.

4.11. Corollary. If one adds to Corollary 4.9 the hypothesis that for every $x \in S$: $\psi(x) \geq a(d(x,E))$ for some function $a \in \mathscr{K}$, then $x(t) \to E$ as $t \to \infty$ for every solution which does not tend to ∞ as $t \to \infty$.

4.12. Let us next examine the case of a solution which cannot be continued up to infinity, or in other words for which ω is finite. As will become apparent from the theorems to follow, we shall be obliged, in compensation, to impose some kind of bound on $f(t,x)$ and further, to content ourselves with weaker conclusions.

4.13. Theorem. Let S be a subset of Ω, closed with respect to Ω; let V and ψ be two functions on $I \times \Omega$ into $\mathscr{R}$, locally lipschitzian in x and continuous; if, for every compact set $C \subset S$, there exist three numbers A, B and $D > 0$ and a function $a \in \mathscr{K}$, such that for every $(t,x) \in I \times C$:

(i) $||f(t,x)|| \leq A$;

(ii) $\psi(t,x) \geq 0$;

(iii) $V(t,x) \geq -B$;

(iv) $D^+V(t,x) \leq -a(\psi(t,x))$;

(v) $D^+\psi(t,x) \geq -D$ (or $D^+\psi(t,x) \leq D$);

then

$$\psi^*(t) = \min\left\{\psi(t,x(t)),\ d(x(t),\partial\Omega),\ \frac{1}{1 + ||x(t)||}\right\} \to 0$$

as $t \to \omega$, for every solution $x(t)$ such that $x(J^+) \subset S$.

Proof. If $\min\left\{d(x(t),\partial\Omega),\ \frac{1}{1 + ||x(t)||}\right\} \to 0$ as $t \to \omega$, there is nothing left to prove. Otherwise, there exists an ε, $0 < \varepsilon < 1$, such that, for some sequence $\{t_n\}$ such that $t_n \to \omega$ as $n \to \infty$, $x(t_n)$ belongs to the compact set

$$\overline{[B(0,\tfrac{1}{\varepsilon} - 1) \setminus B(\partial\Omega,\varepsilon)]} \cap S.$$

One shows as in Theorem 4.6 that $V(t)$ is bounded from below. It remains to verify Hypotheses (iii) and (iv) of the fundamental Lemma 4.3. Now $\psi^*(t)$ is greater than or equal to some given ε, only if $x(t)$ belongs to some compact set as above. One knows that there exists for this set a function $a \in \mathscr{K}$ such that, for every (t,x) in the set: $D^+V(t,x) \leq -a(\psi(t,x))$, and a fortiori: $D^+V(t,x) \leq -a(\psi^*(t,x))$, where the meaning of ψ^* is obvious. At last, again on the same compact set, $f(t,x)$ is bounded. The same is true for $D^+(d(x(t),\partial\Omega))$ and $D^+\left[\frac{1}{1 + ||x(t)||}\right]$. Therefore $D^+\psi^*(t)$ is also bounded, either above or below. The fundamental lemma can thus be applied and the proof is complete. Q.E.D.

4.14. The following corollary parallels Corollary 4.9, in that it introduces a function ψ independent of t and thereby leads to a situation considered by J. P. LaSalle [1968].

4.15. <u>Corollary</u>. Let S be a subset of Ω, closed with respect to Ω; let V be a function on $I \times \Omega$ into $\mathscr{R}$, locally lipschitzian in x and continuous; let ψ be a continuous function on Ω into $\mathscr{R}$; if, for every compact set $C \subset S$, there exist two numbers A and $B > 0$ such that, for every $(t,x) \in I \times C$:

(i) $||f(t,x)|| \leq A$;

(ii) $V(t,x) \geq -B$;

(iii) $D^+V(t,x) \leq -\psi(x) \leq 0$;

then $\Lambda^+ \cap \Omega \subset E = \{x \in S: \psi(x) = 0\}$, for every solution x such that $x(J^+) \subset S$.

<u>Proof</u>. For every compact set $C \subset S$, there exists obviously a function $a \in \mathscr{K}$ such that, for every $x \in C$: $\psi(x) \geq a(d(x,E))$. If $d(x,E)$ is identified with the function $\psi(t,x)$ of Theorem 4.13, Hypothesis (iv) of this theorem is satisfied. On the other hand, since f is bounded on $I \times C$, Hypothesis (v) of this theorem is verified for the same choice of $\psi(t,x)$. Putting $\partial S \cap \partial\Omega = M$, one concludes that

$$\min\left[d(x(t),E),\ d(x(t),M),\ \frac{1}{1 + ||x(t)||}\right] \to 0$$

as $t \to \omega$, for every solution x such that $x(J^+) \subset S$. It follows that $\Lambda^+ \cap \Omega \subset E$, for otherwise there would exist a sequence $\{t_n\} \subset J^+$, $t_n \to \omega$, and a point $y \in S \setminus E$ such that $x(t_n) \to y$, and the lim inf of each of the following sequences:

$$\{d(x(t_n),E)\},\quad \{d(x(t_n),M)\},\quad \left\{\frac{1}{1 + ||x(t_n)||}\right\},$$

would be strictly positive, which is absurd. Q.E.D.

4.16. As in Section 4.10, it follows from the inclusion $\Lambda^+ \cap \Omega \subset E$, that $\Lambda^+ \subset E \cup M$, and if $\overline{S} \subset \Omega$, that $\Lambda^+ \subset E$. This is the situation described in Theorem 1 (a) of J. P. LaSalle [1968]. More precise forms of asymptotic behavior can be proved at the expense of reinforcing the boundedness hypothesis on $f(t,x)$. This is the object of the following theorem, which is presented as an exercise, because its proof is not very different from that of Theorem 4.13.

4.17. Exercise. Let S be any subset of Ω; let V and ψ be two functions on $I \times \Omega$ into $\mathscr{R}$, locally lipschitzian in x and continuous; if, for every $\rho > 0$, there exists three numbers $A, B, D > 0$ and a function $a \in \mathscr{K}$ such that, for every $(t,x) \in I \times [S \setminus B(M,\rho)]$:

(i) $||f(t,x)|| \leq A$;

(ii) $\psi(t,x) \geq 0$;

(iii) $V(t,x) \geq -B$;

(iv) $D^+V(t,x) \leq -a(\psi(t,x))$;

(v) $D^+\psi(t,x) \geq -D$ (or $D^+\psi(t,x) \leq D$);

then

$$\psi^*(t) = \min\{\psi(t,x(t)), d(x(t),M)\} \to 0$$

as $t \to \omega$, for every solution $x(t)$ such that $x(J^+) \subset S$.

The next two statements, again left as exercises, yield sufficient conditions for a solution to approach some set as $t \to \omega$.

4.18. Exercise. If one adds to the hypotheses of Theorem 4.17 that for every $(t,x) \in I \times \Omega$ and some set $E^* \subset \Omega$: $\psi(t,x) \geq d(x,E^*)$, then $x(t) \to E^* \cup M$ as $t \to \omega$.

4.19. Exercise. Let S be any subset of Ω; let us put $M = \partial S \cap \partial\Omega$; let $V: I \times \Omega \to \mathscr{R}$ be a function which is locally lipschitzian in x and continuous; if, for every $\rho > 0$, there exist two numbers A and $B > 0$ and a function $a \in \mathscr{K}$ such that, for every $(t,x) \in I \times [S \setminus B(M,\rho)]$:

(i) $||f(t,x)|| \leq A$;

(ii) $V(t,x) \geq -B$;

(iii) $D^+V(t,x) \leq -a(d(x,M))$;

then $x(t) \to M$ as $t \to \omega$ for every solution $x(t)$ such that $x(J^+) \subset S$.

5. The Invariance Principle for Asymptotically Autonomous and Related Equations

5.1. As has been shown in Chapter VII, a solution x of an autonomous differential equation approaches the largest invariant set contained in the set E where the derivative $\dot{V}(x)$ of the auxiliary function vanishes. No such property exists for non autonomous equations, because in this case, the limit sets are not invariant. But there are special classes of non autonomous equations, for instance periodic, asymptotically autonomous, almost periodic, etc. for which the limit sets possess some easily recognizable property which we shall term here, for convenience, "pseudo-invariance". In these cases, a conclusion of the following type is obtained: any solution x approaches the largest pseudo-invariant subset of E.

A good way to establish these pseudo-invariance results is to prove first a regularity theorem for the solutions of the differential equation. This theorem is also interesting in itself. We state and prove it hereafter for differential equations of the Carathéodory type, a kind of equation frequently encountered in control theory. On the fundamental theory of such equations, cf. E. A. Coddington and N. Levinson [1955].

5.2. General Hypotheses. Let Ψ be a domain of $\mathscr{R} \times \mathscr{R}^n$ and consider the space $\mathscr{F}$ of functions $f(t,x)$ on Ψ into $\mathscr{R}^n$, with the following properties:

(i) f is Lebesgue measurable in t for fixed x;

(ii) f is continuous in x for fixed t;

(iii) for any compact subset K of Ψ, if we put $a = \inf \{t: (t,x) \in K\}$ and $b = \sup \{t: (t,x) \in K\}$, there exists a real function $m_K(t)$ on $[a,b]$ such that $m_K(t)$ is bounded almost everywhere and Lebesgue measurable over $[a,b]$, or, for some $p \in]1,\infty[$, $m_K(t)^p$ is Lebesgue integrable over $[a,b]$, and further, for every $(t,x) \in K$: $||f(t,x)|| \leq m_K(t)$.

As is customary, we shall not distinguish between two functions of $\mathscr{F}$ when they are equivalent, i.e. when, for every fixed x, they differ on a subset of measure zero of the appropriate set of t values. Let us now introduce a topology on $\mathscr{F}$. Suppose f and g are functions of $\mathscr{F}$ and K is a compact subset of Ψ. If Z_K is the family of continuous functions $z(t)$ on some interval $J_z \subset \mathscr{R}$ into $\mathscr{R}^n$ and such that their graph is in K, we define

$$d_K(f,g) = \sup_{z \in Z_K} \left[\int_{J_z} ||f(\tau,z(\tau)) - g(\tau,z(\tau))||^p d\tau\right]^{1/p}.$$

As is apparent, d_K is a family of semi-distances on $\mathcal{F}$ and yields the desired topology. This topology is clearly Hausdorff and therefore metrisable. Let $d(f,g)$ be a distance associated with it. Obviously, if $\{f_i\} \subset \mathcal{F}$ is some sequence, stating that "$f_i \to g$ as $i \to \infty$" is but a short cut for "whatever the compact subset K of ψ, $d_K(f_i,g) \to 0$ as $i \to \infty$".

We shall consider, for some point $(t_0,x_0) \in \Psi$ and some function $f \in \mathcal{F}$, the Cauchy problem

$$\dot{x} = f(t,x), \quad x(t_0) = x_0. \tag{5.1}$$

Further, there will be a sequence of such problems, i.e.

$$\dot{x}_i = f_i(t,x_i) \quad x_i(t_{0i}) = x_{0i}, \quad i = 1, 2, \ldots \tag{5.2}$$

Conditions (i) to (iii) guarantee that, through (t_0,x_0) (resp. (t_{0i},x_{0i})), there passes at least one Carathéodory solution of the corresponding problem (5.1) (resp. (5.2)). All solutions mentioned below will be understood in the sense of Carathéodory.

In the lemma below, all trajectories will be confined to some compact cylindrical subset T of Ψ, defined as $T = J \times B$, where $J = [t_0 - l, t_0 + l]$ and $B = \{x \in \mathcal{R}^n: ||x - x_0|| \leq r\}$ for some quantities $l > 0$ and $r > 0$. As the only function $m_K(t)$ of Hypothesis (iii) to be mentioned explicitly will be associated with T and the function f, there will be no possible misunderstanding if we write it simply $m(t)$.

5.3. Lemma. In these general hypotheses, if $(t_{0i}, x_{0i}) \to (t_0, x_0)$ and $f_i \to f$ as $i \to \infty$ and if $\{x_i: J \to B\}$ is a sequence of solutions of the corresponding problems (5.2), then

(a) there exists a subsequence $\{x_{i(k)}: k = 1, 2, \ldots\}$ and a function $x: J \to B$ such that $x_{i(k)}(t) \to x(t)$ as $k \to \infty$, uniformly for $t \in J$;

(b) $x(t)$ is a solution of problem (5.1);

(c) if there exists no other solution of problem (5.1), $x_i(t) \to x(t)$ as $i \to \infty$, uniformly for $t \in J$.

Proof. (a) The x_i are uniformly bounded, since their trajectories are in T. Let us show that they are equicontinuous. If t_1, t_2 are any two points of J, one gets for every i, $p > 1$ and $q = p/(p-1)$, and using Hölder's inequality

$$||x_i(t_2) - x_i(t_1)|| \leq \left|\int_{t_1}^{t_2} ||f_i(\tau, x_i(\tau)) - f(\tau, x_i(\tau))|| d\tau\right|$$

$$+ \left|\int_{t_1}^{t_2} ||f(\tau, x_i(\tau))|| d\tau\right| \leq [d_T(f_i, f) + [\int_J m(\tau)^p d\tau]^{1/p}] |t_2 - t_1|^{1/q}.$$

If $p = 1$ and M is a bound on $m(t)$, one obtains as easily that

$$||x_i(t_2) - x_i(t_1)|| \leq d_T(f_i, f) + M|t_2 - t_1|.$$

The equicontinuity of the x_i follows in both cases, since $f_i \to f$. This result may not be immediately evident for $p = 1$. But observe that the equicontinuity obtains for the subfamily of the x_i corresponding to every i greater than some sufficiently large n. On the other hand, the equicontinuity is obvious for the finite family $\{x_1, \ldots, x_n\}$.

(b) We write henceforth $\{x_i\}$ instead of $\{x_{i(k)}\}$

for the uniformly convergent subsequence which we know to exist owing to the classical theorem of Arzelà-Ascoli. One gets easily that

$$x_i(t) = x_{i0} + \int_{t_{0i}}^{t} f(\tau,x(\tau))d\tau + \int_{t_{0i}}^{t} [f(\tau,x_i(\tau))-f(\tau,x(\tau))]d\tau$$

$$+ \int_{t_{0i}}^{t} [f_i(\tau,x_i(\tau)) - f(\tau,x_i(\tau))]d\tau. \quad (5.3)$$

But, for $p > 1$,

$$||\int_{t_{0i}}^{t} [f(\tau,x_i(\tau)) - f(\tau,x(\tau))]d\tau||$$

$$\leq [\int_J ||f(\tau,x_i(\tau)) - f(\tau,x(\tau))||^p d\tau]^{1/p}|t-t_{0i}|^{1/q},$$

and, by the dominated convergence theorem, the second member approaches zero as $i \to \infty$. Further

$$||\int_{t_{0i}}^{t} [f_i(\tau,x_i(\tau)) - f(\tau,x_i(\tau))d\tau|| \leq d_T(f_i,f)|t-t_{0i}|^{1/q},$$

and again, the right member approaches zero. Passing to the limit for $i \to \infty$ in both members of (5.3) proves that $x(t)$ is a solution of the Cauchy problem (5.1). The case $p = 1$ can be taken care of even more simply.

(c) This part of the thesis is obvious. Q.E.D.

With a view to extending the conclusions of this theorem in some way to non continuable solutions, we first prove the following lemma, for which the general hypotheses remain unchanged.

5.4. <u>Lemma</u>. Let Ψ_1, Ψ_2 be two bounded open sets of Ψ, with $\overline{\Psi}_1 \subset \Psi_2 \subset \overline{\Psi}_2 \subset \Psi$. There exist two quantities $l > 0$ and $r > 0$ such that, for every $(t_0,x_0) \in \Psi_1$,

(a) the compact cylinder of length $2l$, radius r and center (t_0, x_0) is contained in Ψ_2;

(b) if $f_i \to f$ as $i \to \infty$, for every sequence $\{(t_{0i}, x_{0i})\} \subset \Psi_1$ with $(t_{0i}, x_{0i}) \to (t_0, x_0)$ as $i \to \infty$, and for i sufficiently large, all solutions of the Cauchy problem (5.1) and (5.2) exist on the interval $[t_0 - l, t_0 + l]$ and have their trajectories contained in the cylinder mentioned under (a).

<u>Proof</u>. Suppose we use on Ψ the distance induced by the norm: $||(t,x)|| = |t| + ||x||$. One chooses $r > 0$ smaller than half the distance from Ψ_1 to the frontier of Ψ_2, and $l' > 0$ such that

$$\int_{t_0-l'}^{t_0+l'} m(\tau)d\tau < \frac{r}{4}$$

where $m(t)$ is associated here with $f(t,x)$ and $\overline{\Psi}_2$. Any compact cylinder of radius r, length $l = \min\{r, l'\}$ and center $(t_0, x_0) \in \Psi_1$ is contained in Ψ_2 and therefore thesis (a) is proved. On the other hand, thesis (b) is obvious for the solutions of problem (5.1). Supposing $p > 1$, consider a sequence (t_{0i}, x_{0i}) approaching some fixed $(t_0, x_0) \in \Psi_1$, and select an integer N such that, for every $i \geq N$,

$$d_{\overline{\Psi}_2}(f_i, f) < \frac{r}{4} \frac{1}{(2l)^{1/q}}$$

and $||x_{0i} - x_0|| < r/4$. Assume that for some $i \geq N$, a solution $x_i(t)$ of $\dot{x}_i = f_i(t, x_i), x_i(t_0) = x_{0i}$, doesn't exist on the whole interval $[t_0 - l, t_0 + l]$. But whenever $x_i(t)$ exists and is contained in Ψ_2

$$x_i(t) = x_{0i} + \int_{t_{0i}}^{t} [f_i(\tau, x_i(\tau)) - f(\tau, x_i(\tau))]d\tau + \int_{t_{0i}}^{t} f(\tau, x_i(\tau))d\tau.$$

One gets that

$$||x_i(t) - x_0|| \leq ||x_{0i} - x_0|| + d_{\overline{\Psi}_2}(f_i,f)(21)^{1/q} + \frac{r}{4}.$$

So, $x_i(t)$ doesn't approach the frontier of Ψ_2 and hence cannot cease to exist on $[t_0 - 1, t_0 + 1]$, which is absurd. That it remains within the afore-said cylinder follows from the above inequalities. The proof for $p = 1$ is left to the reader. Q.E.D.

5.5. Theorem. In the general hypotheses above, let $f_i \to f$ as $i \to \infty$; let $\{(t_{0i}, x_{0i})\} \subset \Psi$ be some sequence such that $(t_{0i}, x_{0i}) \to (t_0, x_0) \in \Psi$ as $i \to \infty$. For $i = 1, 2, \ldots$, let $x_i\colon]\alpha_i, \omega_i[\to \mathscr{R}^n$ be a non continuable solution of $\dot{x} = f_i(t,x)$, $x(t_{0i}) = x_{0i}$. Then there exist a non continuable solution $x\colon]\alpha,\omega[\to \mathscr{R}^n$ of problem (5.1) and an increasing subsequence $\{i(k)\colon k = 1, 2, \ldots\}$ such that, for every t_1, t_2 with $\alpha < t_1 < t_2 < \omega$, one gets for k large, that $\alpha_{i(k)} < t_1 < t_2 < \omega_{i(k)}$ and $x_{i(k)}(t) \to x(t)$ uniformly on $[t_1, t_2]$.

Proof. A similar theorem is proved in P. Hartman [1964], in a less general setting. The proof, which is almost the same in both cases (the only significant differences are in the lemmas), is outlined here for completeness. We show the existence of x on a right maximal interval only, but the reasoning runs alike for left and right. Let $\Psi_1, \Psi_2, \ldots$ be a sequence of bounded open sets such that $\overline{\Psi}_i \subset \Psi_{i+1}$ for every i and

$$\Psi = \bigcup_{1 \leq i < \infty} \Psi_i.$$

Suppose, without loss of generality, that $(t_0, x_0) \in \Psi_1$ and let $2l_1$ be the length of the compact cylinders associated with Ψ_1 in Lemma 1. Let (t_{0i},x_{0i}) in Ψ tend to (t_0,x_0). Lemma 1 shows that there exists a subsequence $x_{i(k)}(t)$, $k = 1, 2,\ldots,$ of solutions of the corresponding problems (5.2), such that $x_{i(k)}(t)$ tends to some $x(t)$ uniformly on $[t_0,t_0 + l_1]$. Either $(t_0 + l_1,x(t_0 + l_1))$ belongs to Ψ_1, or it does not. If it does, we start from this point as a new initial point to prove, by the same argument, the existence of a new subsequence, again written $x_{i(k)}(t)$, with the same convergence property, but this time on $[t_0,t_0 + 2l_1]$. Repeating this process proves either the existence of a subsequence $x_{i(k)}(t)$ converging to $x(t)$ over $[t_0,\infty[$ uniformly on every finite interval, or allows one to reach a point outside Ψ_1. But this point will be in Ψ_r for some $r > 1$, and we can repeat in Ψ_r what we have done in Ψ_1, choosing of course a new length l_r. The rest of the proof is obvious. Q.E.D.

5.6. Invariance properties of limit sets. The differential equation to be considered from now on, namely

$$\dot{x} = f(t,x) \tag{5.4}$$

will have its second member f defined on a set $\Psi = I \times \Omega$ where $I =]\tau,\infty[$ for some $\tau \in \mathscr{R}$ or $\tau = -\infty$, and Ω is an open subset of $\mathscr{R}^n$, while its range is still in $\mathscr{R}^n$. The reason for particularizing in this way the set Ψ of the previous section will become apparent when we introduce below the translates of f. Further, f is still supposed to verify Hypotheses (i) to (iii) above.

Let us recall that if $x:]\alpha,\omega[\to \mathscr{R}^n$ is a non continuable solution of Equation (5.4), a point x^* of $\bar{\Omega}$ is said to be a <u>positive limit point</u> of x if there exists a sequence $\{t_i\}$ of time-values such that $t_i \to \omega$ and $x(t_i) \to x^*$ as $i \to \infty$. The set of all positive limit points of x is called the <u>positive limit set</u> of x and is represented by $\Lambda^+(x)$. Amongst the properties of the limit sets, only their invariance will be studied here. In this context, we shall need the following proposition: if $\Lambda^+(x) \cap \Omega$ is not empty, then $\omega = \infty$.

The <u>translate</u> of the function f by a given amount $a > 0$ is the function defined thus

$$f_a: \Psi \to \mathscr{R}^n, \ (t,x) \to f_a(t,x) = f(t + a,x).$$

The following two hypotheses concerning f will be used successively below.

(A) There exists an $f^* \in \mathscr{F}$ such that $f_a \to f^*$ as $a \to \infty$.

(B) For every sequence $\{t_i\}$ such that $t_i \to \infty$ as $i \to \infty$, there is a subsequence $\{t_{i(k)}: k = 1, 2,\ldots\}$ and a function $f^* \in \mathscr{F}$ such that $f_{t_{i(k)}} \to f^*$ as $k \to \infty$.

Let us designate by $S(f,B)$ the set of all functions f^* obtainable in this way. Any such function f^* will be called a <u>limit function</u> and $\dot{x} = f^*(t,x)$ a <u>limit equation</u>.

Several simple remarks are appropriate here:

(1) (A) implies (B).

(2) $\mathscr{F}$ being a vector space for the usual addition of functions and product of a function by a scalar, the subset of functions of $\mathscr{F}$ possessing property (A) is a linear subspace of $\mathscr{F}$.

(3) The same is true for (B).

(4) If f possesses property (A), for any $a' > 0$, $f_{a+a'} \to f^*$ as $a \to \infty$, but also $f_{a+a'} \to f^*_{a'}$ as $a \to \infty$. Therefore f^* is actually a constant with respect to t and the limit equation is autonomous. In this sense, an equation satisfying property (A) is asymptotically autonomous.

(5) A rephrasing of (B) is as follows: for every sequence $\{t_i\}$ such that $t_i \to \infty$ as $i \to \infty$, the family $\{f_{t_i}\}$ of translates of f is relatively compact in $\mathscr{F}$.

(6) If f possesses property (B) and is continuous, and if one uses for f the topology of uniform compact convergence, then it is known that $f(t,x)$ is the sum of two functions $g(t,x)$ and $h(t,x)$ such that, for fixed x, g is almost periodic in the sense of Bohr and $h(t,x) \to 0$ as $t \to \infty$. One may then say that $f(t,x)$ is asymptotically almost periodic, since the limit function is almost periodic. Without the continuity assumption on f, the class of functions satisfying (B) becomes larger of course, and its extent might well deserve some further exploration.

A set $F \subset \Omega$ is said to be semi-invariant with respect to equation (5.4) whose second member is supposed to possess property (A), if, for every $(t_0,x_0) \in I \times F$, there is at least one non continuable solution $x^*:]\alpha,\omega[\to \mathscr{R}^n$ of the limit equation $\dot{x} = f^*(t,x)$ with $x^*(t_0) = x_0$, such that $x^*(t) \in F$ for every $t \in]\alpha,\omega[$.

A set $F \subset \Omega$ is said to be quasi-invariant with respect to equation (5.4) whose second member is supposed to possess property (B) if, for every $(t_0,x_0) \in I \times F$, there exists a function $f^* \in S(f,B)$ and at least one non continuable solution $x^*:]\alpha,\omega[\to \mathscr{R}^n$ of $\dot{x} = f^*(t,x)$ with $x^*(t_0) = x_0$ such that $x^*(t) \in F$ for every $t \in]\alpha,\omega[$.

As is well known (see e.g. P. Hartman [1964]), for any solution of an autonomous differential equation with continuous right member, $\Lambda^+(x) \cap \Omega$ is semi-invariant. If uniqueness of the solutions is assumed, this set is even invariant. Let us now prove two simple but important theorems, first of semi-invariance for equations whose second member satisfies Hypothesis (A), and then of quasi-invariance for the case of property (B).

5.7. <u>Theorem</u>. For every solution x of Equation (5.4) whose second member possesses property (A), $\Lambda^+(x) \cap \Omega$ is semi-invariant.

<u>Proof</u>. Let $x_0^* \in \Lambda^+(x) \cap \Omega$. As observed above, $\omega = \infty$. Let $\{t_i\}$ be such that $t_i \to \infty$ and $x(t_i) \to x_0^*$ as $i \to \infty$. Put $x(t_i) = x_{0i}$. If $f^*(t,x)$ is the limit function of f, then for any $t_0 \in I$ and starting with i large enough for $t_i - t_0$ to belong to I, one may write that:

$$d(f_{t_i - t_0}, f^*) \to 0 \quad \text{as} \quad i \to \infty.$$

Further, $x(t + t_i - t_0)$ is a solution of $\dot{x} = f_{t_i - t_0}(t,x)$, $x(t_0) = x_{0i}$. By Theorem 5.5, an appropriate subsequence of the $x(t + t_i - t_0)$ tends to some solution $x^*(t)$, with $x^*(t_0) = x_0^*$ of the equation $\dot{x} = f^*(t,x)$. But the limit of this subsequence belongs of course to $\Lambda^+(x) \cap \Omega$. Hence the thesis. Q.E.D.

5.8. Theorem. For every solution x of Equation (5.4) whose second member possesses property (B), $\Lambda^+(x) \cap \Omega$ is quasi-invariant.

The proof is a kind of obvious paraphrase of the preceding one.

5.9. Extensions of the invariance principle. Suppose now for simplicity that we come back to the general hypotheses described in Sections 1 and 4.1, which are a particular case of the Carathéodory conditions of the present section. Assume further that the hypotheses of Corollary 4.9 are satisfied. As has been observed already (see 4.10), if $\bar{S} \subset \Omega$, then $\Lambda^+ \subset E$. If one knows that Λ^+ is compact, and therefore that $x(t) \to \Lambda^+$ as $t \to \infty$, and that $f(t,x)$ satisfies property (A) (or (B)), it follows immediately from Theorem 5.7 (or Theorem 5.8) that $x(t)$ approaches the largest semi-invariant (or quasi-invariant) subset of E. A similar conclusion holds of course in correspondence with Corollary 4.15.

6. Dissipative Periodic Systems

6.1. Consider the Cauchy problem

$$\dot{x} = f(t,x), \tag{6.1}$$

$$x(0) = x_0, \tag{6.2}$$

where $f(t,x)$ is supposed to be defined and continuous on $\mathscr{R} \times \mathscr{R}^n$, sufficiently regular to ensure uniqueness of the solutions, and ω-periodic, i.e. $f(t+\omega,x) = f(t,x)$ for some $\omega > 0$ and any $(t,x) \in \mathscr{R} \times \mathscr{R}^n$.

6.2. Suppose that all solutions can be continued to $+\infty$. We define the translation operator

$$T\colon \mathscr{R}^n \to \mathscr{R}^n,\ x_0 \to Tx_0 = x(\omega;0,x_0)$$

and construct what is known as a semi-flow (see, for instance, N. P. Bhatia and O. Hajek [1969]), a function

$$\mathcal{T}: \mathcal{N}^+ \times \mathcal{R}^n \to \mathcal{R}^n, \quad (k, x_0) \to T^k x_0,$$

where $\mathcal{N}^+ = \{0, 1, 2, \ldots\}$, verifying the following properties:

(i) $T^0 x_0 = x_0$;

(ii) $T^k(T^l x_0) = T^{k+l} x_0$ for any k and l in $\mathcal{N}^+$;

(iii) $T^k x_0$ is continuous with respect to x_0, for any k in $\mathcal{N}^+$.

Further, we define the solution of the semi-flow T through $x_0 \in \Omega$ as

$$\tilde{x}: \mathcal{N}^+ \to \mathcal{R}^n, \quad k \to T^k x_0,$$

its positive orbit

$$\tilde{\gamma}^+(\tilde{x}) = \{x: (\exists k \geq 0)\ T^k x_0 = x\},$$

and the corresponding positive limit set

$$\tilde{\Lambda}^+(\tilde{x}) = \{x: (\exists \{k_i\})\ k_i \to \infty \text{ and } T^{k_i} x_0 \to x \text{ as } i \to \infty\}.$$

6.3. A set $S \subset \mathcal{R}^n$ will be said to be invariant if, for every $x_0 \in S$ and every $n' \in \mathcal{Z}$: $x(n'\omega; 0, x_0)$ is defined and $x(n'\omega; 0, x_0) \in S$ ($\mathcal{Z}$ is the set of all positive and negative integers). Paraphrasing what has been done for autonomous ordinary differential equations, one proves easily the following proposition.

Proposition. If $\tilde{\gamma}^+(\tilde{x})$ is bounded, $\tilde{\Lambda}^+(\tilde{x})$ is nonempty and compact, $\tilde{\Lambda}^+(\tilde{x}) \cap \Omega$ is invariant and

$$T^k x_0 \to \tilde{\Lambda}^+(\tilde{x}) \quad \text{as} \quad k \to \infty.$$

6.4. Similarly, one can transpose LaSalle's Theorem VII.3.2.

Theorem. Let S be a compact subset of $\mathscr{R}^n$ and $\tilde{x}$ a solution through x_0 such that $\tilde{\gamma}^+(\tilde{x}) \subset S$. Suppose $V: S \to \mathscr{R}$ is a continuous function such that, for every $k > 0$: $V(T^k x_0) \leq V(T^{k-1} x_0)$. If M is the largest invariant subset of

$$\tilde{E} = \{x \in S: Tx \in S \text{ and } V(Tx) = V(x)\},$$

then $\tilde{x}(k) \to M$ as $k \to \infty$.

The proof is a mere repetition of the proof of Theorem VII.3.2.

6.5. We are now ready to obtain a dissipativity condition for Equation (6.1).

Theorem. Suppose all solutions of (6.1) can be continued to infinity and there exists a positive constant R and a continuous function $V: \mathscr{R}^n \setminus B_R \to \mathscr{R}$ such that:

(i) $V(x) \geq \phi(||x||)$, with $\phi: [R,\infty[\to \mathscr{R}$ a continuous increasing function such that $\phi(r) \to \infty$ as $r \to \infty$;

(ii) for any x such that $x \notin B_R$ and $Tx \notin B_R$: $V(Tx) < V(x)$;

then the system is dissipative.

Proof. Due to the observation stated as Exercise VI.6.19, the only thing to prove is that, for each $x \in \mathscr{R}^n$, there exists a $t \geq 0$ such that $x(t;0,x_0) \in B_R$. If this were wrong, there would exist an $x_0 \in \mathscr{R}^n \setminus B_R$ such that for every $t \geq 0$: $x(t;0,x_0) \in \mathscr{R}^n \setminus B_R$. Then, from assumptions

(i) and (ii), the solution $\tilde{x}$ of the flow $\mathcal{T}$ through x_0 is such that

$$\tilde{\gamma}^+(\tilde{x}) \subset S = \{x \in \mathcal{R}^n \setminus B_R: V(x) \leq V(x_0)\}$$

and S is compact. Further, (ii) implies that the set

$$\tilde{E} = \{x \in S: Tx \in S \quad \text{and} \quad V(Tx) = V(x)\}$$

is empty. Hence a contradiction, by Theorem 6.4. Q.E.D.

Notice that V. A. Pliss [1964] proves that the conditions of Theorem 6.5 are also necessary. A condition such as (ii) is easy to verify using Dini derivatives.

6.6. Exercise. Replace, in Theorem 6.5, assumption (ii) by (ii-1) for any x such that $x \notin B_R$ and $Tx \notin B_R$: $V(Tx) \leq V(x)$; (ii-2) for any $x \notin B_R$, there exists a $k > 0$ such that either $T^k x \in B_R$ or $V(T^k x) < V(x)$.

6.7. Theorem. Suppose that, for some $R > 0$, there exists a function V: $\mathcal{R} \times (\mathcal{R}^n \setminus B_R)$, locally lipschitzian in x and continuous, such that:

(i) $V(t + \omega, x) = V(t,x)$;

(ii) $V(t,x) \geq \phi(||x||)$ with ϕ: $[R,\infty[\to \mathcal{R}$ a continuous increasing function such that $\phi(r) \to \infty$ as $r \to \infty$;

(iii) for each (t,x) in $\mathcal{R} \times (\mathcal{R}^n \setminus B_R)$, $D^+V(t,x) < 0$;

then the system is dissipative.

Proof. Again the only thing to prove is the existence of an $R' > 0$ such that for every $x_0 \in \mathcal{R}^n$ there is a $t \geq 0$ such that $x(t;0,x_0) \in B_{R'}$. It is clear from assumptions (ii) and (iii) that all solutions can be continued to ∞. Let us choose R' such that

$$\phi(R') > \sup\{V(t,x): t \in [0,\omega],\ ||x|| = R\}.$$

It follows then from assumptions (ii) and (iii) that for every $x \in B_R$ and $t \geq 0$; $x(t;0,x_0) \in B_{R'}$. Hence if $x \notin B_{R'}$ and $Tx \notin B_{R'}$, then $x(t;0,x) \notin B_R$ for every $t \in [0,\omega]$. Using (iii), we get that $V(0,Tx) = V(\omega,Tx) < V(0,x)$. All assumptions of Theorem 6.5 are satisfied with $V(0,x)$ as auxiliary function. This proves the theorem. Q.E.D.

6.8. Exercise. Replace assumption (iii) in Theorem 6.7 by: (iii-1) for each (t,x) in $\mathscr{R} \times (\mathscr{R}^n \setminus \overline{B}_R)$, $D^+V(t,x) \leq \psi(x) \leq 0$, where ψ is a locally lipschitzian function from $\mathscr{R}^n \setminus \overline{B}_R$ into $\mathscr{R}$; (iii-2) for each $x \in \mathscr{R}^n \setminus \overline{B}_R$ such that $\psi(x) = 0$, $D^+\psi(t,x) \neq 0$.

6.9. Exercise. Replace assumption (iii) in Theorem 6.7 by: (iii') for each (t,x) in $\mathscr{R} \times (\mathscr{R}^n \setminus \overline{B}_R)$, $D^+V(t,x) \leq 0$ and there exists a $t_1 \in [0,\omega]$ with $D^+V(t_1,x) < 0$.

6.10. Various examples of second and third order dissipative equations appear in V. A. Pliss [1964]. As a simple illustration, let us consider the class of electrical networks introduced in Section VII.6 and described by the equations

$$C\,\frac{dv}{dt} = f,$$

$$L\,\frac{di}{dt} = g,$$

with f and g defined in Section VII.6.3. Suppose, using again the notations of the same section, that

(α) R, L and C are constant, symmetric, positive definite matrices;

(β) $a = a(t)$ is ω-periodic for some $\omega > 0$.

<u>Theorem</u>. If $v^T \frac{\partial G}{\partial v} > 0$ for large enough $||v||$, the system is dissipative.

<u>Proof</u>. The only thing to prove is that for some $R > 0$ and every $x_0 \in \mathscr{R}^n$, there exists a $t \geq 0$ such that $x(t;0,x_0) \in B_R$. Consider the auxiliary function

$$V = \frac{1}{2} v^T C v + \frac{1}{2} i^T L i$$

whose derivative

$$\dot{V} = v^T f + i^T g = -v^T \frac{\partial G}{\partial v}(v) - i^T R i - i^T a$$

is negative for $||(v,i)||$ large enough. Theorem 6.7 applies. Q.E.D.

7. <u>Bibliographical Note</u>

The one-parameter families of auxiliary functions have been introduced by L. Salvadori [1969] and then used in several subsequent papers, for instance L. Salvadori [1971], P. Fergola and V. Moauro [1970], L. Gambardella and L. Salvadori [1971], L. Gambardella and C. Tenneriello [1971], A. D'Anna [1973]. Concerning this method, see also W. Hahn [1971]. Theorem 2.5 on weak attractivity which has been chosen to illustrate this method is adapted from a theorem of V. M. Matrosov $[1962]_1$ (cf. Section II.2 of the present book) on asymptotic stability. In his paper of [1971], L. Salvadori showed how the bound on the second member of the differential equation could be dispensed with: this result appears here in Corollary 2.9. The comparison of the two methods of proof of Theorem 2.6 is borrowed from J. L. Corne [1973]. The latest and most general paper on one-parameter families of auxiliary functions is L. Salvadori [1974].

The use of a vectorial auxiliary function meant to compensate for the fact that $\dot{V}(t,x)$ is not negative definite appears in N. Rouche [1968]. In connection with Section 3, one will remember that such a function is used to expel the solutions from some compact set. This result has been extended successively by N. Rouche [1971], M. Laloy $[1974]_2$ and N. Rouche [1974] and inspired Lemma 3.2.

Section 4 is adapted from J. L. Corne and N. Rouche [1973]. It has already been mentioned in the text that the important Corollaries 4.9 and 4.15 are due to J. P. LaSalle [1968], who however limits his study to solutions contained in some closed subset of Ω.

As for Section 5, two papers opened the way towards rather general regularity theorems and the problem of asymptotically autonomous equations: they are L. Markus [1956] and Z. Opial [1960]. The generalizations of the invariance principle for periodic equations appears in J. P. LaSalle [1962], for asymptotically autonomous equations in T. Yoshizawa [1963] (see also T. Yoshizawa [1966]) and for asymptotically autonomous equations in R. K. Miller [1965]. A regularity theorem concerning equations with continuous second members and the topology of uniform convergence will be found in P. Hartman [1964]. For a topology akin to the one adopted in the present text, but in connection with Volterra integral equations, and therefore encompassing ordinary differential equations, see R. K. Miller and G. R. Sell [1968], [1970]. Similar regularity theorems along with pseudo-invariance properties studied in the setting of dynamical systems associated with non autonomous differential equations appear in R. K. Miller and G. R. Sell [1970] and G. R. Sell [1971]. Theorems 5.7 and 5.8 as they appear here come from N. Rouche [1976].

Concerning the dissipative systems dealt with in Section 6, cf. V. A. Pliss [1964], which contains an extensive bibliography, as well as C. Corduneanu [1957], V. M. Gerstein [1969], V. V. Nemytskii [1965] and, at a more abstract level, J. E. Billotti and J. P. LaSalle [1971] and J. K. Hale, J. P. LaSalle and M. Slemrod [1972]. The comparison method has been applied to dissipative equations by V. M. Matrosov [1969], a paper which also gives an extensive bibliography, and N. Pavel [1971]. Various applications appear in A. de Castro [1953], B. Manfredi [1956] as well as in the book of J. M. Skowronski [1969] and in several papers of the same author. Cf. also S. Ziemba [1961].

CHAPTER IX

THE COMPARISON METHOD

1. Introduction

The role of the comparison method, as already studied in Section II.3, might be roughly characterized as follows: the solutions of the comparison equation push on before themselves those of the original equation. Following V. M. Matrosov [1973], this can be viewed as an interesting extension of the notion of a mathematical model. In the usual sense, a mathematical model operates approximately like the system it represents, whereas the solutions of the comparison equation fit approximately those of the original one, but while remaining "on the same side" at any time. One might speak of a one-sided model. The comparison method is sufficiently important in itself to justify the presence here of a complete chapter devoted to this subject. Further, it introduces to some typical applications.

In Section II.3, we considered a scalar comparison

equation. After some preliminaries on differential inequalities, we generalize here this theory to the case of vectorial equations and apply it thereafter to two practical problems: the stability of composite, inter-connected systems, and a problem of market stability in a Walrasian economy. A last section shows how the comparison method can be applied, besides stability and attractivity, to a wide variety of qualitative concepts.

2. Differential Inequalities

2.1. Let Ψ be an open subset of $\mathscr{R}^m$ for some integer m, and consider the differential equation

$$\dot{u} = F(t,u) \tag{2.1}$$

associated with a function $F: I \times \Psi \to \mathscr{R}^m$, $(t,u) \to F(t,u)$.

We shall need on $\mathscr{R}^m$ some kind of partial order defined in the following natural way: for u and v in $\mathscr{R}^m$, we write $u \geq v$ if $u_i \geq v_i$ for every $i = 1,\ldots,m$. Similarly, $u > v$ will mean that $u_i > v_i$ for $i = 1,\ldots,m$. The function F will be said to be quasi-monotone increasing if, for every pair of points (t,u) and (t,v) in $I \times \Psi$ and every $i = 1,\ldots,m$, one gets $F_i(t,u) \geq F_i(t,v)$ whenever $u_i = v_i$ and $u \geq v$.

2.2. Most of this section will be devoted to explaining and proving the following important property: if F is continuous and quasi-monotone increasing, all solutions of (2.1) issuing from an initial point $(t_0,u_0) \in I \times \Psi$ are bracketed by two particular solutions: a maximum and a minimum solution. A solution $u^+: [t_0,\omega[\to \mathscr{R}^m$ with initial conditions (t_0,u_0) is said to be a <u>(right) maximum solution</u> if every other solution $u: [t_0,\tilde{\omega}[\to \mathscr{R}^m$, passing through (t_0,u_0), is such that $u^+(t) \geq u(t)$ for every $t \in [t_0,\omega[\cap [t_0,\tilde{\omega}[$. Similarly, a solution $u^-(t)$ defined on $[t_0,\omega[$, with initial conditions (t_0,u_0) is called a <u>(right) minimum solution</u> if, for any other solution $u(t)$ issuing from $[t_0,u_0[$ and defined on some interval $[t_0,\tilde{\omega}[$, one has $u^-(t) \leq u(t)$ for every $t \in [t_0,\omega[\cap [t_0,\tilde{\omega}[$. To prove the existence of maximum and minimum solutions, we shall need the following lemmas.

2.3. <u>Lemma</u>. Let $G: I \times \Psi \to \mathscr{R}^m$ be a continuous quasi-monotone increasing function and $v: [t_0,\omega[\to \mathscr{R}^m$, for some ω finite or infinite, be a solution of $\dot{v} = G(t,v)$. If $w: [t_0,\omega[\to \Psi$ is a continuous,function such that

(i) $w(t_0) \leq v(t_0)$;

((ii) $D^+w(t_0) < G(t_0,w(t_0))$;

(iii) $D^-w(t) < G(t,w(t))$ for $t \in]t_0,\omega[$;

then $w(t) < v(t)$ for every $t \in]t_0,\omega[$.

<u>Proof</u>. a) Let us prove first that, for ε small enough, $w(t) < v(t)$ for $t \in]t_0,t_0 + \varepsilon[$. This is obvious, by reason of continuity, if $w(t_0) < v(t_0)$. Suppose then that for some i, $w_i(t_0) = v_i(t_0)$. One has

$$D^+w_i(t_0) < G_i(t_0,w(t_0)) \leq G_i(t_0,v(t_0)) = \frac{dv_i}{dt}(t_0).$$

Hence

$$D^+(w_i - v_i)(t_0) < 0 \quad \text{and} \quad w_i(t_0) = v_i(t_0),$$

which implies that, for ε small enough and $t \in]t_0,t_0 + \varepsilon[$,

$w_i(t) < v_i(t)$.

b) Assume now that there exist values of $t \in]t_0 + \varepsilon,\omega[$ such that $w(t) \nless v(t)$ and let τ be the infimum of these t. Notice by the way that, for a partial order, $\nless$ is not equivalent to $\geq$! Then $\tau \in [t_0 + \varepsilon,\omega[$, $w(t) < v(t)$ for $t \in]t_0,\tau[$ and $w(\tau) \nless v(\tau)$. For one i at least, one gets $w_i(\tau) = v_i(\tau)$. Therefore, as above

$$D^- w_i(\tau) < G_i(\tau,w(\tau)) \leq G_i(\tau,v(\tau)) = \frac{dv_i}{dt}(\tau),$$

and there exists an $\eta > 0$ such that for some $t \in [\tau - \eta,\tau[$: $w_i(t) > v_i(t)$, which contradicts the definition of τ.

Q.E.D.

2.4. Lemma. If we replace assumptions (i), (ii) and (iii) of Lemma 2.3 by

(i) $w(t_0) \geq v(t_0)$;

(ii) $D_+ w(t_0) > G(t_0,w(t_0))$;

(iii) $D_- w(t) > G(t,w(t))$ for $t \in]t_0,\omega[$;

then $w(t) > v(t)$ for $t \in]t_0,\omega[$.

The proof is similar to the one of Lemma 2.3. We are now ready to prove an existence theorem for maximum and minimum solutions.

2.5. Theorem. If F is a continuous quasi-monotone increasing function, there exist through every point $(t_0,u_0) \in I \times \Psi$ a unique non continuable maximum solution and a unique non continuable minimum solution.

Proof. Let us prove first the existence of the maximum solution. Consider a cylinder

$$T = \{(t,u) \in \mathscr{R}^{m+1}: \ |t - t_0| \leq a, \ \ ||u - u_0|| \leq b\},$$

where a and b are chosen such that

(i) $T \subset I \times \Psi$;

(ii) $b = Ma$ where $M > \sup_{(t,u) \in T} ||F(t,u)||$.

It is known that M can be chosen large enough in order that, for any integer $\nu \geq 1$, the Cauchy problem

$$\dot{v}_i = F_i(t,v) + \frac{1}{\nu} = G_i^{\nu}(t,v), \quad v_i(t_0) = u_{0i}, \quad i = 1,\ldots,m$$

has at least one solution $v^{(\nu)}: [t_0,t_0 + a] \to B[u_0,b]$. Further, any solution u of (2.1) through (t_0,u_0) is defined on $[t_0,t_0 + a]$ and

$$\dot{u}_i = F_i(t,u) < F_i(t,u) + \frac{1}{\nu} = G_i^{\nu}(t,u)$$

$$\dot{v}_i^{(\nu-1)} = F_i(t,v^{(\nu-1)}) + \frac{1}{\nu - 1} > G_i^{\nu}(t,v^{(\nu-1)})$$

for $t \in [t_0,t_0 + a]$ and $\nu \geq 2$. It follows from Lemmas 2.3 and 2.4 that

$$u(t) < v^{(\nu)}(t) < v^{(\nu-1)}(t) \quad \text{for} \quad t \in]t_0,t_0+a] \quad (2.2)$$

and the sequence of functions $v^{(\nu)}(t)$ converges pointwise to a function $u^+(t)$. But this convergence is uniform since the functions $v^{(\nu)}$ are equicontinuous on the compact interval $[t_0,t_0 + a]$ (cf. J. Dieudonné [1960, p. 136]). Further, u^+ is a solution of (2.1) as results from passing to the limit in the equation

$$v^{(\nu)}(t) = u_0 + \int_{t_0}^{t} [F(\sigma,v^{(\nu)}(\sigma)) + \frac{1}{\nu}]d\sigma.$$

It is a maximum solution, as follows from (2.2). By a standard argument, this solution can be continued to the right, until it approaches the boundary of Ψ (see e.g. N. Rouche and J. Mawhin [1973, Vol. I, p. 82]). At last, the uniqueness of the maximum solution is obvious. The existence of the minimum solution is proved in the same way. Q.E.D.

The following result is basic for the comparison m method.

2.6. Comparison Lemma. Let F be a continuous, quasi-monotone increasing function and $u^+ : [t_0,\omega[\to \mathscr{R}^m$ the maximum solution through some point $(t_0,u_0) \in I \times \Psi$. Assume v: $[t_0,\tilde{\omega}[\to \mathscr{R}^m$, $\tilde{\omega} \leq \omega$, is a continuous function such that $(t,v(t)) \in I \times \Psi$ and

(i) $v(t_0) \leq u_0$;

(ii) $Dv(t) \leq F(t,v(t))$ for $t \in]t_0,\tilde{\omega}[$,

where $Dv(t)$ is any derivative of $v(t)$. Then $v(t) \leq u^+(t)$ for any t in $[t_0,\tilde{\omega}[$.

Proof. a) Consider the function

$$V(t) = v(t) - \int_{t_0}^{t} F(\tau,v(\tau))d\tau.$$

It follows from (ii) that V is a decreasing function on $]t_0,\tilde{\omega}[$ and, by continuity, on $[t_0,\tilde{\omega}[$. Hence

$$D^+V(t_0) \leq 0 \quad \text{and} \quad D^-V(t) \leq 0 \quad \text{for} \quad t \in]t_0,\tilde{\omega}[.$$

Therefore, for any integer $\nu \geq 1$, one may write:

$$D^+v_i(t_0) < F_i(t_0,v(t_0)) + \frac{1}{\nu}, \quad i = 1,\ldots,m;$$

$$D^-v_i(t) < F_i(t,v(t)) + \frac{1}{\nu}, \quad t \in]t_0,\tilde{\omega}[, \quad i = 1,\ldots,m.$$

As in Theorem 2.5, let $v^{(\nu)} : [t_0,t_0 + a] \to \mathscr{R}^m$ be a solution of

$$\dot{v}_i^{(\nu)}(t) = F_i(t, v^{(\nu)}(t)) + \frac{1}{\nu} = G_i^\nu(t, v^{(\nu)}(t)), \; v^{(\nu)}(t_0) = u_0.$$

It was proved in Theorem 2.5 that for a small enough, the functions $v^{(\nu)}$ converge uniformly to the maximum solution u^+. On the other hand, it follows from Lemma 2.3 that

$$v(t) < v^{(\nu)}(t) \quad \text{for} \quad t \in]t_0, t_0 + a[.$$

By passing to the limit as $\nu \to \infty$, one gets that

$$v(t) \leq u^+(t) \quad \text{for} \quad t \in [t_0, t_0 + a]. \tag{2.3}$$

b) It remains to prove that (2.3) holds for any $t \in [t_0, \tilde{\omega}[$. If it were not true, let τ be the infimum of all $t \in]t_0 + a, \tilde{\omega}[$ such that $v(t) \leq u^+(t)$. By continuity $v(\tau) \leq u^+(\tau)$. The reasoning under a) with τ replacing t_0 proves that, for a' small enough and $t \in]\tau, \tau + a'[$: $v(t) \leq u^+(t)$, which contradicts the definition of τ. Q.E.D.

The same type of proof yields the following result.

2.7. <u>Lemma</u>. Let F be a continuous, quasi-monotone increasing function and $u^-: [t_0, \omega[\to \mathscr{R}^m$ the minimum solution through some point $(t_0, u_0) \in I \times \Psi$. Assume $v: [t_0, \tilde{\omega}[\to \mathscr{R}^m$, $\tilde{\omega} \leq \omega$, is a continuous function such that $(t, v(t)) \in I \times \Psi$ and

(i) $v(t_0) \geq u_0$;

(ii) $Dv(t) \geq F(t, v(t))$ for $t \in]t_0, \tilde{\omega}[$,

where $Dv(t)$ is any Dini derivative of $v(t)$. Then $v(t) \geq u^-(t)$ for any $t \in [t_0, \tilde{\omega}[$.

3. A Vectorial Comparison Equation in Stability Theory

3.1. The general idea of this section is to generalize the comparison method as introduced in II.3 to the use of a vector comparison equation like (2.1). More precisely, consider the equation

$$\dot{x} = f(t,x) \tag{3.1}$$

along with the general hypotheses of Section I.2.2 and a vector comparison equation

$$\dot{u} = F(t,u) \tag{3.2}$$

with a critical point at the origin $u = 0$. Assume further that $F: I \times \Psi \to \mathscr{R}^m$ is continuous and quasi-monotone increasing. Given some initial conditions $(t_0,u_0) \in I \times \Psi$, we shall write $u^+: K = [t_0,\omega[\to \mathscr{R}^m$, $t \to u^+(t)$ for the maximum solution of (3.2) through (t_0,u_0). Similarly, $u^-: L \to \mathscr{R}^m$, $t \to u^-(t)$ represents the minimum solution through (t_0,u_0). Let $e \in \mathscr{R}^m$ be the vector $e = (1, 1, \ldots, 1)$ and suppose $\Psi \supset \{u: 0 \leq u \leq e\}$.

3.2. As already noticed in II.3.2, the only solutions u of (3.2) to be considered here are such that $u(t) \geq 0$. Therefore the relevant stability concepts for (3.2) are the following straightforward modifications of our previous ones. The origin $u = 0$ will be called <u>stable</u> if

$$(\forall\varepsilon > 0)(\forall t_0 \in I)(\exists\delta > 0)(\forall u_0: 0 \leq u_0 \leq \delta e)(\forall t \geq t_0, t \in K)$$
$$u^+(t) < \varepsilon e,$$

and <u>uniformly stable</u> if

$$(\forall\varepsilon > 0)(\exists\delta > 0)(\forall t_0 \in I)(\forall u_0: 0 \leq u_0 \leq \delta e)(\forall t \geq t_0, t \in K)$$
$$u^+(t) < \varepsilon e.$$

Provided $\varepsilon \leq 1$, it is clear that for the maximum solutions mentioned in these definitions, $K = [t_0,\infty[$.

3.3. Similarly, one defines uniform attractivity as

$(\exists\delta > 0)(\forall\varepsilon > 0)(\exists\sigma > 0)(\forall t_0 \in I)(\forall u_0\colon 0 \leq u_0 \leq \delta e)\ t_0 + \sigma \in K$ and $(\forall t \geq t_0 + \sigma, t \in K)\ u^+(t) < \varepsilon e$.

In Chapter VI, we studied six variants of this definition. Here, due to the quasi-monotonicity of F, only three of them are distinct, as will be shown now.

Proposition. In the language of Chapter VI, let W_1 be a word obtained by combining some of the expressions $(\exists\delta > 0)$, $(\forall\varepsilon > 0)$ and $(\forall t_0 \in I)$, and let W_2 be either void or identical with $(\forall t_0 \in I)$. Then for Equation (3.2), the concepts

$$p\colon W_1(\exists\sigma > 0)(\forall u_0\colon 0 \leq u_0 \leq \delta e)\ W_2(t_0 + \sigma \in K) \text{ and } (\forall t \geq t_0 + \sigma, t \in K)\ u^+(t) < \varepsilon e,$$

$$q\colon W_1(\forall u_0\colon 0 \leq u_0 \leq \delta e)(\exists\sigma > 0)\ W_2(t_0 + \sigma \in K) \text{ and } (\forall t \geq t_0 + \sigma, t \in K)\ u^+(t) < \varepsilon e,$$

are equivalent.

Proof. Obviously $p \Longrightarrow q$. Suppose now q is true. There is no loss of generality to assume that $\delta \leq 1$, $\varepsilon \leq 1$, in which case $K(t_0,\delta e) = [t_0,\infty[$. Let us choose σ in p corresponding to $u_0 = \delta e$ in q. From the Comparison Lemma 2.6 and for any $u_0\colon 0 \leq u_0 \leq e$, one has

$$0 \leq u^+(t;t_0,u_0) \leq u^+(t;t_0,\delta e)$$

which implies that $K(t_0,u_0) = K(t_0,\delta e) = [t_0,\infty[$ and proposition p. Q.E.D.

Among the three distinct attractivity concepts for (3.2), equi-attractivity turns out to be particularly useful. Its definition reads:

$$(\forall t_0 \in I)(\exists \delta > 0)(\forall \varepsilon > 0)(\exists \sigma > 0)(\forall u_0\colon 0 \le u_0 \le \delta e)\ t_0 + \sigma \in K$$

and

$$(\forall t \ge t_0 + \sigma, t \in K)\ u^+(t) < \varepsilon e.$$

The proposition above implies that it is equivalent to attractivity for Equation (3.2).

3.4. The instability concept relevant for Equation (3.2) is:

$$(\exists \varepsilon > 0)(\exists t_0 \in I)(\forall \delta > 0)(\exists u_0\colon 0 \le u_0 \le \delta e)(\exists t \ge t_0, t \in L)\ u^-(t) \nless \varepsilon e.$$

3.5. Theorem (V. M. Matrosov [1962]$_2$). Suppose there exists a function F as described in 3.1 and a continuous function $V\colon I \times \Omega \to \mathscr{R}^m$ which is locally lipschitzian in x and such that, for some function $a \in \mathscr{K}$ and every $(t,x) \in I \times \Omega$:

(i) $\max\limits_i V_i(t,x) \ge a(||x||)$; $V(t,0) = 0$;

(ii) $D^+V(t,x) \le F(t,V(t,x))$;

then

(a) stability of $u = 0$ implies stability of $x = 0$;

(b) equi-attractivity of $u = 0$ implies equi-attractivity of $x = 0$;

if moreover, for some function $b \in \mathscr{K}$ and every $(t,x) \in I \times \Omega$

(iii) $\max\limits_i V_i(t,x) \le b(||x||)$,

then

(c) uniform stability of $u = 0$ implies uniform stability of $x = 0$;

(d) uniform attractivity of $u = 0$ implies uniform attractivity of $x = 0$.

Proof. Because of (i) and (ii) and the Comparison Lemma 2.6, we have, for any $(t_0,x_0) \in I \times \Omega$ and any $t \geq t_0$, $t \in J(t_0,x_0)$ where $u^+(t;t_0,V(t_0,x_0))$ is defined, that

$$a(||x||) \leq \max_i V_i(t,x) \leq \max_i u_i^+(t;t_0,V(t_0,x_0)). \tag{3.3}$$

The origin being stable for (3.2), there exists a $\delta^*(t_0,\varepsilon) > 0$ such that, for arbitrary u_0: $0 \leq u_0 \leq \delta^* e$ and $t \geq t_0$: $u^+(t;t_0,u_0) < a(\varepsilon)e$. Notice in passing that $u^+(t;t_0,u_0)$ is defined over $[t_0,\infty[$. By continuity, there exists a $\delta(t_0,\varepsilon)$ such that $||x_0|| < \delta$ implies $u_0 = V(t_0,x_0) \leq \delta^* e$. But it follows then from (3.3) that $||x(t;t_0,x_0)|| < \varepsilon$ for any $t \geq t_0$. Thus $x = 0$ is stable, and Thesis (a) is proved. This proof was a kind of repetition of the corresponding proof in Theorem II.3.4. The same is true for (b), (c) and (d). Q.E.D.

3.6. Example. Let us try, for the system of equations,

$$\dot{x} = e^{-t}x + y \sin t - (x^3 + xy^2)\sin^2 t,$$
$$\dot{y} = x \sin t + e^{-t}y - (x^2y + y^3)\sin^2 t,$$

to use the auxiliary function

$$V = Ax^2 + 2Bxy + Cy^2, \qquad (A > 0,\ B^2 - AC < 0)$$

in connection with Theorem I.4.2. One computes

$$\dot{V} = 2(e^{-t} - (x^2 + y^2)\sin^2 t)V + 2(Bx^2 + (A+C)xy + By^2)\sin t,$$

an expression which is positive for any $t \in \{k\pi,\ k$ an integer$\}$. Hence Theorem I.4.2 cannot be applied. One can however use the quadratic vector function

$$V = ((x + y)^2, (x - y)^2)^T$$

which is such that

$$\dot{V} \leq 2 \begin{pmatrix} e^{-t} + \sin t & 0 \\ 0 & e^{-t} - \sin t \end{pmatrix} V,$$

whereas, for each of the equations

$$\dot{u} = 2(e^{-t} \pm \sin t)u,$$

the origin is uniformly stable, as can be shown by direct integration:

$$|u| = |u_0| \exp \{2[e^{-t_0} - e^{-t} \pm \cos t_0 \mp \cos t]\}$$

$$\leq |u_0| e^4 \exp (2e^{-t_0}).$$

3.7. Exercise (N. P. Bhatia and V. Lakshmikantham [1965]). Suppose there exist a function F as described in 3.1, a continuous function $V: I \times \Omega \to \mathscr{R}^m$ which is locally lipschitzian in x, and a continuous function $k: I \to \mathscr{R}$ such that, for some function $a \in \mathscr{K}$ and every $(t,x) \in I \times \Omega$:

(i) $\max_i V_i(t,x) \geq a(||x||)$, $V(t,0) = 0$;

(ii) $D^+(k(t)V(t,x)) \lesssim F(t,k(t)V(t,x))$;

(iii) $k(t) > 0$, $k(t) \to \infty$ as $t \to \infty$;

then stability of $u = 0$ implies equi-asymptotic stability of $x = 0$.

Notice that assumptions (i) and (ii) can be replaced by

(i') $\max_i V_i(t,x) \geq k(t)a(||x||)$, $V(t,0) = 0$;

(ii') $D^+V(t,x) \leq F(t,V(t,x))$.

These results should be compared with Theorem I.6.31 and Exercise II.3.9.

3.8. Exercise. With assumptions (i), (ii) and (iii) of Theorem 3.5 and if $\Psi \supset \{u: u \geq 0\}$, uniform global attractivity of $u = 0$, i.e.

$$(\forall\delta > 0)(\forall\varepsilon > 0)(\exists\sigma > 0)(\forall t_0 \in I)(\forall u_0: 0 \leq u_0 \leq \delta e)\ t_0+\sigma \in K$$

and

$$(\forall t \geq t_0 + \sigma, t \in K)\ u^+(t) < \varepsilon e$$

implies uniform global attractivity of $x = 0$.

3.9. Theorem. Suppose there exist a function F as described in 3.1 and a continuous function $V: I \times \Omega \to \mathscr{R}^m$ which is locally lipschitzian in x and such that, for some function $a \in \mathscr{K}$ and every $(t,x) \in I \times \Omega$:

(i) $\max_i V_i(t,x) \leq a(||x||)$;

(ii) for any $\delta > 0$ and $t_0 \in I$, there exists an $x_0 \in B(0,\delta_0)$ such that $V(t_0,x_0) > 0$;

(iii) $D^+V(t,x) \geq F(t,V(t,x))$;

then instability of $u = 0$ implies instability of $x = 0$.

Proof. The origin being unstable for (3.2), there exist $\varepsilon^* > 0$ and $t_0 \in I$ such that $(\forall\delta^* > 0)(\exists u_0: 0 \leq u \leq \delta^* e)$ $(\exists t \geq t_0,\ t \in L)\ u^-(t) \nleq \varepsilon^* e$. Let ε be such that $a(\varepsilon) \leq \varepsilon^*$.

For a given $\delta > 0$, let us choose $\delta^* > 0$ such that

$$(\forall u_0: 0 \leq u_0 \leq \delta^* e)(\exists x_0: ||x_0|| < \delta)\ u_0 \leq V(t_0,x_0). \quad (3.4)$$

This is possible, because of (ii). Let us now fix u_0: $0 \leq u_0 \leq \delta^* e$ and $t \geq t_0$, $t \in L$ such that $u^-(t) \nleq \varepsilon^* e$. If x_0 is chosen according to (3.4) and $t \notin J(t_0,x_0)$, the theorem is proved, since a solution cannot cease to exist without leaving $B(0,\varepsilon)$. If $t \in J(t_0,x_0)$, we can apply Lemma 2.7 and obtain

$$a(||x(t)||) \geq \max_i V_i(t,x(t)) \geq \max_i u_i^-(t;t_0,u_0) \geq a(\varepsilon).$$

Hence $||x(t)|| \geq \varepsilon$. Q.E.D.

There exist several variants of this theorem. Let us give two of them.

3.10. <u>Theorem</u> (V. M. Matrosov [1962]$_2$). Suppose there exist a function F as described in 3.1 and a continuous function $V: I \times \Omega \to \mathscr{R}^m$, which is locally lipschitzian in x and such that, for some constant M and every $(t,x) \in I \times \Omega$:

(i) $V_1(t,x) < M$;

(ii) $D^+V(t,x) \geq F(t,V(t,x))$;

then, the following "instability" concept

$$(\exists t_0 \in I)(\forall \delta > 0)(\exists x_0 \in B(0,\delta))(\exists t \geq t_0, t \in L)$$
$$u_1^-(t;t_0,V(t_0,x_0)) \geq M$$

implies instability of $x = 0$.

3.11. Exercise (V. M. Matrosov $[1962]_2$). Suppose there exists a $\mathscr{C}^2$ function $V: I \times \Omega \to \mathscr{R}$ such that, for every $(t,x) \in I \times \Omega$:

(i) $V(t,x)$ is bounded from above;

(ii) $\ddot{V} \geq aV + 2b\dot{V}$, where $a \geq 0$, and $b \geq 0$ if $a = 0$;

(iii) for any $t \in I$, the function

$$[\sqrt{b^2 + a} - b]\, V(t,x) + \dot{V}(t,x)$$

takes positive values in any neighborhood of $x = 0$;

then the critical point $x = 0$ is unstable.

A wealth of related results appear in V. M. Matrosov $[1962]_2$.

4. Stability of Composite Systems

4.1. The difficulty of studying the stability properties of a system increases usually with its dimension. One way to get over the difficulty of building a Liapunov function for a large system might be described as follows: one considers the system as constituted by the interconnection into a complex whole of several smaller subsystems, each of which is simpler to study. The complete system will be called composite.

4.2. More precisely, suppose the basic building block of our system is a so-called transfer system, described by the equations

$$\begin{aligned} \dot{x}_i &= f_i(t,x_i) + D_i u_i, \\ y_i &= H_i x_i, \end{aligned} \qquad i = 1, \ldots, m,$$

where $x_i \in \mathscr{R}^{n_i}$ is the <u>state vector</u>, $u_i \in \mathscr{R}^{p_i}$ is the <u>input</u> and $y_i \in \mathscr{R}^{q_i}$ the <u>output</u>. We assume that the $f_i\colon I \times \mathscr{R}^{n_i} \to \mathscr{R}^{n_i}$ satisfy the general hypotheses I.2.2 and that the matrices D_i and H_i, being rectangular with the appropriate number of rows and columns, are constant. An example of such a transfer system is the tansistor model given in II.6.5, where $x_i = y_i = (v_1, v_2)$ and $u_i = (i_1, i_2)$.

The m transfer systems are interrelated by the following equations:

$$u_i = \sum_{1 \le j \le m} B_{ij} y_i, \qquad i = 1,\ldots,m,$$

where the B_{ij} are appropriate constant matrices. Then the system as a whole is described by the equations

$$\dot{x}_i = f_i(t,x_i) + \sum_{1 \le j \le m} D_i B_{ij} H_j x_j, \; i = 1,\ldots,m. \tag{4.1}$$

4.3. <u>Theorem</u> (F. N. Bailey [1965]). Suppose $B_{ii} = 0$ for $i = 1,\ldots,m$ in System (4.1). Suppose for $i = 1,\ldots,m$, there exist $\mathscr{C}^1$-functions $V_i\colon I \times \mathscr{R}^{n_i} \to \mathscr{R}$ and positive constants c_{1i}, c_{2i}, c_{3i} and c_{4i} such that:

(i) $c_{1i}||x_i||^2 \le V_i(t,x_i) \le c_{2i}||x_i||^2$;

(ii) $\dfrac{\partial V_i}{\partial t} + \dfrac{\partial V_i}{\partial x_i}(t,x_i)f(t,x_i) \le -c_{3i}||x_i||^2$;

(iii) $||\dfrac{\partial V_i}{\partial x_i}(t,x_i)|| \le c_{4i}||x_i||$;

assume at last that the eigenvalues of the matrix A whose elements are defined by

$$a_{ii} = - c_{3i} / 2c_{2i},$$

$$a_{ij} = c_{4i}^2 \sum_{1 \le k \le m} ||D_i B_{ik} H_k||^2 / 2c_{3i}\, c_{1j}, \quad i \neq j,$$

have strictly negative real parts;

then the critical point $x = 0$ of System (4.1) is uniformly, globally, asymptotically stable.

Proof. Considering the comparison equation

$$\dot{u} = Au \tag{4.2}$$

the only thing one has to verify, owing to Theorem 3.5 and Exercise 3.8 is that

$$\begin{bmatrix} \dot{V}_1 \\ \cdot \\ \cdot \\ \cdot \\ \dot{V}_m \end{bmatrix} \le A \begin{bmatrix} V_1 \\ \cdot \\ \cdot \\ \cdot \\ V_m \end{bmatrix}.$$

But one computes easily that

$$\dot{V}_i = \frac{\partial V_i}{\partial t} + \frac{\partial V_i}{\partial x_i}(t,x_i) f_i(t,x_i) + \frac{\partial V_i}{\partial x_i}(t,x_i) \sum_{1 \le j \le m} D_i B_{ij} H_j x_j$$

$$\le -c_{3i}||x_i||^2 + c_{4i}\Big(\sum_{1 \le j \le m} ||D_i B_{ij} H_j||\ \ ||x_j||\Big)||x_i||$$

$$\le -\frac{c_{3i}}{2}||x_i||^2 + \frac{c_{4i}^2}{2c_{3i}}\Big(\sum_{\substack{1 \le j \le m \\ i \neq j}} ||D_i B_{ij} H_j||\ ||x_j||\Big)^2$$

$$\le -\frac{c_{3i}}{2}||x_i||^2 + \frac{c_{4i}^2}{2c_{3i}} \sum_{1 \le k \le m} ||D_i B_{ik} H_k||^2 \sum_{\substack{1 \le j \le m \\ j \neq i}} ||x_j||^2$$

$$\le \sum_{1 \le j \le m} a_{ij} V_j .$$

Q.E.D.

4.4. Exercise. In Theorem 4.3, one may also define A by

$$a_{ii} = - c_{3i} / 2c_{2i}$$

$$a_{ij} = (m-1)c_{4i}^2 ||D_i B_{ij} H_j||^2 / 2c_{3i}c_{1j}, \quad i \neq j.$$

4.5. A simple example of a composite system is a closed loop, as shown on Fig. 9.1. The input of every subsystem is the output of the preceding one. In that case, $p = q_{i-1}$ and the matrix B built up from the submatrices B_{ij} is

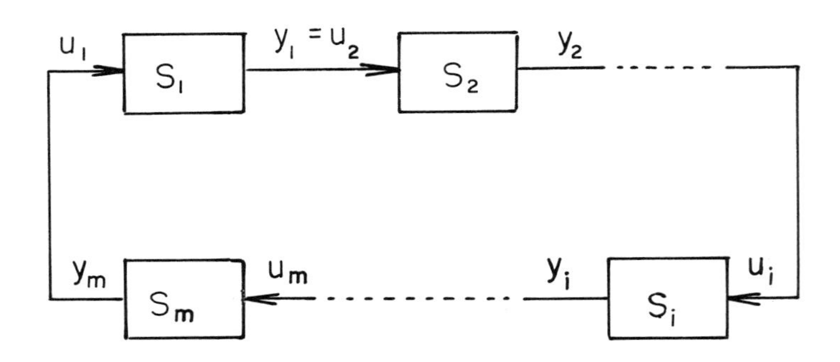

Fig. 9.1. Closed loop system.

$$B = \begin{pmatrix} 0 & \cdots\cdots & 0 & I_{p_1} \\ I_{p_2} & & 0 & 0 \\ 0 & & \vdots & \vdots \\ \vdots & & \vdots & \vdots \\ \vdots & & \vdots & \vdots \\ 0 & \cdots\cdots 0 & I_{p_m} & 0 \end{pmatrix}$$

where I_{p_i} is a $p_i \times p_i$ unit matrix. To investigate the stability of such a system, we shall need the following lemma.

4.6. Lemma. The solution $u = 0$ of the differential

system

$$\begin{aligned} \dot{u}_1 &= -a_1u_1 + b_1u_m, \\ \dot{u}_i &= -a_iu_i + b_iu_{i-1}, \quad i = 2, \ldots, m, \end{aligned} \tag{4.3}$$

with $a_i > 0$ and $b_i > 0$ is globally asymptotically stable if

$$\Pi \frac{a_i}{b_i} > 1. \tag{4.4}$$

Proof. The characteristic equation for (4.3) is

$$\prod_{1 \leq i \leq m} (\lambda + a_i) - \prod_{1 \leq i \leq m} b_i = 0.$$

But for $\operatorname{Re} \lambda \geq 0$,

$$\prod_{1 \leq i \leq m} b_i = \left| \prod_{1 \leq i \leq m} (\lambda + a_i) \right| = \prod_{1 \leq i \leq m} |\lambda + a_i|$$

$$\geq \prod_{1 \leq i \leq m} (\operatorname{Re} \lambda + a_i) \geq \prod_{1 \leq i \leq m} a_i,$$

which contradicts (4.4). Hence $\operatorname{Re} \lambda < 0$. Q.E.D.

4.7. Theorem. Consider System (4.1) with B defined as in 4.5. Suppose there exist m functions V_i and $4m$ positive constants c_{1i}, c_{2i}, c_{3i} and c_{4i} satisfying assumptions (i), (ii) and (iii) of Theorem 4.3. If the gain estimates

$$\eta_i = \frac{c_{4i}}{c_{3i}} \left(\frac{c_{2i}}{c_{1i}} \right)^{1/2} ||D_i|| \;\; ||H_i||$$

of the subsystems are such that

$$\prod_{1 \leq i \leq m} \eta_i^2 < 1,$$

then the zero solution of the composite system (4.1) is uniformly globally asymptotically stable.

Proof. The comparison equation (4.2) is of the form (4.3), with

$$a_i = c_{3i} / 2c_{2i}, \qquad i = 1, \ldots, m,$$

$$b_1 = c_{4i}^2 \|D_1 H_m\|^2 / 2c_{3i} c_{1m};$$

$$b_i = c_{4i}^2 \|D_i H_{i-1}\|^2 / 2c_{3i} c_{1,i-1}, \quad i = 2, \ldots, m.$$

But owing to Lemma 4.6, $u = 0$ is globally asymptotically stable for (4.2) since

$$\prod_{1 \le i \le m} \frac{a_i}{b_i} = \prod_{1 \le i \le m} \frac{c_{3i}^2}{c_{4i}^2} \frac{c_{1i}}{c_{2i}} \frac{1}{\|D_i H_{i-1}\|^2} \ge \prod_{1 \le i \le m} \frac{1}{\eta_i^2} > 1,$$

where H_0 has to be interpreted as another symbol for H_m.

Q.E.D.

5. An Example from Economics

We show in this section how the comparison method can be used to prove that, under some suitable conditions, a market tends to some given evolution independent of initial conditions. In the Walrasian approach to price evolution, the price of any commodity, be it services or goods, is supposed to increase when demand exceeds supply and to decrease otherwise. On the other hand, the supply is an increasing function of the price, whereas the opposite occurs for the demand (see e.g. J. R. Hicks [1939] or P. A. Samuelson [1947]). The model discussed below comes, after simplification, from D. D. Siljak [1973].

Suppose we divide the market into n groups of commodities, the ith group being composed of k_i items. The subscripts i and j, running from i to n, will designate the groups, while l running from 1 to k_i will label the commodities in one group. For instance p_{il} will be the price of the lth item of the ith group, p_i a column vector formed by the prices in i^{th} group, and p a column vector formed from the p_i's. On the other hand, the capitals D and S with proper subscripts will designate demand and supply respectively, whereas $g = D - S$ is called the <u>excess demand</u>.

The equations for the prices in the Walrasian approach are

$$\dot{p}_{il} = h_{il}[t,(D_{il}(t,p) - S_{il}(t,p))]$$

where $h_{il}(t,0) = 0$ and

$$\frac{\partial h_{il}}{\partial x}(t,x) > 0, \quad \frac{\partial D_{il}}{\partial p_{il}} \leq 0, \quad \frac{\partial S_{il}}{\partial p_{il}} > 0.$$

All the functions introduced up to here are defined on some appropriate domain (for instance most of them are positive) which we shall not need to specify any further. Anyhow, the equations of the problem may be written under the general form

$$\dot{p}_{il}(t,p) = g_{il}(t,p) \quad \text{with} \quad \frac{\partial g_{il}}{\partial p_{il}} < 0. \tag{5.1}$$

It is natural to ask under what conditions all solutions of (5.1) approach some particular solution p_0. Writing $P = p - p_0$, we get equations of the form

$$\dot{P} = g(t,p) - g(t,p_0(t)) = f(t,P) \tag{5.2}$$

with $\quad f(t,0) = 0, \ \dfrac{\partial f_{il}}{\partial P_{il}}(t,P) = \dfrac{\partial g_{il}}{\partial p_{il}}(t,p_0 + P) < 0.$

These conditions are consistent with the following further

hypotheses. Assume there exist constants α_i and β_{ij} such that, for every $i, j = 1, \ldots, n$,

(i) $0 \leq \beta_{ij}$;

(ii) $0 \leq \beta_{ii} < \alpha_i$;

(iii) $f_{il}(t,P) = a_{il}(t,P_i) + b_{il}(t,P)$;

(iv) $\sum_l a_{il}(t,P_i)P_{il} \equiv P_i^T a_i \leq -\alpha_i P_i^T P_i$;

(v) $\sum_l P_{il} b_{il}(t,P) \equiv P_i^T b_i \leq \sum_{1 \leq j \leq n} \beta_{ij} ||P_i||\ ||P_j||$

where $||P_i|| = \sqrt{\sum_l P_{i,l}^2}$.

Equation (iii) can be viewed as a way to consider separately the evolution of prices in one group and the interactions between groups, including readjustments in one group due to variations in the others; (iv) refers to the adjustments of prices in one group and (v) is a bound on the reactions between groups. Notice that (v) is implied by the following simple condition

$$\sum_l |b_{il}| \leq \sum_j \beta_{ij} ||P_j||.$$

Indeed

$$\sum_l P_{il} b_{il} \leq \sum_l |P_{il}||b_{il}| < \sum_l ||P_i|||b_{il}|.$$

Condition (iv) is sufficient for exponential stability of the origin for the "decoupled" system

$$\dot{P}_i = a_i(t,P_i),$$

as is shown by using the Liapunov function $V_i = ||P_i||$. This suggests using the vector Liapunov function

$$V = (V_1,\ldots,V_n) = ((P_1^TP_1)^{1/2},\ldots,(P_i^TP_i)^{1/2},\ldots,(P_n^TP_n)^{1/2})$$

and the comparison equation

$$\dot{u}_i = -\alpha_i u_i + \sum_j \beta_{ij}u_j = F_i(u) \tag{5.3}$$

to study the stability properties of the origin for Equation (5.2) under assumptions (i) to (v). But since (5.3) is linear with constant coefficients, the origin will be globally asymptotically (exponentially) stable provided all eigenvalues of the matrix C, whose elements C_{ij} are given by

$$C_{ii} = -\alpha_i + \beta_{ii}, \quad C_{ij} = \beta_{ij} \quad (i \neq j),$$

have negative real parts; (observe that by hypothesis, $C_{ii} < 0$ and $C_{ij} \geq 0$ for $i \neq j$). L. Metzler [1945] has proved that this is the case if and only if the so-called Hicks conditions, familiar to economists, are satisfied: the principal determinants of C must alternate in sign:

$$\text{(vi)}\quad C_{11} < 0, \begin{vmatrix} C_{11} C_{12} \\ C_{21} C_{22} \end{vmatrix} > 0,\ldots,(-1)^j \begin{vmatrix} C_{11} & \cdots & C_{1j} \\ \cdots & \cdots & \cdots \\ C_{j1} & \cdots & C_{jj} \end{vmatrix} > 0,$$

$$1 \leq j \leq n.$$

An extensive list of equivalent criteria appears in the review article by M. Fiedler and V. Ptak [1960].

Let us now show how (i) to (vi) imply global asymptotic stability for the origin in Equation (5.2) provided all solutions of this equation can be continued to infinity. Indeed, putting $V_i = ||P_i|| = \sqrt{P_i^TP_i}$, we compute, if $V_i \neq 0$,

$$\begin{aligned}\dot{V}_i = V_i^{-1}P_i^T\dot{P}_i &= V_i^{-1}P_i^T[a_i(t,P_i) + b_i(t,P)] \\ &\leq -\alpha_i V_i^{-1}P_i^TP_i + \sum_j V_i^{-1}\beta_{ij}||P_j||\ \ ||P_i|| \\ &\leq -\alpha_i V_i + \sum_j \beta_{ij}V_j.\end{aligned}$$

If $V_i(t) = 0$, then for every $h > 0$

$$\begin{aligned}V_i(t+h) - V_i(t) &\leq h \sup_{\substack{\tau \in [t,t+h] \\ V_i(\tau) \neq 0}} \dot{V}_i(\tau) \\ &\leq h \sup_{\substack{\tau \in [t,t+h] \\ V_i(\tau) \neq 0}} [-\alpha_i V_i(\tau) + \sum_j \beta_{ij}V_j(\tau)].\end{aligned}$$

Since the V's are continuous functions, dividing by h and taking the limit $h \to 0^+$, one gets

$$D^+V_i(t) \leq -\alpha_i V_i(t) + \sum_j \beta_{ij}V_j(t).$$

As is readily verified, all the hypotheses of Theorem 3.5 are satisfied, and this proves the announced property.

6. A General Comparison Principle

6.1. The comparison method, which was applied in Section 3 to the concepts of stability and attractivity, can of course be also used successfully in connection with most concepts introduced in Chapter VI. It shows how a qualitative property relating to some supposedly simple differential equation entails a similar property for another, usually more complex, equation. In this chapter, we analyse from a theoretical point of view, this kind of transfer of a property from one equation to another: of the property itself, we shall retain just enough for the transfer to take place. In other words, a comparison principle will be derived, which will apply to a whole class of concepts.

We consider again Equations (3.1) and (3.2) with the general hypotheses 3.1, and, in order to make things easier, we assume that <u>all solutions under consideration are unique and continuable to</u> $+\infty$.

6.2. Using the symbolism of Chapter VI and given any qualitative concept

$$C: W,\gamma * (\lambda,\beta)$$

relating to Equation (3.1), we define a similar concept

$$C^0: W^0,\gamma^0 * (\lambda^0,\beta^0)$$

for equation (3.2) by substituting:

(1) $u_0 \in \mathscr{A}^0$ to $x_0 \in \mathscr{A}$;

(2) $K(t_0,u_0)$ to $J(t_0,x_0)$; notice that, owing to our continuability hypothesis, both intervals equal $[t_0,\infty[$;

(3) proposition $\beta^0 \equiv [u(t;t_0,u_0) \in \mathscr{B}^0]$ to β.

The other quantities appearing in $W,\gamma * (\lambda,\beta)$ remain unchanged. We assume that the word W^0 thus obtained is admissible. The new concept C^0 will be said to be a <u>comparison concept</u> with respect to C and some auxiliary function $V(t,x)$ if the sets $\mathscr{A}^0$ or $\mathscr{B}^0$ are such that:

(i) if x_0 is universal and defined after t_0

$$(\forall t_0 \in I)(\forall x_0 \in \mathscr{A})(\exists u_0 \in \mathscr{A}^0) \quad V(t_0,x_0) \leq u_0;$$

(ii) if x_0 is universal and defined before t_0

$$(\forall x_0 \in \mathscr{A})(\exists u_0 \in \mathscr{A}^0)(\forall t_0 \in I) \quad V(t_0,x_0) \le u_0;$$

(iii) if x_0 is existential and defined after t_0

$$(\forall t_0 \in I)(\forall u_0 \in \mathscr{A}^0)(\exists x_0 \in \mathscr{A}) \quad V(t_0,x_0) \le u_0;$$

(iv) if x_0 is existential and defined before t_0

$$(\forall u_0 \in \mathscr{A}^0)(\exists x_0 \in \mathscr{A})(\forall t_0 \in I) \quad V(t_0,x_0) \le u_0;$$

(v) $(\forall t \in I)(\forall u \in \mathscr{B}^0)(\forall x \in \Omega\colon V(t,x) \le u) \quad x \in \mathscr{B}$.

Notice that these requirements don't define $\mathscr{A}^0$ nor $\mathscr{B}^0$ completely. This definition of a comparison concept will perhaps appear puzzling at first sight. Its meaning becomes clear in the sequel. Anyhow, it is fairly easy to build such a concept, as is shown by the following two lemmas.

6.3. Lemma. Suppose there exists a continuous function $V\colon I \times \Omega \to \mathscr{R}^m$ such that, for some function $a \in \mathscr{K}$ and every $(t,x) \in I \times \Omega$: $\max_i V_i(t,x) \ge a(||x||)$, $V(t,0) = 0$. Consider a concept C with x_0 universal and defined after t_0, $\mathscr{A} = B(0,\delta) \cap \Omega$ and $\mathscr{B} = B(0,\varepsilon) \cap \Omega$. Then a comparison concept is defined by the sets

$$\mathscr{A}^0 = \{u\colon u \ge 0,\ u_i \le \sup_{x \in \mathscr{A}} V_i(t_0,x)\},$$

$$\mathscr{B}^0 = \{u\colon \max_i u_i < a(\varepsilon)\}.$$

Proof. As x_0 is universal and defined after t_0, we can choose $u_0 = V(t_0,x_0)$. Further, for any $t \in I$, $u \in \mathscr{B}^0$ and $x \in \Omega$ such that $V(t,x) \le u$,

$$a(||x||) \le \max_i V_i(t,x) \le \max_i u_i < a(\varepsilon).$$

Hence $x \in B(0,\varepsilon) \cap \Omega$. Q.E.D.

Similarly, if x_0 is defined before t_0, one can prove the following lemma.

6.4. Lemma. Suppose there exists a function $V: I \times \Omega \to \mathscr{R}^m$ and that for some functions $a \in \mathscr{K}$ and $b \in \mathscr{K}$ and every $(t,x) \in I \times \Omega$: $a(||x||) \leq \max_i V_i(t,x) \leq b(||x||)$. Consider a concept C with x_0 universal and defined before t_0, $\mathscr{A} = B(0,\delta) \cap \Omega$ and $\mathscr{B} = B(0,\varepsilon) \cap \Omega$. Then a comparison concept is defined by the sets

$$\mathscr{A}^0 = \{u\colon u \geq 0, \max_i u_i \leq b(\delta)\},$$

$$\mathscr{B}^0 = \{u\colon \max_i u_i < a(\varepsilon)\}.$$

In most cases, the definitions of the sets $\mathscr{A}^0$ and $\mathscr{B}^0$ can be simplified to yield a new concept, which we shall call auxiliary concept and which implies the comparison concept.

6.5. Lemma. Let C^0 be a comparison concept as defined by Lemma 6.3 or 6.4 and suppose the corresponding assumptions are satisfied. Assume further that

(i) if δ is defined before t_0, there exists a function $b' \in \mathscr{K}$ such that for any $(t,x) \in I \times \Omega$

$$\max_i V_i(t,x) \leq b'(||x||);$$

(ii) if ε is existential: $a(r) \to \infty$ as $r \to \infty$;

then the auxiliary concept C^* obtained from C^0 by substituting

$$\mathscr{A}^* = \{u \in \mathscr{R}^m\colon 0 \le u \le \delta e\},$$

$$\mathscr{B}^* = \{u \in \mathscr{R}^m, u < \varepsilon e\},$$

to $\mathscr{A}^0$ and $\mathscr{B}^0$, implies the comparison concept C^0.

<u>Proof</u>. Let δ^*, ε^* be the values of δ, ε in C^*. In what follows, W_1^0, W_2^0 and W_3^0 are parts of a word chosen in such a way that the appearances of δ are left explicit.

If δ is universal, one has the following implications:

$$W_1^0(\forall\delta^* > 0)W_2^0(\forall u_0 \in \{u\colon 0 \le u \le \delta^* e\})W_3^0, \gamma^0 * (\lambda^0, \beta^0)$$

$$\Longrightarrow W_1^0(\forall\delta > 0)(\exists\delta^* > 0\colon \mathscr{A}^0 \subset \{u\colon 0 \le u \le \delta^* e\})W_2^0$$
$$(\forall u_0 \in \{u\colon 0 \le u \le \delta^* e\})W_3^0, \gamma^0 * (\lambda^0, \beta^0)$$

$$\Longrightarrow W_1^0(\forall\delta > 0)W_2^0(\forall u_0 \in \mathscr{A}^0)W_3^0, \gamma^0 * (\lambda^0, \beta^0) \equiv C^0.$$

Consider next an existential δ and the concept

$$W_1^0(\exists\delta^* > 0)W_2^0(\forall u_0 \in \{u\colon 0 \le u \le \delta^* e\})W_3^0, \gamma^0 * (\lambda^0, \beta^0).$$

Given $\delta^* > 0$, there exist $\delta > 0$ such that $\mathscr{A}^0 \subset \mathscr{A}^* = \{u\colon 0 \le u \le \delta^* e\}$, in which case

$$C^0 \equiv W_1^0(\exists\delta > 0)W_2^0(\forall u_0 \in \mathscr{A}^0 \subset \mathscr{A}^*)W_3^0, \gamma^0 * (\lambda^0, \beta^0)$$

is satisfied.

One proves in the same way that

$$\ldots\ \varepsilon^* \ldots\ \gamma^0 * (\lambda^0, u(t) \in \{u \in \mathscr{R}^m, u \le \varepsilon^* e\})$$

implies

$$\ldots\ \varepsilon \ldots\ \gamma^0 * (\lambda^0, u(t) \in \{u \in \mathscr{R}^m, u \le a(\varepsilon) e\}).$$

Indeed, given any one of the quantities ε, ε^*, one can always choose the other one in such a way that $\varepsilon^* \leq a(\varepsilon)$.

At last, combining the above arguments proves that $C^* \Longrightarrow C^0$. Q.E.D.

6.6. General Comparison Theorem. Assume there exists a function F as described in 3.1 and 6.1, and a continuous function $V: I \times \Omega \to \mathscr{R}^m$ which is locally lipschitzian in x and such that, for every $(t,x) \in I \times \Omega$: $D^+V(t,x) \leq F(t,V(t,x))$. Then a qualitative concept C will be satisfied whenever the corresponding comparison concept C^0 is satisfied.

Proof. Let $C = W, \gamma * (\lambda,\beta)$ be the concept under consideration and $C^0 = W^0, \gamma^0 * (\lambda^0,\beta^0)$ be the corresponding comparison concept. The problem is, knowing that C^0 is true, to prove that the existential variables can be chosen in W such that $\gamma * (\lambda,\beta)$ is true. Let us choose them by the following recurrence. The first k variables being fixed in W and W^0, let us look at the next one.

1) If it is universal, we fix it in W, and, except for x_0, we choose the same value in W^0; if this variable is x_0, let us choose $u_0 \in \mathscr{A}^0$ such that $V(t_0,x_0) \leq u_0$. Notice that, if t_0 is not fixed yet, we can choose u_0 such that for every $t_0 \in I$: $V(t_0,x_0) \leq u_0$.

2) If the next variable is existential, consider the corresponding variable in W^0 such that C^0 is satisfied. Except for x_0, we adopt the same value for the variable in W. For x_0, we choose $x_0 \in \mathscr{A}$ such that $V(t_0,x_0) \leq u_0$.

Let us now show that $\gamma * (\lambda,\beta)$ is true. First γ and γ^0 are equivalent since $J = K = [t_0,\infty[$. The variable t in λ^0 or λ is chosen according to the rules 1) or 2) used for the variables in W. From Lemma 2.6 we deduce that

$$V(t,x(t)) \leq u(t;t_0,u_0)$$

and from the definition of the comparison concept (cf. Conditions (v))

$$u(t;t_0,u_0) \in \mathscr{B}^0 \Longrightarrow x(t) \in \mathscr{B}. \qquad \text{Q.E.D.}$$

7. Bibliographical Note

The comparison method has already been used, in substantially the same form as here, by R. Conti [1956] in a theorem on the continuability of solutions using a scalar comparison equation. It was applied by C. Corduneanu [1960] to the study of stability and asymptotic stability. V. M. Matrosov $[1962]_2$ and R. Bellman [1962] both extended the results of Corduneanu to the case of vectorial comparison equations. These extensions are based on the work of T. Wazewski [1950] on differential inequalities, an account of which appears in several books; for instance W. Walter [1964], J. Szarski [1967] or V. Lakshmikantham and S. Leela [1969]. The last one contains many original contributions of the authors in several domains, as for instance conditional stability. A fairly complete survey of the results obtainable by the comparison method, not only in stability theory, but in the study of existence and uniqueness of solutions, continuability, convergence of approximations, etc. is given in a series of papers by V. M. Matrosov [1968-1969].

Most results of Section 3 come from V. M. Matrosov $[1962]_2$. Theorem 3.5 generalizes the corresponding result of Matrosov in the sense that negative V_i functions can be used here, whereas in the original paper, all these functions had to be positive and their sum had to be positive definite. Exercise 3.7 comes from N. P. Bhatia and V. Lakshmikantham [1965]. Theorem 3.9 is a variant of a result of V. M. Matrosov, where one used a comparison concept independent of the equation in x. Theorem 3.10 is an example of Matrosov's results on instability.

Section 4 is based on a paper by F. N. Bailey [1965]. An application to a control system describing the longitudinal motion of a plane appears in A. A. Piontkovskii and L. D. Rutkovskaya [1967]. Economic problems akin to the one treated in Section 5 appear in D. D. Siljak [1973]. On interconnected systems, see also A. N. Michel [1974].

The comparison principle presented in Section 6 comes from P. Habets and K. Peiffer [1973]. Some generalizations to comparison equations depending not only of the auxiliary variable u, but also on x, are due to V. M. Matrosov [1963] and P. Habets and K. Peiffer [1975]. On the other hand, with a view to generalize the monotonicity hypothesis, V. Lakshmikantham, A. R. Mitchell and R. W. Mitchell [1975] have established some comparison results for differential equations on cones.

Amongst the questions directly related to the subject matter of this chapter, but which have not been treated here for lack of space, let us cite the combined use of the comparison method and the theory of sectors by V. M. Matrosov [1965], partial stability in C. Corduneanu [1964], K. Peiffer and N. Rouche [1969], and C. Risito [1972], and at last several results of Dang Chau Phien [1968] on stability and asymptotic stability, where the hypothesis on $\max_i V_i(t,x)$ is substantially weakened with respect to the theorems of the present chapter.

APPENDIX I

DINI DERIVATIVES AND MONOTONIC FUNCTIONS

1. The Dini Derivatives

1.1. Let a, b, $a < b$, be two real numbers and consider a function $f:]a,b[\to \mathcal{R}$, $t \to f(t)$ and a point $t_0 \in]a,b[$. We assume that the reader is familiar with the notions of

$$\limsup_{t \to t_0} f(t) \quad \text{and} \quad \liminf_{t \to t_0} f(t).$$

The same concepts of lim sup and inf, but for $t \to t_0 +$ mean simply that one considers, in the limiting processes, only the values of $t > t_0$. A similar meaning is attached to $t \to t_0 -$. Remember that, without regularity assumptions on f, any lim sup or inf exists if we accept the possible values $+\infty$ or $-\infty$. The extended real line will be designated hereafter by $\overline{\mathcal{R}}$. Thus $\overline{\mathcal{R}} = \mathcal{R} \cup \{-\infty\} \cup \{+\infty\}$.

The four Dini derivatives of f at t_0 are now defined by the following equations:

$$D^+f(t_0) = \limsup_{t \to t_0+} \frac{f(t) - f(t_0)}{t - t_0},$$

$$D_+f(t_0) = \liminf_{t \to t_0+} \frac{f(t) - f(t_0)}{t - t_0},$$

$$D^-f(t_0) = \limsup_{t \to t_0-} \frac{f(t) - f(t_0)}{t - t_0},$$

$$D_-f(t_0) = \liminf_{t \to t_0-} \frac{f(t) - f(t_0)}{t - t_0} .$$

They are called respectively the <u>upper right, lower right,</u> <u>upper left</u> and <u>lower left derivatives of</u> f <u>at</u> t_0. Further, the function $t \to D^+f(t)$ on $]a,b[$ into $\overline{\mathscr{R}}$ is called the <u>upper right derivative of</u> f <u>on the interval</u> $]a,b[$, and similarly for D_+, D^- and D_-.

1.2. <u>Remarks</u>. a) It is clear that, in the absence of regularity assumptions on f, any Dini derivative may equal $-\infty$ or $+\infty$.

b) However, if there is a Lipschitz condition for f on some neighborhood of t_0, then all four derivatives are finite.

c) The four Dini derivatives of f at some point $t_0 \in]a,b[$ are equal if and only if f has a derivative at t_0. This derivative is then of course equal to the common value of the Dini derivatives.

d) The well known properties of lim sup and lim inf yield the elementary rules of calculus applicable to the Dini derivatives. For example, if f_1 and f_2 are two real functions defined on $]a,b[$, one gets for any $t \in]a,b[$ that

$$D^+[f_1(t) + f_2(t)] \leq D^+f_1(t) + D^+f_2(t),$$

and

$$D^+[f_1(t) + f_2(t)] \geq D^+f_1(t) + D_+f_2(t),$$

as long as the additions are possible $[(+\infty) + (-\infty)$ is an example of an addition which is not possible$]$.

e) Another important property is that if f is continuous and g is $\mathscr{C}^1$, then

(1) if for some t: $g(t) \geq 0$, one has $D^+(fg)(t) = f(t)g'(t) + g(t)D^+f(t)$

(2) if for some t: $g(t) \leq 0$, one has $D^+(fg)(t) = f(t)g'(t) + g(t)D_+f(t)$,

where $g'(t)$ is the ordinary derivative.

The proof is as follows:

$$D^+(fg)(t) = \limsup_{h \to 0+} \frac{(fg)(t+h) - (fg)(t)}{h}$$

$$= \limsup_{h \to 0+} [f(t+h) \frac{g(t+h) - g(t)}{h} + g(t) \frac{f(t+h) - f(t)}{h}]$$

$$= \lim_{h \to 0+} f(t+h) \frac{g(t+h) - g(t)}{h} + \limsup_{h \to 0+} g(t) \frac{f(t+h) - f(t)}{h}.$$

Hence the expected result. On the properties of the lim sup and lim inf which enable one to write the above equalities, as well as on other rules of calculus for Dini derivatives, we refer to E. J. McShane [1944].

2. Continuous Monotonic Functions

2.1. Theorem. Suppose f is continuous on $]a,b[$. Then f is increasing on $]a,b[$ if and only if $D^+f(t) \geq 0$ for every $t \in]a,b[$.

Remember that, in this book, f is called increasing on $]a,b[$ if, for any $t_1, t_2 \in]a,b[$, $t_1 < t_2$, one has $f(t_1) \leq f(t_2)$.

Proof. The condition is obviously necessary. Let us prove that it is sufficient.

a) Assume first that $D^+f(t) > 0$ on $]a,b[$. If there exist two points $\alpha, \beta \in]a,b[$, $\alpha < \beta$, with $f(\alpha) > f(\beta)$, then there exist a μ with $f(\alpha) > \mu > f(\beta)$ and some points $t \in [\alpha,\beta]$ such that $f(t) > \mu$. Let ξ be the sup of these points. Of course, ξ is an interior point of $[\alpha,\beta]$, and, due to the continuity of f: $f(\xi) = \mu$. Therefore, for every $t \in]\xi,\beta[$:

$$\frac{f(t) - f(\xi)}{t - \xi} < 0$$

and $D^+f(\xi) \leq 0$, which is absurd.

b) Assume now, as in the statement of the theorem, that $D^+f(t) \geq 0$ on $]a,b[$. For any $\varepsilon > 0$, one gets

$$D^+[f(t) + \varepsilon t] = D^+f(t) + \varepsilon \geq \varepsilon > 0.$$

Hence $f(t) + \varepsilon t$ is increasing on $]a,b[$. And since this is true for any ε, $f(t)$ is also increasing on $]a,b[$.

Q.E.D.

2.2. Remarks. a) This theorem remains true if one replaces the inequality $D^+f(t) \geq 0$ by $D_+f(t) \geq 0$, because the latter implies the former.

b) One proves similarly that $D^+f(t) \geq 0$ can also be replaced by $D^-f(t) \geq 0$: it suffices to substitute the inf of the points t where $f(t) < \mu$ to the sup of the points t where $f(t) > \mu$.

c) In the new theorem thus obtained, D^- may be replaced by D_-. As a consequence, we get the following statement.

2.3. Theorem. Suppose f is continuous on $]a,b[$. Then f is increasing on $]a,b[$ if and only if any of the four Dini derivatives of f is ≥ 0 on $]a,b[$.

2.4. Corollary. If any Dini derivative of the continuous function f is ≥ 0 on $]a,b[$, the same is true of the other three.

2.5. Remark. Analogous monotonicity properties can be established using less than the continuity of f (cf. E. J. McShane [1944]).

2.6. Functions with a bounded Dini derivative. The following theorem is used to estimate the average rate of decrease of a function possessing a Dini derivative bounded from below. It is a straightforward consequence of Theorem 2.3. Hereafter, the symbol D^*f represents any of the four Dini derivatives of f.

Theorem. Let $f: [a,b] \to \mathcal{R}$ be a continuous function such that for any $t \in]a,b[$ and some $A > 0$:

$$D^*f(t) \geq -A. \tag{2.1}$$

Then

$$\frac{f(a) - f(b)}{b - a} \leq A.$$

Proof. One deduces from (2.1) that $D^*(f(t) + At) \geq 0$, and therefore, using Theorem 2.3, that $f(t) + At$ is increasing on $]a,b[$. Therefore $f(b) + Ab \geq f(a) + Aa$. Q.E.D.

2.7. Dini derivative of the maximum of two functions. Concerning three functions f, g, h such that $h(t) = \max (f(t), g(t))$, the following theorem gives an estimation of a Dini derivative of $h(t)$ in terms of the corresponding derivatives of f and g.

Theorem. Let f, g, and h be three continuous functions on $[a,b]$ into $\mathscr{R}$, such that $h(t) = \max(f(t),g(t))$. If $D^+f(t) \leq 0$ and $D^+g(t) \leq 0$ for $t \in]a,b[$, then $D^+h(t) \leq 0$ for $t \in]a,b[$.

Proof. Otherwise, one would have, by Theorem 2.1, for two points a', b' with $a \leq a' < b' \leq b$, that $\max(f(b'), g(b')) > \max(f(a'),g(a'))$, and therefore either $f(b') > f(a')$ or $g(b') > g(a')$. But, using Theorem 2.1 again, this contradicts either $D^+f \leq 0$ or $D^+g \leq 0$. Q.E.D.

3. The Derivative of a Monotonic Function

3.1. The theorem stated (without proof) in this section is a key theorem for Liapunov's direct method. It mentions the Lebesgue integral of a function f on an interval $[a,b]$, which will be written

$$\int_a^b f(\tau)d\tau$$

On this concept, we refer to E. J. McShane [1944] or to A. N. Kolmogorov and S. V. Fomin [1961]. Only one of its elementary properties will be recalled below, in order to clear the statement of the theorem.

3.2. A subset E of the real line $\mathscr{R}$ is said to have measure zero if there exists, for every $\varepsilon > 0$, a finite or countable collection $I_1, I_2, \ldots$ of open intervals such that $\cup I_i \supset E$ and $\Sigma \Delta I_i < \varepsilon$, where ΔI_i is the length of I_i. When a property is verified at each point of some interval $[a,b] \in \mathscr{R}$, except at the points of a set of measure zero, one says that the property is true almost everywhere on $[a,b]$ or for almost all $t \in [a,b]$.

If a function $f: [a,b] \to \mathscr{R}$ is Lebesgue integrable over $[a,b]$, then any function $g: [a,b] \to \mathscr{R}$ which is equal to f almost everywhere on $[a,b]$ is also Lebesgue integrable on $[a,b]$, and the integral of g equals the integral of f. Therefore, it makes sense to speak of the integral over $[a,b]$ of a function which is defined only almost everywhere on $[a,b]$: it can be extended to the whole of $[a,b]$ by choosing arbitrary values at the points where it was originally undefined.

3.3. Theorem. If $f: [a,b] \to \mathscr{R}$ is an increasing function, f has a finite derivative $f'(t)$ almost everywhere on $[a,b]$; this derivative is Lebesgue integrable and one has, for any $t \in [a,b]$,

$$f(t) = \int_a^t f'(\tau)d\tau + h(t)$$

where h is an increasing function and $h'(t)$ vanishes almost everywhere on $[a,b]$.

For a proof, see E. J. McShane [1944] or H. L. Royden [1963].

3.4. Corollary. In the hypotheses of Theorem 3.3,

$$f(b) - f(a) \geq \int_a^b f'(\tau)d\tau. \tag{3.1}$$

3.5. Remarks. a) This inequality becomes an equality if one adds the hypothesis that f is absolutely continuous on $[a,b]$. On this point, cf. the reference books already mentioned. It does exist an example of a function f on $[a,b]$ into $\mathscr{R}$, which is increasing, uniformly continuous, whose derivative vanishes almost everywhere, and such that

$f(b) > f(a)$. Of course, for this function, which is not absolutely continuous

$$f(b) - f(a) > \int_a^b f'(\tau)d\tau = 0.$$

Cf. K. Kuratowski [1961], p. 187.

b) Since the derivative of f, when it exists, equals all four Dini derivatives, the inequality (3.1) can also be written under the form

$$\int_a^b D^+f(\tau)d\tau \leq f(b) - f(a),$$

or similarly while replacing D^+ by D_+, D^- or D_-. Remember that it is valid when f is increasing.

4. Dini Derivative of a Function along the Solutions of a Differential Equation

4.1. For some τ, $-\infty \leq \tau < \infty$ and some open subset $\Omega \subset \mathscr{R}^n$, consider a continuous function

$$f\colon]\tau,\infty[\times \Omega \to \mathscr{R}^n, \ (t,x) \to f(t,x)$$

and the associated differential equation $\dot{x} = f(t,x)$. Further, let $V\colon]\tau,\infty[\times \Omega \to \mathscr{R}$ be a continuous function, satisfying a local Lipschitz condition for x, uniformly with respect to t.

4.2. One has often to verify that a function like $V(t,x)$ is, so to say, <u>decreasing along the solutions of the differential equation</u>. This means that for any solution $x\colon J \to \mathscr{R}^n$, J an open interval, of the equation $\dot{x} = f(t,x)$, the function $\tilde{V}\colon J \to \mathscr{R}^n$, $t \to \tilde{V}(t) = V(t,x(t))$ is decreasing. The following theorem is crucial, for it enables one to check this property <u>without any knowledge of the solutions</u>.

4.3. <u>Theorem</u> (T. Yoshizawa [1966]). In these general hypotheses, let $x: J \to \mathscr{R}^n$ be any solution and let $t^* \in J$. Putting $x(t^*) = x^*$, one gets

$$D^+\tilde{V}(t^*) = \limsup_{h \to 0+} \frac{V(t^* + h, x^* + hf(t^*,x^*)) - V(t^*,x^*)}{h}. \tag{4.1}$$

<u>Proof</u>. One has, for $h > 0$ small,

$$\begin{aligned} &V(t^* + h, x(t^* + h)) - V(t^*, x(t^*)) = \\ &V[t^* + h, x^* + hf(t^*,x^*) + h\varepsilon(t^*,x^*,h)] - V(t^*,x^*) \\ &\leq V(t^* + h, x^* + hf(t^*,x^*)) + kh||\varepsilon(t^*,x^*,h)|| - V(t^*,x^*), \end{aligned}$$

where $\varepsilon \to 0$ with h and k is a Lipschitz constant on some neighborhood of x^*. Therefore

$$\begin{aligned} D^+\tilde{V}(t^*) &= \limsup_{h \to 0+} \frac{V(t^* + h, x(t^* + h)) - V(t^*, x(t^*))}{h} \\ &\leq \limsup_{h \to 0+} \frac{V(t^* + h, x^* + hf(t^*,x^*)) - V(t^*,x^*)}{h}. \end{aligned}$$

One obtains similarly for $h > 0$ small, that

$$\begin{aligned} &V(t^* + h, x(t^* + h)) - V(t^*, x(t^*)) \geq \\ &\qquad V(t^* + h, x^* + hf(t^*,x^*)) - kh||\varepsilon(t^*,x^*,h)|| - V(t^*,x^*), \end{aligned}$$

whence

$$D^+\tilde{V}(t^*) \geq \limsup_{h \to 0+} \frac{V(t^* + h, x^* + hf(t^*,x^*)) - V(t^*,x^*)}{h}.$$

Q.E.D.

4.4. <u>Remarks</u>. a) We shall admit the symbol $D^+V(t^*,x^*)$ to represent the second member of (4.1), and this quantity will be called occasionally the upper right Dini derivative of

$V(t,x)$ (along the solutions of the differential equation).

b) It is a simple consequence of Theorems 2.1 and 4.3 that if $D^+V(t,x) \geq 0$ on $]\tau,\infty[\times \Omega$, then $V(t,x)$ is increasing along the solutions of the differential equation. The analogous statement for $V(t,x)$ decreasing is obvious.

c) There is a theorem similar to 4.3 for any other Dini derivative D_+, D^- and D_-.

d) It is noticeable that no uniqueness property has been assumed for the solutions of the differential equation.

e) If, for some $\varepsilon > 0$ and every $(t,x) \in]\tau,\infty[\times \Omega$, one has $D^+V(t,x) \geq -\varepsilon$, then for any solution $x: J \to \mathcal{R}^n$ and any points $a, b \in J$, $a < b$,

$$\frac{\tilde{V}(a) - \tilde{V}(b)}{b - a} \leq \varepsilon.$$

This is a consequence of Theorems 2.6 and 4.3.

APPENDIX II

THE EQUATIONS OF MECHANICAL SYSTEMS

Assuming some knowledge of analytical mechanics, we gather here a few precise definitions concerning Lagrangian and Hamiltonian systems and recall their most fundamental properties.

1. To a mechanical holonomic system with n degrees of freedom with generalized (or Lagrangian) coordinates $q \in \mathscr{R}^n$ and generalized velocities $\dot{q} \in \mathscr{R}^n$, there will correspond a <u>kinetic energy</u> of the form

$$T: I \times \Omega \times \mathscr{R}^n \to \mathscr{R}, \quad (t,q,\dot{q}) \to T(t,q,\dot{q}),$$

where Ω is some domain (connected open set) of $\mathscr{R}^n$ and $I =]\tau,\infty[$ for some $\tau \in \mathscr{R}$. With respect to $\dot{q}$, T is a polynomial of the second degree. It will often be written thus:

$$T(t,q,\dot{q}) = \frac{1}{2}\dot{q}^T A(t,q)\dot{q} + b(t,q)^T\dot{q} + d(t,q)$$

where A is an $n \times n$ matrix, b is an $n \times 1$ matrix and d is a scalar, all defined on $I \times \Omega$. Further, A, b and d will be supposed to be $\mathscr{C}^1$ functions.

2. The <u>potential energy</u> will be designated by Π. In this book, we are prevented from choosing the usual letter

V to represent the potential function, because in the domain of Liapunov's direct method, V is the ritual symbol for auxiliary functions. Π will be supposed to be a $\mathscr{C}^1$ function

$$\Pi : I \times \Omega \to \mathscr{R}, \ (t,q) \to \Pi(t,q).$$

3. We shall also consider <u>Lagrangian forces</u> which do not derive from a potential function. They will form on n-vector depending on velocities, coordinates and time:

$$Q : I \times \Omega \times \mathscr{R}^n \to \mathscr{R}, \ (t,q,\dot{q}) \to Q(t,q,\dot{q}).$$

We assume Q to be continuous.

4. The <u>Lagrangian equations of motion</u> are, in the usual notations,

$$\frac{d}{dt}\frac{\partial L}{\partial \dot{q}} - \frac{\partial L}{\partial q} = Q \tag{4.1}$$

where $L = T - \Pi$. The question arises immediately whether Equation (4.1) can be solved with respect to $\ddot{q}$, and therefore be brought to normal form $\dot{x} = f(t,x)$. This can be done around every point (t,q) where $A(t,q)$ is regular.

5. A situation frequently encountered is when the matrices B and C vanish identically and A is independent of t. In this case we write the latter $A(q)$. Such a compact form is obtained for T for instance in case of <u>time independent constraints</u>. Then Equation (4.1) can be solved with respect to $\ddot{q}$ in the neighborhood of some point $(t,q,\dot{q})$ if $\det A(q) \neq 0$. But $A(q)$ is everywhere positive semi-definite and therefore $\det A(q) \neq 0$ if and only if $A(q)$ is positive definite. The following useful property of the kinetic

energy is easy to prove.

Proposition. For $q_0 \in \Omega$, there exists a neighborhood N of q_0, $N \subset \Omega$ and a function $a \in \mathcal{K}$ such that, for every $(q,\dot{q}) \in N \times \mathcal{R}^n$, $\dot{q}^T A(q)\dot{q} \geq a(||\dot{q}||)$, if and only if $A(q_0)$ is positive definite.

(Hint: choose N compact, remember A is continuous and can be diagonalized by an orthogonal change of coordinates).

6. The dissipative forces are an important family of generalized forces: Q is called dissipative if, for every $(t,q,\dot{q}) \in I \times \Omega \times \mathcal{R}^n$: $Q^T(t,q,\dot{q})\dot{q} \leq 0$. Of course $Q^T\dot{q}$ is the power supplied by the forces Q. The case of complete dissipation is when there exists a function $a \in \mathcal{K}$ such that, for every $(t,q,\dot{q}) \in I \times \Omega \times \mathcal{R}^n$: $Q^T(t,q,\dot{q})\dot{q} \leq -a(||\dot{q}||)$. Dissipative forces which are linear with respect to $\dot{q}$ are called viscous friction forces. As observed by A. I. Lur'e [1968], they are not the only ones to be derivable from a Rayleigh dissipation function (on this notion, see for instance H. Goldstein [1950]). Non energic forces (in the terminology of G. D. Birkhoff [1927]) are particular dissipative forces, namely those for which $Q^T(t,q,\dot{q})\dot{q} = 0$ identically. Gyroscopic forces are non energic forces which are linear with respect to $\dot{q}$.

7. Proposition. If $Q(t,q,\dot{q})$ is dissipative, $Q(t,q,0) = 0$ for every $(t,q) \in I \times \Omega$.

Proof. Suppose on the contrary that for some $(t^*,q^*) \in I \times \Omega$ and some i, $1 \leq i \leq n$: $Q_i(t^*,q^*,0) \neq 0$ or, more specifically, that $Q_i(t^*,q^*,0) > 0$ (the case < 0 would be treated alike). By continuity, one would get $Q_i(t^*,q^*,\dot{q}) > 0$ in some neighborhood of $\dot{q} = 0$. Let us, in this neighborhood, choose $\dot{q}^* = (0,0,\ldots,\dot{q}_i^*, 0,\ldots,0)$, with $\dot{q}_i^* > 0$. One would obtain $Q^T(t^*,q^*,\dot{q}^*)\dot{q}^* > 0$, which is excluded. Q.E.D.

8. This proposition introduces the following statement: if T is quadratic in $\dot{q}$ and Q is dissipative, the Lagrange equations (1) have an equilibrium at every point where $\partial\Pi/\partial q = 0$. This covers obviously the case where Q vanishes identically.

9. When T is independent of t (in this case we write $T(q,\dot{q})$), i.e. when A, b and d are time-independent, and when further $Q \equiv 0$, the Lagrange equations admit of a first integral called the <u>Painlevé integral</u>. It reads

$$E(q,\dot{q}) = \dot{q}^T \frac{\partial L}{\partial \dot{q}} - L = \frac{1}{2}\dot{q}^T A(q)\dot{q} - d(q) + \Pi(q).$$

If $Q \neq 0$, E is no more a first integral, and one computes easily that $\dot{E}(t,q,\dot{q}) = Q(t,q,\dot{q})^T\dot{q}$, where $\dot{E}$ is the derivative of E along the solutions of (4.1). If further T is quadratic with respect to $\dot{q}$, i.e. if B and C vanish identically, then $E(q,\dot{q}) = T(q,\dot{q}) + \Pi(q)$ is the total energy and, as a first integral, is called the <u>energy integral</u>. As already mentioned, this situation arises mainly when the constraints are time independent.

10. Other first integrals are the so-called <u>integrals of conjugate momenta</u>. When L does not depend on some coordinate q_k and the corresponding force Q_k vanishes identically,

$$\frac{\partial L}{\partial \dot{q}_k}(t,q,\dot{q})$$

is a first integral of the motion.

11. Ignorable coordinates and equations of Routh. Consider the mechanical system described in Sections 1 to 4. Suppose the number n of degrees of freedom is ≥ 2, and for some m, $1 \leq m \leq n$

$$\frac{\partial T}{\partial q_i} = 0 \qquad i = n - m + 1,\ldots,n.$$

Henceforth, we shall write $q = (q_1,\ldots,q_{n-m})$ for the first $n - m$ Lagrangian coordinates and $r = (r_1,\ldots,r_m)$ for the last m. Observe therefore that, by doing this, we change the assignment of the symbol q! The coordinates $r_1,\ldots,r_m$ are said to be ignorable if one has further that

(1) the potential does not depend on r: we shall write it $\Pi(t,q)$;

(2) the only non vanishing generalized forces correspond to the first $n - m$ degrees of freedom and do not depend on r and $\dot{r}$: we shall write them $Q = (Q_1,\ldots,Q_{n-m})$ with $Q = Q(t,q,\dot{q})$.

The kinetic energy may be written under the form

$$T(t,q,\dot{q},r) = T_2^*(t,q,\dot{q}) + U(t,q,\dot{q},\dot{r}) + T_2^{**}(t,q,\dot{r}) + T_1^*(t,q,\dot{q}) + T_1^{**}(t,q,\dot{r}) + T_0(t,q)$$

where T_2^* is quadratic in $\dot{q}$, U is bilinear in $(\dot{q},\dot{r})$, T_2^{**} is quadratic in $\dot{r}$, T_1^* is linear in $\dot{q}$ and T_1^{**} is linear in $\dot{r}$.

The regularity assumptions for all functions T, Π

and Q remain those of Sections 1 to 3. The Lagrange equations of motion read

$$\frac{d}{dt}\frac{\partial T}{\partial \dot{q}} = \frac{\partial}{\partial q}(T - \Pi) + Q \tag{11.1}$$

$$\frac{\partial T}{\partial \dot{r}} = c \tag{11.2}$$

where c is a constant of integration. Assuming that for some $\bar{q}$, $T_2^{**}(t,\bar{q},\dot{r})$ is positive definite with respect to $\dot{r}$, Equation (11.2) can be solved for $\dot{r}$ in some appropriate neighborhood of $\bar{q}$, yielding a function $\dot{r} = \dot{r}(t,q,\dot{q},c)$. In this case, $\dot{r}$ can be eliminated from the equations of motion. An elegant procedure to achieve this is by using Routh's function, obtained by substituting $\dot{r}$ in terms of $t,q,\dot{q},c$ in $T - c^T\dot{r}$. One gets readily for this function

$$R(t,q,\dot{q},c) = [T_2^* - T_2^{**} + T_1^* + T_0]_{\dot{r}=\dot{r}(t,q,\dot{q},c)} - \Pi(t,q),$$

the detailed computation appearing in A. I. Lur'e [1968]. For our theoretical purposes, we only need to observe here that, with respect to $\dot{q}$, R is a polynomial of the second degree, and in this polynomial, the terms of the second degree do not depend on c. This enables us to write

$$R(t,q,\dot{q},c) = R_2(t,q,\dot{q}) + R_1(t,q,\dot{q},c) + R_0(t,q,c) - \Pi(t,q), \tag{11.3}$$

where R_2 and R_1 are respectively quadratic and linear in $\dot{q}$. By the way $R_2 = T_2^*$.

Using R, one gets the equations of motion under the convenient form, called <u>Routh's equations</u>,

$$\frac{d}{dt}\frac{\partial R}{\partial \dot{q}} - \frac{\partial R}{\partial q} = Q,$$
$$\dot{c} = 0. \tag{11.4}$$

In case of time-independent constraints, the formulas above become simpler: the argument t disappears everywhere and further, all functions with subscript 1 or 0 vanish identically. As is well known, there exist time-dependent constraints such that t disappears everywhere, but the functions R_1 and R_0 do not vanish.

12. Stationary motions. Consider an equilibrium $q = \bar{q}$, $\dot{q} = 0$, $c = \bar{c}$ of Equations (11.4). A *generalized stationary motion* (C. Risito [1972]) is a motion of the original system corresponding to such an equilibrium, with $\dot{r}$ of course given as a function of t by Equation (11.2). If the equations of motion are autonomous, $\dot{r}$ is a constant and the corresponding motion is called *stationary*. Another adjective sometimes used instead of stationary is *merostatic*.

13. For the sake of references, let us now describe a type of system most frequently encountered in the applications. The kinetic energy $T(q,\dot{q})$ is defined and $\mathscr{C}^1$ on $\Omega \times \mathscr{R}^n$ and is, for every q, a positive definite quadratic form with respect to $\dot{q}$. The potential function $\Pi(q)$ is defined and $\mathscr{C}^1$ on Ω. When one has to study an equilibrium of such a system, it is convenient to locate it at the point $q = 0$, i.e. to assume $\frac{\partial \Pi}{\partial q}(0) = 0$, and to adjust Π in such a way that $\Pi(0) = 0$. This of course requires that the origin be a point of Ω.

14. For the system just described, with $T(q,\dot{q}) = \frac{1}{2}\dot{q}^T A(q)\dot{q}$, the equation

$$p = \frac{\partial T}{\partial \dot{q}} = A(q)\dot{q}$$

can be uniquely solved for $\dot{q}$. The kinetic energy expressed as a function of p and q reads

$$\tilde{T}(p,q) = \frac{1}{2} p^T B(q) p$$

where $B(q) = A^{-1}(q)$. Of course $A(q)$ and $B(q)$ are simultaneously positive definite. The <u>Hamiltonian function</u>

$$H(p,q) = \tilde{T}(p,q) + \Pi(q)$$

is equal to the total energy expressed as a function of p and q, and the Hamilton's equations of motion are

$$\dot{p} = -\frac{\partial H}{\partial q}, \quad \dot{q} = \frac{\partial H}{\partial p}.$$

APPENDIX III

LIMIT SETS

1. Let $I =]\tau,\infty[$ for some $\tau \in \mathscr{R}$ and Ω an open set of $\mathscr{R}^n$. We consider the differential equation

$$\dot{x} = f(t,x) \tag{1}$$

where $f: I \times \Omega \to \mathscr{R}^n$ is some continuous function. Let $x: J =]\alpha,\omega[\to \mathscr{R}^n$ be a non-continuable solution of this equation. A point $y \in \overline{\Omega}$ is called a <u>positive limit point</u> of x if there exists a sequence $\{t_n\}$ of time-values, such that $\{t_n\} \subset J$, $t_n \to \omega$ and $x(t_n) \to y$ as $n \to \infty$. The <u>positive limit set</u> of the solution x is the set of its positive limit points. It is designated by $\Lambda^+(x)$. <u>Negative limit points</u> and <u>negative limit sets</u> can be defined alike. Let us recall that the <u>positive semi-orbit</u> of x corresponding to some $t_0 \in J$ is the set $\gamma^+(x,t_0) = \{x(t): t \in [t_0,\omega[\}$.

2. <u>Theorem</u>. For every $t_0 \in J$: $\overline{\gamma^+(x,t_0)} = \gamma^+(x,t_0) \cup \Lambda^+(x)$.

<u>Proof</u>. Since $\Lambda^+(x) \subset \overline{\gamma^+(x,t_0)}$, it is clear that

$$\overline{\gamma^+(x,t_0)} \supset \gamma^+(x,t_0) \cup \Lambda^+(x).$$

Suppose now that some point $y \in \overline{\gamma^+(x,t_0)}$. There exists a sequence $\{t_i\} \subset J$, $t_i \geq t_0$ such that $x(t_i) \to y$. If $t_i \to \omega$ as $i \to \infty$, then $y \in \Lambda^+(x)$. Otherwise, there exists an infinite subsequence $\{t_i'\}$ which is bounded from above by some $t' < \omega$. The t_i' are such that $x(t_i') \to y$ as $i \to \infty$. But since they are infinite in number and bounded, there exists a subsequence $\{t_i''\}$ approaching some $t'' < \omega$. Hence, by continuity, $x(t_i'') \to x(t'') = y$ as $i \to \infty$, and therefore $y \in \gamma^+(x,t_0)$. Q.E.D.

3. <u>Theorem</u>. $\Lambda^+(x)$ is a closed set.

<u>Proof</u>. Let $\{y_i\} \subset \Lambda^+(x)$ be some sequence such that $y_i \to y$ when $i \to \infty$, and let us prove that $y \in \Lambda^+(x)$. For every i, there exists a sequence $\{t_{ij}\}$ such that $t_{ij} \to \omega$ and $x(t_{ij}) \to y_i$ when $j \to \infty$. We shall now construct some particular sequences $\{t_{ij}\}$. Suppose first of all that $\{t_{1j}\}$ be given. Let us choose $t_{22} > t_{12}$ and such that $d(x(t_{22}),y_2) < 1/2$. Choose further $t_{33} > t_{13}$ and such that $d(x(t_{33}), y_3) < 1/3$, and so on for $i = 4, 5, \ldots$ Of course $t_{ii} \to \omega$ when $i \to \infty$ and $d(x(t_{ii}),y_i) < 1/i$ for every i. On the other hand, we impose no restriction whatsoever to the t_{ij} for $i \neq j$. From the two propositions

$$(\forall\varepsilon > 0)(\exists N > 0)(\forall i \geq N) \quad d(x(t_{ii}),y_i) < \varepsilon/2,$$
$$(\forall\varepsilon > 0)(\exists N' > 0)(\forall i \geq N') \quad d(y_i,y) < \varepsilon/2,$$

we deduce easily that $x(t_{ii}) \to y$ as $i \to \infty$, and therefore that $y \in \Lambda^+(x)$. Q.E.D.

4. <u>Theorem</u>. If $\gamma^+(x,t_0)$ is bounded, $\Lambda^+(x)$ is non empty, compact and connected.

Proof. Since $\overline{\gamma^+(x,t_0)}$ is compact, $\Lambda^+(x)$ is not empty. Further, $\Lambda^+(x)$ is compact, since it is a closed subset of a compact set. Let us now prove that $\Lambda^+(x)$ is connected. In the opposite case, $\Lambda^+(x)$ would be the union of two compact disjoint sets Λ_1 and Λ_2, such therefore that $d(\Lambda_1,\Lambda_2) = \delta$ for some $\delta > 0$. There would exist a sequence $t_0 < t_1 < t_1'$ $< t_2 < t_2' < \dots$ such that $t_i \to \omega$, $d(x(t_i),\Lambda_1) \to 0$ and $d(x(t_i'),\Lambda_2) \to 0$ as $i \to \infty$. Therefore, for every large enough i, there would exist a t_i^* such that $t_i < t_i^* < t_i'$ and $d(x(t_i^*), \Lambda_1 \cup \Lambda_2) > \delta/4$. But the $x(t_i^*)$ would admit a cluster point, since $\gamma^+(x,t_0)$ is bounded. This point would belong to Λ^+, but also it would be apart from Λ^+ by at least the distance $\delta/4$, which is a contradiction. Q.E.D.

5. Theorem. If $\Lambda^+(x)$ is bounded, $x(t) \to \Lambda^+(x)$ as $t \to \omega$.

Proof. If it were not the case, there would exist an $\varepsilon > 0$ and a sequence of time-values $t_0, t_1, \dots$ such that $t_i \to \omega$ when $i \to \infty$ and $d(x(t_i),\Lambda^+(x)) > \varepsilon$. But there would exist a subsequence of the $x(t_i)$ approaching a limit point. This point would belong to $\Lambda^+(x)$ and be at a distance of $\Lambda^+(x)$ larger than ε, which is impossible. Q.E.D.

6. Exercise. Prove that $\omega = \infty$ as soon as there exists a positive limit point in Ω.

7. Exercise. If x is a periodic solution, one has for every t_0 that $\Lambda^+(x) = \gamma^+(x,t_0) = \{x(t)\colon t \in J\}$.

8. Theorem. If the differential equation is autonomous and if the solutions possess the uniqueness property, then

$$\Lambda^+(x) = \bigcap_{t_0 \in J} \overline{\gamma^+(x,t_0)}.$$

Proof. If x is periodic, the theorem is a direct consequence of Exercise 7. If it is not, let us first deduce from Theorem 2 that

$$\bigcap_{t_0 \in J} \overline{\gamma^+(x,t_0)} = \left(\bigcap_{t_0 \in J} \gamma^+(x,t_0)\right) \cup \Lambda^+(x).$$

The theorem follows from the fact that the intersection in the second member is empty. Q.E.D.

9. For autonomous equations, the most important property of the limit sets is that they are semi-invariant. To define this notion, let

$$\dot{x} = f(x) \tag{2}$$

be the differential equation, with f a continuous function defined on some open subset Ω of $\mathscr{R}^n$. Uniqueness of the solutions is not required. First of all, a set $F \subset \Omega$ is said to be invariant if, for every $x_0 \in F$ and for all non-continuable solutions $x(t)$ of Equation (2), defined on some interval J and such that $x(t_0) = x_0$, one has $x(t) \in F$ for every $t \in J$. Further, $F \subset \Omega$ is said to be semi-invariant if, for every $x_0 \in F$, there is one such non-continuable solution with the same property. If uniqueness of the solutions is assumed, semi-invariance is of course equivalent to invariance.

10. The following regularity theorem will be used to prove the semi-invariance of the limit sets.

Theorem. Let $\{x_{0i}\} \subset \Omega$ be a sequence of points such that $x_{0i} \to x_0$ as $i \to \infty$, for some $x_0 \in \Omega$. Let x_i:

$]\alpha_i, \omega_i[\to \mathscr{R}^n$ be a non-continuable solution of the Cauchy problem $\dot{x} = f(x)$, $x(0) = x_{0i}$. Then there exist a non-continuable solution $x\colon]\alpha,\omega[\to \mathscr{R}^n$ of the Cauchy problem $\dot{x} = f(x)$, $x(0) = x_0$ and an increasing subsequence $\{i(k)\colon k = 1, 2, \ldots\}$ such that for every t_1, t_2 with $\alpha < t_1 < t_2 < \omega$, one gets, for k large, that $\alpha_{i(k)} < t_1 < t_2 < \omega_{i(k)}$ and $x_{i(k)}(t) \to x(t)$ uniformly on $[t_1,t_2]$ as $k \to \infty$.

For the proof, cf. P. Hartman [1964].

11. Theorem. If the differential equation is autonomous, $\Lambda^+(x) \cap \Omega$ is semi-invariant.

Proof. Let $x_0^* \in \Lambda^+(x) \cap \Omega$. We know that $\omega = \infty$ (see Exercise 6). Let $\{t_i\}$ be such that $t_i \to \infty$ and $x(t_i) \to x_0^*$ as $i \to \infty$. Put $x(t_i) = x_{0i}$. Then $x(t + t_i)$ is a solution of the Cauchy problem $\dot{x} = f(x)$, $x(0) = x_{0i}$. It follows from Theorem 10 that there exists an appropriate subsequence of the $x(t + t_i)$ approaching some solution $x^*(t)$ of the Cauchy problem $\dot{x} = f(x)$, $x(0) = x_0^*$, uniformly on every compact subinterval of the interval J^* of definition of x^*. Therefore, $x^*(t) \in \Lambda^+(x)$ for every $t \in J^*$. Q.E.D.

LIST OF EXAMPLES

BIBLIOGRAPHY

Amaldi, U., see Levi-Civita, T.

Amundson, N. R., see Warden, R. B.

Antosiewicz, H. A. [1958], A survey of Liapunov's second method, Contributions to the theory of non linear oscillations, vol. IV, ed. by S. Lefschetz, Princeton Univ. Press, 141-166.

Appell, P. [1932], Traité de mécanique rationnelle, tome IV, fascicule I, Gauthier-Villars, Paris, 235.

Appell, P. [1953], Traité de mécanique rationnelle, tome II, 2d edition, Gauthier-Villars, Paris, 339.

Aris, R., see Warden, R. B.

Arnold, V. I. [1961], The stability of the equilibrium position of a Hamiltonian system of ordinary differential equations in the general elliptic case, Dokl. Akad. Nauk. SSSR, 137, 255-257, Soviet Math. Dokl., 2, 247-249.

Avdonin, L. N. [1971], A theorem on the instability of equilibrium, Prikl. Mat. Meh., 35, 1089-1090, J. Appl. Math. Mech., 35, 1036-1038.

Avramescu, C. [1973], Quelques remarques sur les concepts qualitatifs attachés à une équation différentielle, Proceedings of the conference on differential equations and their applications, Iasi, Romania, October, 24-27.

Bailey, F. N. [1965], The application of Liapunov's second method to interconnected systems, SIAM J. Control, ser. A, 3, 443-462.

Barbashin, E. A. [1967], Introduction to the theory of stability, Wolters-Noordhoff, Groningen, The Netherlands, 1970; translation of the Russian edition, Moskow, 1967.

Barbashin, E. A., and Krasovski, N. N. [1952], On the stability of motion in the large (Russian), Dokl. Akad. Nauk. SSSR, 86, 453-456.

Beletskii, V. V. [1966], Motion of an artificial satellite about its center of mass, Israel program for scientific translations, Jerusalem.

Bellman, R. [1953], Stability theory of differential equations, McGraw-Hill, New York.

Bellman, R. [1962], Vector Liapunov Functions, SIAM J. Control, ser. A, 1, 32-34.

Bellman, R. [1970], Introduction to matrix analysis, McGraw-Hill, New York.

Bhatia, N. P. [1966], Weak attractors in dynamical systems, Bol. Soc. Mat. Mexicana, 11, 56-64.

Bhatia, N. P., and Hajek, O. [1969], Local semi-dynamical systems, Lecture Notes in Mathematics, 90, Springer Verlag, Berlin - Heidelberg - New York.

Bhatia, N. P., and Lakshmikantham, V. [1965], An extension of Liapunov's direct method, Mich. Math. J., 12, 183-191.

Bhatia, N. P., and Szegö, G. P. [1967], Dynamical systems: stability theory and applications, Lecture Notes in Mathematics, 35, Springer Verlag, Berlin - Heidelberg - New York.

Billotti, J. E., and LaSalle, J.P. [1971], Dissipative periodic processes, Bull. Amer. Math. Soc., 77, 1082-1088.

Birkhoff, G. D. [1927], Dynamical systems, American Mathematical Society Colloquium Publications, vol. IX, Providence, Rhode Island.

Brayton, R. K., and Moser, J. K. [1964], A theory of nonlinear network, Quart. Appl. Math., 22, 1-33, 81-104.

Bushaw, D. [1969], Stabilities of Liapunov and Poisson types, SIAM Rev., 11, 214-225.

Cabannes, H. [1966], Cours de mécanique générale, Dunod, Paris.

Cesari, L. [1959], Asymptotic behavior and stability problems in ordinary differential equations, Springer Verlag, Berlin - Heidelberg - New York.

Chernous'ko, F. L. [1964], On the stability of regular precession of a satellite, Prikl. Mat. Meh., 28, 155-157, J. Appl. Math. Mech., 28, 181-184.

Chetaev, N. G. [1934], A theorem on instability (Russian), Dokl. Akad. Nauk. SSSR, 1, 529-531.

Chetaev, N. G. [1952], On the instability of equilibrium in some cases where the force function is not maximum (Russian), Prikl. Mat. Meh., 16, 89-93.

Chetaev, N. G. [1955], The stability of motion, Pergamon Press, New York, 1961; translation of the Russian edition, Moskow, 1955.

Coddington, E. A., and Levinson, N. [1955], Theory of ordinary differential equations, McGraw-Hill, New York.

Conti, R. [1956], Sulla prolungabilità delle soluzioni di un sistema di equazioni differenziali ordinarie, Boll. Un. Mat. Ital., 11 (3), 510-514.

Conti, R., see Reissig, R.

Coppel, W. A. [1965], Stability and asymptotic behavior of differential equations, D. C. Heath and Company, Boston.

Corduneanu, C. [1957], Systèmes différentiels admettant des solutions bornées, C. R. Acad. Sci. Paris, ser. A-B, 245, 21-24.

Corduneanu, C. [1960], Applications of differential inequalities to stability theory (Russian), An. Sti. Univ. "Al I. Cuza", Iasi Sect. I a Mat., 6, 47-58.

Corduneanu, C. [1964], Sur la stabilité partielle, Revue Roumaine de Math. Pures et Appl., 9, 229-236.

Corduneanu, C. [1971], Principles of differential and integral equations, Allyn and Bacon Inc., Massachusetts.

Corne, J. L. [1973], L'attractivité ... mais c'est très simple, Thèse de doctorat, Louvain-la-Neuve.

Corne, J. L., and Rouche, N. [1973], Attractivity of closed sets proved by using a family of Liapunov functions, J. Differential Equations, 13, 231-246.

Dana, M. [1972], Conditions for Liapunov stability, J. Differential Equations, 12, 590-609.

Dang Chau Phien [1973], Stabilité d'ensembles et fonctions vectorielles de Liapunov, Séminaires de mathématique appliquée et mécanique, 58, Louvain University.

Dang Chau Phien, and Rouche, N. [1970], Stabilité d'ensembles pour des équations différentielles ordinaires, Riv. Mat. Univ. Parma, 11, (2), 1-15.

D'Anna, A. [1973], Asymptotic stability proved by using vector Liapunov functions, Ann. Soc. Sci. Bruxelles, sér. I, 87 (2), 119-139.

De Castro, A. [1953], Sulle oscillazioni non-lineari dei sistemi in uno o più gradi di libertà, Rend. Sem. Mat. Univ. Padova, 22, 294-304.

Deprit, A., and Deprit-Bartholomé, A. [1967], Stability of the triangular lagrangian points, Astronom. J., 72, 173-179.

Dieudonné, J. [1960], Foundations of modern analysis, Academic Press, New York - London.

D'Onofrio, B., Sarno, R., Laloy, M. [1974], Unboundedness properties of solutions of nth order differential equations, Boll. Un. Mat. Ital., (4), 10, 451-459.

Duhem, P. [1902], Sur les conditions nécessaires pour la stabilité de l'équilibre d'un système visqueux, C. R. Acad. Sci. Paris, 135, 939-941.

Etkin, B. [1959], Dynamics of flight: stability and control, Wiley, New York.

Fergola, P., and Moauro, V. [1970], On partial stability, Ricerche Mat., 19, 185-207.

Fiedler, M., and Ptak, V. [1960], On matrices with non positive off diagonal elements and positive principal minors, Czechoslovak Math. J., 12 (87), 382-400.

Fomin, S. V., see Kolmogorov, A. N.

Gambardella, L., and Salvadori, L. [1971], On the asymptotic stability of sets, Ricerche Mat., 20, 143-154.

Gambardella, L., and Tenneriello, C. [1971], On a theorem of N. Rouche, Rend. Accad. Sci. Fis. Mat. Napoli, (4), 38, 145-150.

Gerstein, V. M. [1969], The dissipativity of a certain two-dimensional system, Differencial'nye Uravnenija, 5, 1438-1444.

Goel, N. S., Maitra, S. C., and Montroll, E. W. [1971], Nonlinear models of interacting populations, Academic Press, New York - London.

Goldstein, H. [1950], Classical mechanics, Addison-Wesley, Reading.

Goodwin, B. [1963], Temporal organization of cells, Academic Press, New York.

Gorsin, S. [1948], On stability of motion under constantly acting disturbances, Izv. Akad. Nauk. Kazah. SSR, 56, Ser. Mat. Meh., 2, 46-73.

Gummel, H. K. [1968], A charge-control transistor model for network analysis programs, Proc. IEEE, 56, 751.

Habets, P., and Peiffer, K. [1973], Classification of stability-like concepts and their study using vector Lyapunov functions, J. Math. Anal. Appl., 43, 537-570.

Habets, P., and Peiffer, K. [1975], Attractivity concepts and vector Lyapunov functions, Nonlinear Vibration Problems, 16, 35-52.

Habets, P., and Risito, C. [1973], Stability criteria for systems with first integrals, generalizing theorems of Routh and Salvadori, Equations différentielles et fonctionnelles non linéaires, éd. P. Janssens, J. Mawhin et N. Rouche, Hermann, Paris, 570-580.

Hadamard, J. [1897], Sur certaines propriétés des trajectoires en dynamique, J. Math. Pures et Appl., ser. V, 3, 331-387.

Haddock, J. R. [1972], A remark on a stability theorem of M. Marachkoff, Proc. Am. Math. Soc., 31, 209-212.

Hagedorn, P. [1971], Die Umkehrung der Stabilitätssätze von Lagrange-Dirichlet und Routh, Arch. Rational Mech. Anal., 42, 281-316.

Hagedorn, P. [1972], Eine zusätzliche Bemerkung zu meiner Arbeit: "Die Umkehrung der Stabilitätssätze von Lagrange-Dirichlet und Routh", Arch. Rational Mech. Anal., 47, 395.

Hagedorn, P. [1975], Über die Instabilität konservativer Systeme mit gyroskopischen Kräften, Arch. Rational Mech. Anal., 58, 1-9.

Hahn, W. [1959], Theory and application of Liapunov's direct method, Prentice-Hall, Englewood Cliffs, 1963; translation of the German edition, Springer Verlag, Berlin, 1959.

Hahn, W. [1967], Stability of motion, Springer Verlag, Berlin.

Hahn, W. [1971], On Salvadori's one-parametric families of Liapunov functions, Ricerche Mat., 20, 193-197.

Hajek, O., see Bhatia, N. P.

Halanay, A. [1963], Differential equations: stability, oscillations, time lags, Academic Press, New York, 1966; translation of the Rumanian edition, Bucharest, 1963.

Hale, J. K. [1969], Ordinary differential equations, Wiley-Interscience, New York.

Hale, J. K., LaSalle, J. P., and Slemrod, M. [1972], Theory of a general class of dissipative processes, J. Math. Anal. Appl., 39, 177-191.

Hamel, G. [1903], Über die Instabilität der Gleichgewichtslage eines Systems von zwei Freiheitsgraden, Math. Ann., 57, 541-553.

Hartman, P. [1964], Ordinary differential equations, John Wiley, New York.

Hicks, J. R. [1939], Value and capital, Oxford University Press, London, Clarendon Press, Oxford, 1945 (2d edition).

Hing, C. So., see Mitra, D.

Hoppensteadt, F. C. [1966], Singular perturbations on the infinite interval, Trans. Amer. Math. Soc., 123, 521-535.

Huaux, A. [1964], Sur la méthode directe de Ljapunov, Université Libre de Bruxelles, Faculté Polytechnique de Mons.

Ibrachev, Kh. I. [1947], On Liapunov's second method, Izv. Akad. Nauk. Kazah. SSR, Ser. Fiz. Mat., 42, 101-110.

Kalman, R. E. [1963], Liapunov functions for the problem of Lurie in automatic controls, Proc. Nat. Acad. Sci. USA, 49, 201-205.

Kappel, F., see Knobloch, H. W.

Kneser, A. [1895-1897], Studien über die Bewegungsvorgänge in der Umgebung instabiler Gleichgewichtslagen, J. Reine Angew. Math., 115, 1895, 308-327 and 118, 1897, 186-223.

Knobloch, H. W., and Kappel, F. [1974], Gewöhnliche Differentialgleichungen, B. G. Teubner, Stuttgart.

Koiter, W. T. [1965], On the instability of equilibrium in the absence of a minimum of the potential energy, Nederl. Akad. Wetensch. Proc. ser. B, 68, 107-113.

Kolmogorov, A. N., and Fomin, S. V. [1960], Measure, Lebesgue integrals and Hilbert space, Academic Press, New York - London, 1961; translation of the Russian edition, Moskow, 1960.

Krasovski, N. N. [1956], The inversion of the theorems of Liapunov's second method and the question of stability of motion using the first approximation, Prikl. Mat. Meh., 20, 255-265.

Krasovski, N. N. [1959], Problems of the theory of stability of motion, Stanford Univ. Press, Stanford, California, 1963; translation of the Russian edition, Moskow, 1959.

Krasovski, N. N., see Barbashin, E. A.

Kuratowski, K. [1961], Introduction to set theory and topology, Pergamon Press, Oxford.

Kurzweil, J. [1955], Reversibility of Liapunov's first theorem on stability of motion (Russian), Czechoslovak Math. J., 5, 382-398.

Lagrange, J. L. [1788], Mécanique analytique, Paris, reed. Mallet - Bachelier, Paris, 1853-1855.

Lakshmikantham, V., and Leela, S. [1969], Differential and integral inequalities, theory and applications, Academic Press, New York - London.

Lakshmikantham, V., Mitchell, A. R., and Mitchell, R. W. [1975], Maximal and minimal solutions and comparison results for differential equations in abstract cones, University of Texas at Arlington, TR 27.

Lakshmikantham, V., see Bhatia, N. P.

Laloy, M. [1973]$_1$, Une extension de la méthode des secteurs pour l'étude de l'instabilité à la Liapunov, Ann. Soc. Sci. Bruxelles, sér. I, 87, 17-49.

Laloy, M. [1973]$_2$, Prolongements de la méthode des secteurs de Persidskii, Ann. Soc. Sci. Bruxelles, sér. I, 87, 141-164.

Laloy, M. [1973]$_3$, Utilisation des intégrales premières dans l'étude de la stabilité, Sém. Math. Appl. Méca., Louvain, 62.

Laloy, M. [1973]$_4$, Stabilité à la Liapunov du bétatron, Equations différentielles et fonctionnelles non linéaires, éd. P. Janssens, J. Mawhin, N. Rouche, Hermann, Paris, 582-595.

Laloy, M. [1974]$_1$, La méthode des secteurs appliquée à l'équation scalaire du n^e ordre $d^nx/dt^n = f(t,x)$, An. Sti. Univ. "Al. I. Cuza", Iasi Sect. I a Mat., 20, 39-51.

Laloy, M. [1974]$_2$, Problèmes d'instabilité pour des équations différentielles ordinaires et fonctionnelles, Thèse de doctorat, Louvain-la-Neuve.

Laloy, M. [1975], On equilibrium instability for conservative and partially dissipative mechanical systems, Sém. Math. Appl. Méca., Louvain, 82.

Laloy, M., see D'Onofrio, B.

Lanchester, F. W. [1908], Aerial flight II: aerodonetics, Constable, London.

Lanczos, B. [1962], Variational principles in mechanics, Handbook of Engineering Mechanics, W. Flügge, ed., McGraw-Hill, New York.

LaSalle, J. P. [1960], Some extensions of Liapunov's second method, IRE Trans. Circuit Theory CT-7, 520-527.

LaSalle, J. P. [1962], Asymptotic stability criterion, Proc. Sympos. Appl. Math., vol. 13, 299-307, Amer. Math. Soc., Providence, R. I.

LaSalle, J. P. [1968], Stability theory for ordinary differential equations, J. Differential Equations, 4, 57-65.

LaSalle, J. P., and Lefschetz, S. [1961], Stability by Liapunov's direct method with applications, Academic Press, New York.

LaSalle, J. P., see Billiotti, J. E.; see Hale, J. K.

Leela, S., see Lakshmikantham, V.

Lefschetz, S. [1965], Stability of nonlinear control systems, Academic Press, New York.

Lefschetz, S., see LaSalle, J. P.

Leimanis, E. [1965], The general problem of the motion of coupled rigid bodies about a fixed point, Springer Verlag, Berlin - Heidelberg - New York, p. 300.

Leipholz, H. [1968], Stability theory, Academic Press, New York, 1970; translation from the German edition, Stuttgart, 1968.

Lejeune-Dirichlet, G. [1846], Über die Stabilität des Gleichgewichts, G. Lejeune- Dirichlet's Werke, Vol. II, Georg Reimer, Berlin, 1897; first published in Abh. Preussische Akad. Wiss., 1846.

Leontovich, A. M. [1962], On the stability of Lagrange's periodic solutions of the restricted three-body problem, Dokl. Akad. Nauk. SSSR, 143, 525-528, Soviet Math. Dokl., 3, 425-430.

Levi-Civita, T., and Amaldi, U. [1922-1927], Lezioni di meccanica razionale, Zanichelli, Bologna, vol. I, 1922, vol. II_1, 1926, vol. II_2, 1927.

Levinson, N. [1944], Transformation theory of non-linear differential equations of second order, Ann. of Math., 45, 723-737.

Levinson, N., see Coddington, E. A.

Liapunov, A. M. [1892], Problème général de la stabilité du mouvement, Photo-reproduction in Annals of Mathematics, Studies n° 17, Princeton Univ. Press, Princeton, 1949, of the 1907 French translation of the fundamental Russian paper of Liapunov published in Comm. Soc. Math., Kharkow, 1892.

Liapunov, A. M. [1897], Sur l'instabilité de l'équilibre dans certains cas où la fonction de forces n'est pas un maximum, J. Math. Pures Appl., ser. V, 3, 81-94.

Lotka, A. J. [1920], Analytical note on certain rhythmic relations in organic systems, Proc. Nat. Acad. Sci. USA, 6, 410-415.

Lur'e, A. I. [1968], Mécanique analytique, Librairie Universitaire, Louvain.

McShane, E. J. [1944], Integration, Princeton University Press, Princeton.

Maitra, S. C., see Goel, N. S.

Malkin, I. G. [1944], Stability in the case of constantly acting disturbances, Prikl. Mat. Meh., 8, 241-245.

Malkin, I. G. [1952], Theorie der Stabilität einer Bewegung, translated by W. Hahn and R. Reissig from the Russian edition, 1952, R. Oldenburg, Munich, 1959.

Malkin, I. G. [1954], On the reversibility of Liapunov's theorem on asymptotic stability (Russian), Prikl. Mat. Meh., 18, 129-138.

Manfredi, B. [1956], Sulla stabilità del moto di sistemi a più gradi di libertà in condizioni non lineari, Boll. Un. Mat. Ital., ser. III, 64-71.

Marachkov, M. [1940], On a theorem on stability (Russian), Bull. Soc. Phys.-Math., Kazan, 12, 171-174.

Markeev, A. P. [1969], On the stability of the triangular libration points in the circular bounded three-body problem, Prikl. Mat. Meh., 33, 112-116, J. Appl. Math. Mech., 33, 105-110.

Markus, L. [1956], Asymptotically autonomous differential systems, Contr. Theor. Nonlinear Oscillations, ed. by S. Lefschetz, vol. 3, Princeton Univ. Press, 17-29.

Markus, L., and Yamabe, H. [1960], Global stability criteria for differential equations, Osaka Math. J., 12, 305-317.

Massera, J. L. [1949], On Liapounoff's conditions of stability, Ann. of Math., 50, 705-721.

Massera, J. L. [1956-1958], Contributions to stability theory, Ann. of Math., 64, 1956, 182-206, Erratum Ann. of Math., 68, 1958, 202.

Matrosov, V. M. [1962]$_1$, On the stability of motion, Prikl. Mat. Meh., 26, 885-895, J. Appl. Math. Mech., 26, 1337-1353.

Matrosov, V. M. [1962]$_2$, On the theory of stability of motion, Prikl. Mat. Mech. 26, 992-1002, J. Appl. Math. Mech., 26, 1506-1522.

Matrosov, V. M. [1963], On the stability of motion II (Russian), Trudy Kazan. Aviacion. Inst., 80, 22-33.

Matrosov, V. M. [1965], Development of the method of Liapunov functions in stability theory, Proceedings second All-Union Conference in theoretical and applied mechanics, L. I. Sedov ed., 1965; English transl. Israël program for scientific translation, Jerusalem, 1968.

Matrosov, V. M. [1968-1969], Comparison principle with vector Liapunov functions I-IV, Differencial'nye Uravnenija, 4, 1968, 1374-1386, 1739-1752, 5, 1969, 1171-1185, 2129-2145.

Matrosov, V. M. [1973], Comparison method in system's dynamics, Equations différentielles et fonctionnelles non linéaires, éd. P. Janssens - J. Mawhin - N. Rouche, Hermann, Paris, 407-445.

Mawhin, J., see Rouche, N.

Metzler, L. [1945], Stability of multiple markets: the Hicks conditions, Econometrica, 13, 277-299.

Michel, A. N. [1969], On the bounds of the trajectories of differential systems, Int. J. Control, 10, 593-600.

Michel, A. N. [1974], Stability analysis of interconnected systems, SIAM J. Control, 12, 554-579.

Miller, R. K. [1965], Asymptotic behavior of solutions of nonlinear differential equations, Trans. Amer. Math. Soc., 115, 400-416.

Miller, R. K., and Sell, G. R. [1968], Existence, uniqueness and continuity of solutions of integral equations, Ann. Mat. Pura Appl., (4), 80, 135-152.

Miller, R. K., and Sell, G. R. [1970], Volterra integral equations and topological dynamics, Mem. Amer. Math. Soc., 102.

Mitchell, A. R., see Lakshmikantham, V.

Mitchell, R. W., see Lakshmikantham, V.

Mitra, D., and Hing C. So. [1972], Existence conditions for L_1 Lyapunov functions for a class of nonautonomous systems, IEEE Trans. Circuit Theory, CT-19, 594-598.

Moauro, V., see Fergola, P.

Montroll, E. W., see Goel, N. S.

Moser, J. K., see Brayton, R. K.

Muller, W. D. [1965], Zum Nachweis der Instabilität der Rubelagen bei gewissen Differentialgleichungen, Z. Angew. Math. Mech., 45, 359-360.

Myshkis, A. D. [1947], Sur un lemma géométrique qui s'applique dans la théorie de stabilité au sens de Liapunov, Dokl. Akad. Nauk. SSSR, 55, 299-302.

Nemytskii, V. V. [1965], Some modern problems in the qualitative theory of ordinary differential equations, Russian Math. Surveys, 20 (IV), 1-34.

Opial, Z. [1960], Sur la dépendance des solutions d'un système d'équations différentielles de leurs seconds membres, application aux systèmes presque autonomes, Ann. Polon. Math., 8, 75-89.

Oziraner, A. S. [1972], On certain theorems of Liapunov's second method, Prikl. Mat. Meh., 36, 396-404, J. Appl. Math. Mech., 36, 373-381.

Painlevé, P. [1897], Sur les positions d'équilibre instable, C. R. Acad. Sci. Paris, sér. A-B, 25 (2), 1021-1024.

Painlevé, P. [1904], Sur la stabilité de l'équilibre, C. R. Acad. Sci. Paris, sér. A-B, 138, 1555-1557.

Pavel, N. [1971], On dissipative systems, Boll. Un. Mat. Ital., 4, 701-707.

Pavel, N. [1972]$_1$, A generalization of ultimately bounded systems, An. Sti. Univ. "Al. I. Cuza" Iasi Sect. I a Mat., 18, 81-86.

Pavel, N. [1972]$_2$, On the boundedness of solutions of a system of differential equations, Tôhoku Math. J., 24, 21-32.

Peiffer, K., and Rouche, N. [1969], Liapunov's second method applied to partial stability, J. Mécanique, 8, 323-334.

Peiffer, K., see Habets, P.; see Rouche, N.

Persidski, K. P. [1933], On the stability of motion in first approximation, Mat. Sb., 40, 284-293.

Persidski, K. P. [1946], On the theory of stability of solutions of differential equations (Russian), Doctoral dissertation, Moskov. Gos. Univ., a summary was published in Uspehi Mat. Nauk., 1, 250-255.

Persidski, K. P. [1947], On Liapunov's second method, Izv. Akad. Nauk. Kazah, SSR, ser. Fiz.-Mat., 42, 48-55.

Persidski, S. K. [1961], On the second method of Liapounov, Prikl. Mat. Meh., 25, 17-23, J. Appl. Math. Mech., 25, 20-28.

Persidski, S. K. [1968], Investigation of stability of solutions of some nonlinear systems of differential equations, Prikl. Mat. Meh., 32, 1122-1125, J. Appl. Math. Mech., 32, 1141-1144.

Persidski, S. K. [1970], Investigating the stability of solutions of systems of differential equations, Prikl. Mat. Meh., 34, 219-226, J. Appl. Math. Mech., 34, 209-215.

Piontkovskii, A. A., and Rutkovskaya, L. D. [1967], Investigation of certain stability theory problems by the vector Lyapunov function method, Avtomat. i Telemeh., 10, 23-31, Automat. Remote Control, 10, 1422-1429.

Pliss, V. A. [1964], Nonlocal problems of the theory of oscillations, Academic Press, New York - London, 1966; translated from the Russian edition, Moscow - Leningrad, 1964.

Pluchino, S. [1971], Osservazioni sulla stabilità dei moti merostatici di un sistema olonomo od anolonomo, Boll. Un. Mat. Ital., 4, 213-219.

Poisson, S. D. [1838], Traité de mécanique, 3e éd., Société belge de Librairie, Bruxelles.

Pontryagin, L. S. [1961], Ordinary differential equations, Addison-Wesley, Reading, Massachusetts, 1962; translation of the Russian edition, Moskow, 1961.

Popov, V. M. [1962], Absolute stability of nonlinear systems of automatic control, Avt. Telemeh., 22, 961-979, Aut. Rem. Control, 22, 857-875.

Popov, V. M. [1966], Hyperstability of control systems, Springer Verlag, Berlin - Heidelberg - New York, 1973; revised translation of the Romanian edition, Bucarest, 1966.

Pozharitskii, G. K. [1958], On the construction of Liapunov functions from the integrals of the equations of the perturbed motion (Russian), Prikl. Mat. Meh., 22, 145-154.

Ptak, V., see Fiedler, M.

Reissig, R., Sansone, G., and Conti, R. [1969], Nichtlineare Differentialgleichungen höherer Ordnung, Edizioni Cremonese, Roma.

Risito, C. [1967], On the Liapunov stability of a system with known first integrals, Meccanica, 2, 197-200.

Risito, C. [1970], Sulla stabilità asintotica parziale, Ann. Mat. Pura Appl., (4), 84, 279-292.

Risito, C. [1971], Some theorems on the stability and the partial asymptotic stability of systems with known first integrals, Comptes rendus des journées nationales du C.B.R.M., Mons, 24-26 mai, 53-56.

Risito, C. [1972], The comparison method applied to the stability of systems with known first integrals, Proc. 6th Int. Conf. on Nonlinear Oscillations, Poznan, 1972, Nonlinear vibration problems, 15, 1974, 25-45.

Risito, C. [1974], Metodi per lo studio dellà stabilità di sistemi con integrali primi noti, Séminaires de mathématique appliquée et mécanique, 74, Louvain University; to appear in Ann. Mat. Pura Appl.

Risito, C. [1975], On the Chetayev method for the construction of a positive definite first integral, Ann. Soc. Sci. Bruxelles, sér, I, 89, 3-10.

Risito, C., see Habets, P.

Rosen, R. [1970], Dynamical system theory in biology, Wiley-Interscience, New York.

Rouche, N. [1968], On the stability of motion, Int. J. Non-Linear Mechanics, 3, 295-306.

Rouche, N. [1969], Quelques critères d'instabilité à la Liapunov, Ann. Soc. Sci. Bruxelles, sér. I, 83, 5-17.

Rouche, N. [1971], Attractivity of certain sets proved by using several Liapunov functions, Symposia Mathematica, 6, 331-343.

Rouche, N. [1974], Théorie de la stabilité dans les équations différentielles ordinaires, Stability Problems, 1st CIME session, 1974, Edizioni Cremonese, Rome, 111-194.

Rouche, N. [1975], The invariance principle applied to non-compact limit set, Boll. Un. Mat. Ital. 11, 306-315.

Rouche, N., and Mawhin J. [1973], Equations différentielles ordinaires, Masson, Paris.

Rouche, N., and Peiffer, K. [1967], Le théorème de Lagrange-Dirichlet et la deuxième méthode de Liapounoff, Ann. Soc. Sci. Bruxelles, Ser. 1, 81, 19-33.

Rouche, N., see Corne, J. L.; see Dang Chau Phien.

Routh, E. J. [1905], The advanced part of a treatise on the dynamics of a system of rigid bodies, Republished by Dover, New York, 1955, from the 6th edition, 1905.

Routh, E. J. [1975], Stability of motion, Republ. of selected papers, Taylor and Francis Ltd., London.

Royden, H. L. [1963], Real analysis, McMillan, New York.

Rubanovskii, V. V., and Stepanov, S. Ia. [1969], On the Routh theorem and the Chetaev method for constructing the Liapunov function from the integrals of the equations of motion, Prikl. Mat. Meh., 33, 904-912, J. Appl. Math. Mech., 33, 882-890.

Rumiantsev, V. V. [1957], On the stability of a motion in a part of variables (Russian), Vestnik Moskov. Univ. Ser. I Mat. Meh., 4, 9-16.

Rumiantsev, V. V. [1966], On the stability of steady motions, Prikl. Mat. Meh., 30, 922-933, J. Appl. Math. Mech., 30, 1090-1103.

Rumiantsev, V. V. [1968], On the stability of steadystate motions, Prikl. Mat. Meh., 32, 504-508, J. Appl. Math. Mech., 32, 517-521.

Rumiantsev, V. V. [1970], On the optimal stabilization of controlled systems, Prikl. Mat. Meh., **34**, 440-456, J. Appl. Math. Mech., **34**, 415-430.

Rumiantsev, V. V. [1971], On the stability with respect to a part of the variables, Symposia Math., **6**, Meccanica non lineare e stabilità, 23-26 Febbraio, 1970, Roma, Ist. Naz. di Alta Matematica, Academic Press, New York, 1971.

Rutkovskaya, L. D., see Piontkovskii, A. A.

Salvadori, L. [1953], Un' osservazione su di un criterio di stabilità del Routh, Rend. Accad. Sci. Fis.-Mat. Napoli, **20**, 269-272.

Salvadori, L. [1966], Sull' estensione ai sistemi dissipativi del criterio di stabilità del Routh, Ricerche Mat., **15**, 162-167.

Salvadori, L. [1968], Sulla stabilità dell' equilibrio nella meccanica dei sistemi olonomi, Boll. Un. Mat. Ital., **4**, 333-344.

Salvadori, L. [1969], Sulla stabilità del movimento, Matematiche, **24**, 218-239.

Salvadori, L. [1971], Famiglie ad un parametro di funzioni di Liapunov nello studio della stabilità, Symposia Math., **6**, 309-330.

Salvadori, L. [1972], Sul problema della stabilità asintotica, Rendiconti dell' Accad. Naz. Lincei, (8), **53**, 35-38.

Salvadori, L. [1974], Some contributions to asymptotic stability theory, Ann. Soc. Sci. Bruxelles, sér, I, **88**, 183-194.

Salvadori, L., see Gambardella, L.

Samuelson, P. A. [1947], Foundations of economic analysis, Harvard University Press, Cambridge, Massachusetts.

Sandberg, I. W. [1969], Some theorems on the dynamic response of nonlinear transistor networks, Bell Syst. Tech. J., **48**, 35-54.

Sansone, G., see Reissig, R.

Sarno, R., see D'Onofrio, B.

Schuur, J. D. [1967], The asymptotic behavior of a solution of the third order linear differential equation, Proc. Amer. Math. Soc., **18**, 391-393.

Sell, G. R. [1971], Topological dynamics and ordinary differential equations, Van Nostrand Reinhold Company, London.

Sell, G. R., see Miller, R. K.

Siljak, D. D. [1973], Competitive economic systems: stability, decomposition and aggregation, Proceeding of the 1973 IEEE Conference on Decision and Control, San Diego, California, December 5-7, 265-275.

Silla, L. [1908], Sulla instabilità dell'equilibrio di un sistema materiale in posizioni non isolate, R. Accad. Lincei, 17, 347-355.

Skowronski, J. M. [1969], Multiple nonlinear lumped systems, Polish scientific publishers, Warsaw.

Slemrod, M., see Hale, J. K.

Smets, H. B. [1961], Stability in the large of heterogeneous power reactors, Acad. Roy. Belg. Bull. Cl. Sci., 47, 382-405.

Stepanov, S. Ia., see Rubanovskii, V. V.

Stern, T. E. [1965], Theory of nonlinear networks and systems, Addison-Wesley, Reading, Massachusetts.

Szarski, J. [1967], Differential inequalities, Polish Scientific Publishers, Warsaw.

Szegö, G. P., see Bhatia, N. P.

Tait, P. G., see Thomson, W. (Lord Kelvin).

Tenneriello, C., see Gambardella, L.

Thomson, W. (Lord Kelvin), and Tait, P. G. [1912], Principles of mechanics and dynamics, Cambridge U. P. (last ed.), 1912, reprinted by Dover Publications, Inc. 2 vol., 1962.

Van Chzhao-Lin [1963], On the converse of Routh's theorem, Prikl. Mat. Meh., 27, 890-893, J. Appl. Math. Mech., 27, 1354-1360.

Verhulst, P. [1845], Recherches mathématiques sur la loi d'accroissement de la population, Mémoires Acad. Roy. Bruxelles, 18, 3-39.

Verhulst, P. [1847], Deuxième mémoire sur la loi d'accroissement de la population, Mémoires Acad. Roy. Bruxelles, 20, 3-32.

Vinograd, R. E. [1957], The inadequacy of the method of characteristic exponents for the study of nonlinear differential equations (Russian), Mat. Sbornik, 41 (83), 431-438.

Volterra, V. [1931], Leçon sur la théorie mathématique de la lutte pour la vie, Gauthier-Villars, Paris.

Vrkoč, I. [1959], Integral stability, Czechoslovak Math. J., 9 (84), 71-129.

Walter, W. [1964], Differential and integral inequalities, Springer Verlag, New York - Heidelberg - Berlin, 1970; translation of the German edition, 1964.

Warden, R. B., Aris, R., and Amundson, N. R. [1964], An analysis of chemical reactor stability and control VIII, Chem. Eng. Sc., 19, 149-172.

Wazewski, T. [1947], Sur un principe topologique de l'examen de l'allure asymptotique des intégrales des équations différentielles ordinaires, Ann. Polon. Math., 20, 279-313.

Wazewski, T. [1950], Systèmes des équations et des inégalités différentielles ordinaires aux deuxièmes membres monotones et leurs applications, Ann. Soc. Polonaise Math., 23, 112-166.

Wintner, A. [1941], The analytical foundation of celestial mechanics, University Press, Princeton.

Yacubovich, V. A. [1962], Solution of certain matrix inequalities occuring in the theory of automatic controls, Dokl. Akad. Nauk. SSSR, 143, 1304-1307.

Yamabe, H., see Markus, L.

Yorke, J. A. [1968], An extension of Chetaev's instability theorem using invariant sets, and an example, Seminar on differential equations and dynamical systems, ed. by G. S. Jones, Springer Verlag, Berlin, 100-106.

Yoshizawa, T. [1959], Liapunov's function and boundedness of solutions, Funkcial. Ekvac., 2, 71-103.

Yoshizawa, T. [1963], Asymptotic behavior of solutions of a system of differential equations, Contributions Differential Equations, 1, 371-387.

Yoshizawa, T. [1966], Stability theory by Liapunov's second method, The Math. Soc. of Japan, Tokyo.

Zhukovski, N. E. [1891], On soaring of birds (Russian), Trudy Otdel. Fiz. Nauk. Obshch. Lyubit. Yestestvz., 4 (2), 29, 1891, Reprinted in Collected Works, 4, Gostekhizdat, 1949, 5.

Ziemba, S. [1961], Some problems of the Warsaw group on the theory of nonlinear oscillations in the last five year programme, Proc. Internat. Symp. Nonlinear Oscillations, T II, Kiev, 161-171.

Zubov, V. I. [1957], The methods of Liapunov and their applications, Noordhoff, Groningen, 1964; translated from the Russian edition, Leningrad, 1957.

AUTHOR INDEX

SUBJECT INDEX